THE HIGH-MOUNTAIN CRYOSPHERE
Environmental Changes and Human Risks

The world's high-mountain cryosphere is undergoing unprecedented changes, with significant implications for both humans and the environment. This edited volume, showcasing cutting-edge research, addresses two primary questions: what are the main drivers of change in high mountains and what are the risks implied by these changes?

Applying an interdisciplinary methodology, this comprehensive book provides a definitive overview of the global drivers of high-mountain cryospheric change, from climate, to economy and culture. From a physical perspective, it examines the complex interplay between climate and the high-mountain cryosphere, with further chapters covering tectonics, volcano–ice interactions, hydrology, slope stability, erosion, ecosystems, and glacier- and snow-related hazards. Societal dimensions, both global and local, of high-mountain cryospheric change are also explored. The book offers unique perspectives on high-mountain cultures, livelihoods, governance, and natural resources management, focusing on how global change influences societies and how people respond to climate-induced cryospheric changes.

An invaluable reference for researchers and professionals in cryospheric science, geomorphology, climatology, environmental studies, and human geography, this volume will also be of interest to practitioners working in global change and risk, including NGOs and policy advisors, as well as to graduate students.

CHRISTIAN HUGGEL is a senior researcher at the Department of Geography, University of Zurich. He has worked in mountain regions and the cryosphere worldwide, including the European Alps, the Andes, Mexico, Alaska, the Caucasus, and the Himalayas. Currently, he leads several projects on climate impacts and adaptation in the tropical Andes, in particular Peru, the Indian Himalayas and the Alps, in collaboration with the Swiss and national governments. He is a lead author of the IPCC Working Group II 5th Assessment Report and has been a science advisor and member of the Swiss delegation at the UNFCCC Conference of Parties (COP).

MARK CAREY is Associate Dean and Associate Professor of History in the Robert D. Clark Honors College at the University of Oregon. His book, *In the Shadow of Melting Glaciers: Climate Change and Andean Society* (Oxford, 2010), won the

Elinor Melville Prize for the best book in Latin American environmental history, awarded by the American Historical Association's Conference on Latin American History. He is a co-founder and co-director of the Transdisciplinary Andean Research Network (TARN) that involves collaborative research with colleagues in the United States, Canada, and South America.

JOHN CLAGUE is Shrum Professor of Science at Simon Fraser University, British Columbia, and is currently the Canada Research Chair in Natural Hazard Research and Director of the Centre for Natural Hazard Research. His research interests include glacial geology, geomorphology, stratigraphy, sedimentology, and natural hazards, and he has consulted for several private-sector firms and government agencies. His other principle professional interest is promoting awareness of earth science to students, teachers, and the general public. Clague is a Fellow of the Royal Society of Canada, former President of the Geological Association of Canada, and recipient of the Geological Society of America Burwell Award.

ANDREAS KÄÄB is Professor for Remote Sensing at the Department of Geo-sciences, University of Oslo. His main research focus is on remote sensing of the cryosphere, in particular glaciers, permafrost, river ice, and related natural hazards in a changing world. He was Chair of the Standing Group on Glacier and Permafrost Hazards in Mountains (GAPHAZ) of the International Association of Cryospheric Sciences (IACS) and of the International Permafrost Association (IPA) for 10 years. Kääb is currently running a number of projects related to observing glaciers and natural hazards from space. In 2008 he was awarded the prize for Excellence in Permafrost Research by the International Permafrost Association.

THE HIGH-MOUNTAIN CRYOSPHERE

Environmental Changes and Human Risks

Edited by

CHRISTIAN HUGGEL
University of Zurich, Switzerland

MARK CAREY
University of Oregon, USA

JOHN J. CLAGUE
Simon Fraser University, Canada

ANDREAS KÄÄB
University of Oslo, Norway

CAMBRIDGE
UNIVERSITY PRESS

CAMBRIDGE
UNIVERSITY PRESS

University Printing House, Cambridge CB2 8BS, United Kingdom

One Liberty Plaza, 20th Floor, New York, NY 10006, USA

477 Williamstown Road, Port Melbourne, VIC 3207, Australia

4843/24, 2nd Floor, Ansari Road, Daryaganj, Delhi - 110002, India

79 Anson Road, #06-04/06, Singapore 079906

Cambridge University Press is part of the University of Cambridge.

It furthers the University's mission by disseminating knowledge in the pursuit of education, learning and research at the highest international levels of excellence.

www.cambridge.org
Information on this title: www.cambridge.org/9781107662759

First published 2015
First paperback edition 2017

A catalogue record for this publication is available from the British Library

Library of Congress Cataloging in Publication data
The high-mountain cryosphere : environmental changes and human risks / edited by
Christian Huggel, University of Zurich, Switzerland, Mark Carey, University of
Oregon, USA, John J. Clague Simon Fraser University, Canada, Andreas Kääb,
University of Oslo, Norway.
pages   cm
ISBN 978-1-107-06584-0 (Hardback)
1. Mountains–Environmental aspects.   2. Glaciers–Environmental aspects.
3. Cryosphere.   4. Climatic changes.   5. Global environmental change.
6. Mountain ecology.   7. Mountain life.   I. Huggel, Christian.
GB501.2.H52 2015
577.5´3–dc23   2015005312

ISBN 978-1-107-06584-0 Hardback
ISBN 978-1-107-66275-9 Paperback

Additional resources for this publication at www.cambridge.org/9781107065840

Contents

Contributors

Simon Allen
Department of Geography, University of Zurich, Winterthurerstr. 190, CH-8057 Zurich, Switzerland, and Institute for Environmental Sciences, University of Geneva Site de Battelle, D, 7 route de Drize, CH-1227, Carouge, Switzerland

Michel Baraer
Département de génie de la construction, École de technologie supérieure, University of Quebec, 1100, rue Notre-Dame Ouest, Montreal, H3C 1K3, Canada

Javiera Barandiarán
Global & International Studies Program, University of California, Santa Barbara, CA 93106-7065, USA

Julie Brugger
Climate Assessment for the Southwest, University of Arizona, PO Box 210156, Tucson, AZ 85721-0156, USA

Jeffrey Bury
Department of Environmental Studies, University of California, Santa Cruz, CA 95064, USA

Mark Carey
Robert D. Clark Honors College, University of Oregon, Eugene, OR 97403, USA

Jonathan Carrivick
School of Geography, University of Leeds, University Road, Leeds LS2 9JT, UK

John J. Clague
Department of Earth Sciences, Simon Fraser University, 8888 University Drive, Burnaby, BC, V5A 1S6, Canada

Hildegard Diemberger
Department of Archaeology and Anthropology, Division of Social Anthropology, Cambridge University, Free School Lane, Cambridge CB2 3RF, UK

Katherine W. Dunbar
Center for Integrative Conservation Studies, University of Georgia, 321 Holmes-Hunter Academic Building, 101 Herty Drive, Athens, GA 30602, USA

Stuart Dunning
Geography Department, Northumbria University, Newcastle-Upon-Tyne, NE1 8ST, UK

Benjamin R. Edwards
Department of Earth Sciences, Dickinson College, Carlisle, PA 17013, USA

Alfonso Fernández
Byrd Polar Research Center, Ohio State University, 108 Scott Hall, 1090 Carmack Road, Columbus, OH 43210, USA, and Department of Geography, Universidad de Concepción, Chile

Andrew G. Fountain
Department of Geology, Portland State University, Portland, OR 97207, USA

Adam French
Energy and Resources Group, University of California, Berkeley, CA 94720, USA

Sven Fuchs
Institute of Mountain Risk Engineering, University of Natural Resources and Life Sciences, Peter-Jordan-Strasse 82, 1190 Vienna, Austria

Christoph Graf
Swiss Federal Research Institute WSL, Mountain Hydrology and Mass Movements, Zuercherstrasse 111, CH-8903 Birmensdorf, Switzerland

Astrid Hovden
Department of Culture Studies and Oriental Languages, University of Oslo, PO Box 1010 Blindern, 0315 Oslo, Norway

Christian Huggel
Department of Geography, University of Zurich, Winterthurerstrasse 190, CH-8057 Zurich, Switzerland

Walter Immerzeel
Faculty of Geosciences, Utrecht University, Heidelberglaan 2, 3584 CS Utrecht, the Netherlands

Matthias Jakob
BGC Engineering, Suite 800, 1045 Howe Street, Vancouver, BC, V6Z 2A9, Canada

Christine Jurt
Department of Geography, Research Group on Environment and Climate: Impacts, Risks and Adaptation, University of Zurich, Winterthurerstrasse 190, CH-8057 Zurich, Switzerland

Andreas Kääb
Department of Geosciences, University of Oslo, PO Box 1047, Blindern, 0316 Oslo, Norway

Margreth Keiler
Institute of Geography, University of Bern, Hallerstrasse 12, 3012 Bern, Switzerland

Oliver Korup
Earth and Environmental Sciences, Potsdam University, Karl-Liebknechtstr. 24 (Hs 27), D-14476 Potsdam, Germany

Michael Krautblatter
Landslide Research, Engineering Geology, Technical University Munich, Arcisstr. 21 80333 Munich, Germany

Kerry Leith
Landslide Research, Engineering Geology, Technical University Munich, Arcisstr. 21 80333 Munich, Germany

Bryan G. Mark
Department of Geography and Byrd Polar Research Center, Ohio State University, 1136 Derby Hall, 154 N Oval Mall, Columbus, OH 43210, USA

Bill McGuire
UCL Hazard Research Centre, Department of Earth Sciences, University College London, Gower Street, London WC1E 6BT, UK

Kerry Milch
Center for Research on Environmental Decisions (CRED), Columbia University, 419 Schermerhorn Hall, 1190 Amsterdam Avenue, New York, NY 10027, USA

James Miller
Department of Marine and Coastal Sciences, Rutgers University, 71 Dudley Road, New Brunswick, NJ 08901, USA

R. Daniel Moore
Departments of Geography and Forest Resources Management, University of British Columbia, 1984 West Mall, Vancouver, BC, V6T 1Z2, Canada

Ben Orlove
School of International and Public Affairs (SIPA), Columbia University, 420 West 118th St., Suite 833, New York, NY 10027, USA

Nick Pepin
Department of Geography, University of Portsmouth, Buckingham Building, Lion Terrace, Portsmouth, PO1 3HE, UK

Duncan Quincey
School of Geography, University of Leeds, University Road, Leeds LS2 9JT, UK

Costanza Rampini
Department of Environmental Studies, University of California, Santa Cruz, Santa Cruz, CA 95064, USA

Imtiaz Rangwala
Western Water Assessment/CIRES & Physical Sciences Division, NOAA ESRL, 325 Broadway, R/PSD NOAA, Boulder, CO, 80305, USA

Dieter Rickenmann
Swiss Federal Research Institute WSL, Mountain Hydrology and Mass Movements, Zuercherstrasse 111, CH-8903 Birmensdorf, Switzerland

Mario Rohrer
Meteodat GmbH, Technoparkstr. 1, CH-8005 Zurich, Switzerland

Nadine Salzmann
Department of Geosciences, University of Fribourg, Chemin de Musée 4, CH-1700 Fribourg, Switzerland

Simon C. Scherrer
Federal Office of Meteorology and Climatology MeteoSwiss, Operation Center 1, CH-8058 Zurich-Airport, Switzerland

Sergey Sokratov
Laboratory of Snow Avalanches and Debris Flows & Arctic Environment Laboratory, Faculty of Geography, Moscow State University, Leninskie gory 1, 119991, Moscow , Russian Federation

Markus Stoffel
Institute of Geological Sciences, University of Berne, Baltzerstrasse 1+3, CH-3012 Berne, Switzerland, and Institute for Environmental Sciences, University of Geneva, 7 route de Drize, CH-1227 Carouge, Switzerland

Mathias Vuille
Department of Atmospheric and Environmental Sciences, University at Albany, NY, 1400 Washington Avenue, Albany, NY 12222, USA

Richard B. Waitt
US Geological Survey, Cascades Volcano Observatory, Vancouver, WA 98683, USA

Rolf Weingartner
Institute of Geography, Oeschger Centre for Climate Change Research, University of Bern, Hallerstrasse 12, CH-3012 Bern, Switzerland

Emily T. Yeh
Department of Geography, University of Colorado at Boulder, Guggenheim 110, 260 UCB, Boulder, CO 80309-0260, USA

Kenneth R. Young
Department of Geography and the Environment, University of Texas at Austin, 305 E. 23rd Street – A3100 – CLA 3.306, Austin, TX 78712-1697, USA

Acknowledgments

This book is a contribution to the Glacier and Permafrost Hazards in Mountains (GAPHAZ) Standing Group of the International Association for the Cryospheric Sciences (IACS) and the International Permafrost Association (IPA). M. Carey acknowledges funding support from the US National Science Foundation under grants 1010132 and 1253779. A. Kääb acknowledges support by the European Research Council under the European Union's Seventh Framework Programme (FP/2007–2013)/ERC grant agreement no. 320816. Special thanks for providing reviews for the individual chapters are due to Marta Chiarle, Pascal Haegeli, and many other anonymous referees.

1

Introduction

Human–environment dynamics in the high-mountain cryosphere

CHRISTIAN HUGGEL, MARK CAREY, JOHN J. CLAGUE, AND ANDREAS KÄÄB

Recent global-scale assessments such as the 5th Assessment Report (AR5) of the Intergovernmental Panel on Climate Change (IPCC) have provided evidence of the rapid changes to the high-mountain cryosphere due to climate change [1,2]. Glaciers, recognized as indicators or 'thermometers' of climate change, have been receding worldwide over the past century, and many glaciers are likely to disappear over the next several decades, leaving behind historically unprecedented landscapes. High mountains are commonly thought of as being remote, but human interactions with this environment are essential for many societies, and rapid biophysical changes can cause societal transformations.

The natural alpine environment, its human dimensions, and their interplay are increasingly being documented. Observation technologies have improved, times series of observations have become longer, and the number of monitoring sites has increased, all of which have increased our understanding of regional changes to the high-mountain cryosphere. Similarly, more research is being conducted on how local people perceive the cryosphere and high mountains, how physical changes affect their livelihoods, and how they respond to such changes.

Nevertheless, there are still substantial gaps in our understanding. For example, until recently regional and local glacier changes in the Himalayas had not been adequately studied, and ground observations there are scarce [3]. In many regions we still lack a comprehensive understanding of how climate and cryosphere changes will affect water resources, slope stability, and vegetation. The most substantial gaps in knowledge, however, are in the human dimensions. The framing of research on cyrospheric change has been around ice and water itself, with little attention to how people perceive, feel, or value those changes, or how distinct social groups living near or far from glaciers are affected by changes to

The High-Mountain Cryosphere, ed. Christian Huggel, Mark Carey, John J. Clague and Andreas Kääb. Published by Cambridge University Press. © Cambridge University Press 2015.

glaciers, snow, and permafrost. Also, the ways in which human activities affect the high-mountain cryosphere, for instance mining, hydropower development, and tourism, have not been well documented. To understand these dynamic intersections between people and the cryosphere, it is crucial to integrate disciplines, to talk across boundaries, and to embrace concepts and methods applicable to coupled natural–human and social–ecological systems.

Risk is where the physical and human world are confronted and interact. Risks in high mountains, as in other regions, are dynamic and not static, which poses enormous challenges to basic and applied research. However, even at a primarily physical level the assessment of cryosphere hazards is often not adequately developed. Over the past years efforts therefore have been made to define standards, notably by GAPHAZ, the Standing Group on Glacier and Permafrost Hazards in Mountains of the International Association of Cryospheric Sciences (IACS) and International Permafrost Association (IPA) [4]. Standards are required to ensure an adequate level of quality and to avoid incorrect assessments with potentially adverse consequences, as experiences in the past have shown. Concepts and terminologies related to hazard and risk assessments must follow recently issued consensus statements such as from the United Nations institutions. GAPHAZ thereby noted on several occasions that there is still a lack of integration of physical/engineering and social aspects of glacier and permafrost hazards; and therefore communication and exchange between natural and social science experts must be strengthened.

This book offers new perspectives on the high-mountain cryosphere to address these gaps by assembling evidence from a broad range of natural and social science disciplines. It examines the intersections among different biophysical and social systems. The contributions provide evidence for interactions among climate and the different components of the cryosphere; among the cryosphere and geological processes; among climate, the cryosphere, and water resources; and among social, cultural, political, and economic processes and the high-mountain environment. As overarching questions, the book asks the following:

- What are the drivers of environmental, social, and economic change?
- How do those drivers interact and impact natural and human systems?
- What types of risks are related to these changes?

The book is divided into three main sections, each of which includes a number of chapters. The focus of the first section is global drivers acting on the high-mountain cryosphere. The second section examines the range of biophysical processes operating in high mountains. The third section considers the consequences of high-mountain change and hazards, and how different societies respond to related risks.

Chapters in the first section of the book provide an integrated perspective on climate, economy, policy, and culture. These global drivers act on different scales of time and space, involve different processes, and yet are connected. The climate perspective is provided by two chapters, authored by Rangwala *et al.* and Salzmann *et al.* These two chapters show how the climate interacts with the cryosphere and the high-mountain environment. Large mountain systems are affected by large-scale circulation patterns, but themselves also affect the local and regional climate, for instance by forming topographic barriers that produce remarkable precipitation gradients, such as in the Andes or the Himalayas (Rangwala *et al.*). These interactions impact the natural environment and the societies living in high-mountain valleys. Climatic extremes have a similarly important effect on the cryosphere and high mountains, for example by producing a range of hazards including landslides, floods, and droughts. Documenting the high-mountain climate is a major challenge, even though an increasing number of ground, airborne, and spaceborne instruments support these efforts (Salzmann *et al.*). Starting with the climate–snow interface, Fuchs *et al.* demonstrate how socio-economic change and dynamics drive avalanche risks, how snow is import-ant in economic terms, and how such risks are addressed in international policy agreements. The economic perspective is extended by Bury, who shows how the confluence of rapid economic growth, deepening integration of global markets, and burgeoning demand for natural resources are increasing mineral and energy extraction and are related to high-mountain and cryosphere changes.

Culture and cultural values are important, but often hidden, elements underpin-ning many social and political processes, and are of interest in high-mountain regions. As demonstrated by Jurt *et al.* in this book, cultural values are typically place-specific and their characteristics therefore need to be studied locally and in depth. At the same time, however, the collection of case studies in this chapter from the Andes, the US North Cascades, and the Alps underscore the transnational characteristics of mountain cultural values. Identity plays a central role in this discussion, and glaciers are part of the identity of people in high-mountain regions. It is notable in this context that the influence of anthropogenic climate change on glaciers over the past decades has recently been quantified [5], raising questions of responsibility for loss of identity and cultural values.

The second section of the book is concerned with a rich and diverse range of processes that act on, and interact with, the high-mountain cryosphere. The chapters in this section offer a biophysical panorama while also highlighting the critical interface with society. Many high mountains coincide with tectonically and seismically active regions. The interplay of the high-mountain cryosphere with tectonic, seismic, and volcanic processes and a warming climate is still not suffi-ciently understood, but has important implications for people living in these areas and beyond, as the analysis by McGuire in this book demonstrates. The three chapters by Korup and Dunning, Krautblatter and Leith, and Rickenmann and Jakob shed light on fundamental aspects of mass turnover in high mountains. The processes act on a range of temporal and spatial scales, from continuous and regional scale to local scale, with potentially dramatic consequences for populations in the reach of such processes.

The search for causes, triggers, and drivers of both continuous processes and rare extreme events, and their changes over time, is a research field that has been stimulated in the context of climate change. The three components of the cryosphere – glaciers, permafrost, and snow – underlie, drive, and interact with a variety of surface processes. Surface and subsurface ice and snow typically are near the melting point and thus are sensitive to changes in boundary conditions, including meteorological and climatic conditions. This sensitivity of snow and ice, in concert with the high gravitational energy and steep topography, make high mountains places where a multitude of slow to very fast mass movements are threats to people. Cumulative effects of climate change on the cryosphere, such as glacier shrinkage, lake formation, permafrost degradation, and exposure of unconsolidated sediment due to glacier retreat, can result in sudden and rapid extreme events.

The sensitivity of the cryosphere to environmental changes also has implications for water resources. Water produced by melting snow and glaciers is a significant element of the hydrologic cycle in all high-mountain regions, and societies and economies directly depend on it, as Mark *et al.* demonstrate in their chapter. Concern is mounting that shrinking glaciers and rising snowlines are negatively affecting agriculture, hydroelectric power generation, and local livelihoods, especially in regions with distinct dry seasons such as the Andes. The transition of ice to water operates at different spatial and temporal scales, and is therefore not easily captured in current models. In several high-mountain and downstream regions, water scarcity and episodic excess are important hazards and risks. Quincey and Carrivick provide insights into some of the Earth's most extreme floods that occur from glacial lakes and can have far-reaching impacts on downstream populations.

The fragility and sensitivity of high-mountain ecosystems are clear from Young's contribution to this book. He focuses on landscape and ecological changes in tropical mountains. A fascinating picture of cryosphere, high-mountain wetland, treeline, and other environmental changes emerges, placed in the context of land use, including animal grazing, burning, mining, deforestation/afforestation, and trekking.

The third and final section of the book substantially extends the social perspective and again integrates the human dimension into the cryosphere and high-mountain processes. Diemberger *et al.* start the section by offering unique insights into how local and indigenous people in the Himalayas of Tibet perceive their environment, how their livelihoods are affected by cryosphere and high-mountain change, and how they respond to those changes. For instance, Tibetans see mountains as the 'owners of the land' that eventually control and enable human livelihoods. Snow and ice are the honor of the mountains, and hence their decline and loss goes far beyond the physical impacts by touching deeply upon beliefs and identity. The local perspective is deepened by learning how these people cope with hazards such as floods and how they make use of a broad range of knowledge to evaluate risks and develop risk-reduction strategies.

The two chapters by Waitt *et al.* and Stoffel and Graf remind us about the impacts and consequences of processes involving interactions among climate, the cryosphere, volcanoes, and geomorphic systems. Volcanic activity and eruptions that interact with snow and ice can be disastrous for humans and can change landscapes, as for instance the cases of Mount St. Helens and the Iceland volcanic eruptions have shown. Floods resulting from the rapid melt of snow and ice on volcanoes can have extreme magnitudes, comparable or even larger than large glacial lake outburst floods (GLOFs). Eruptions from Nevado del Huila volcano in the Cordillera Central in Colombia in 2007 and 2008 produced extremely large volcanic debris flows (lahars) that reached as far as 150 km downstream, destroying infrastructure and population centers [6]. In the mid 1990s, similar mass flows killed as many as 1000 people. In the case of the 2007 and 2008 events, the death toll was limited to a few people, demonstrating the efficacy of early warning and disaster preparedness efforts that have been undertaken in recent years.

The debris flows in an alpine valley in Switzerland described by Stoffel and Graf are much smaller than the Colombian examples mentioned above, but nevertheless have consequences for vital transportation and tourism infrastructure. Their study provides insights into the effects of local-scale climatic changes, including heavy precipitation and thawing permafrost, on debris flow activity and shows the complexity and diversity of processes that eventually have implications for the local economy.

The consequences and responses of the convergence of the effects of climate, the cryosphere, and local and international economies are highlighted in this book by French *et al.* They analyze how social conflicts over water emerge, evolve, and are resolved. They build on the contributions by Bury, Mark *et al.*, and Diemberger *et al.* by putting water resources in their fundamental local and regional social contexts. In Peru, for example, a large number of social conflicts are related to water resources, underlining the importance of developing and applying more integrated approaches to intersecting cryospheric and global changes [7].

The sections and individual chapters of this book deepen understanding of the wide range of processes, both natural and human, that impact or interact with the high-mountain cryosphere. They should stimulate thinking about the drivers of change and how they converge locally or regionally and produce patterns of risk or opportunities. The need and the challenge to adapt to the pace of change in mountains clearly emerge as a critical recommendation from this book. In AR5, the IPCC emphasizes adaptation in relation to climate change, as it enables societies to reduce and manage risks [1]. The ability of high-mountain social–ecological systems to adjust to the pace and extent of cryospheric change will be tested in the next several decades. This book demonstrates that sustainable adaptation and risk-reduction strategies require a profound understanding of the complex interplay of physical processes and society. A thorough understanding of dynamic environmental, social, cultural, economic, and political forces is imperative both to expand knowledge and adapt to global change.

References

1. IPCC. *Climate Change 2014: Impacts, Adaptation, and Vulnerability. Part A: Global and Sectoral Aspects. Contribution of Working Group II to the Fifth Assessment Report of the Intergovernmental Panel on Climate Change.* Cambridge and New York: Cambridge University Press; 2014.
2. IPCC. *Climate Change 2013: The Physical Science Basis. Contribution of Working Group I to the Fifth Assessment Report of the Intergovernmental Panel on Climate Change.* Cambridge and New York: Cambridge University Press; 2013.
3. T Bolch, A Kulkarni, A Kääb, C Huggel, F Paul, JG Cogley, *et al.* The state and fate of Himalayan glaciers. *Science* 2012;**336**:310–314. doi:10.1126/science.1215828.
4. C Huggel, C Burn, JJ Clague, K Hewitt, A Kääb, M Krautblatter, *et al.* GAPHAZ: improving knowledge management of glacier and permafrost hazards and risks in mountains, *European Geoscience Union, Geophysical Research Abstracts* 2014 16: EGU2014-16568.
5. B Marzeion, JG Cogley, K Richter, D Parkes. Attribution of global glacier mass loss to anthropogenic and natural causes. *Science* 2014;**345**:919–921. doi:10.1126/science.1254702.
6. R Worni, C Huggel, M Stoffel, B Pulgarín. Challenges of modeling current very large lahars at Nevado del Huila Volcano, Colombia. *Bulletin of Volcanology* 2012;**74**:309–324. doi:10.1007/s00445-011-0522-8.
7. M Carey, M Baraer, BG Mark, A French, J Bury, KR Young, *et al.* Toward hydro-social modeling: merging human variables and the social sciences with climate–glacier runoff models (Santa River, Peru). *Journal of Hydrology* 2014;**518**, Part A:60–70. doi:10.1016/j.jhydrol.2013.11.006.

Part I
Global drivers

2

Influence of climate variability and large-scale circulation on the mountain cryosphere

IMTIAZ RANGWALA, NICK PEPIN, MATHIAS VUILLE, AND JAMES MILLER

2.1 Introduction

Climate variability and its association with large-scale atmospheric circulation has a major influence on the transport of moisture and heat from the oceans to the continents on interannual to decadal timescales, thereby affecting snow and ice accumulation in mountains on these same timescales. This internal variability of the Earth's climate system is modulated primarily by processes related to ocean circulation that produce variations in the ocean surface conditions on interannual to interdecadal timescales. These, in turn, are communicated to the overlying atmosphere and help shape the large-scale atmospheric patterns, also referred to as atmospheric teleconnections, that eventually affect the characteristics of large-scale atmospheric transport of moisture and heat into a region.

The climate of a particular mountain region may be sensitive to a specific mode of climate variability. At any given time, however, it could also be influenced by different modes of climate variability that are operating at different phases and strengths. Understanding these relationships is useful for assessing short-term regional climate tendencies, which could help in facilitating short-term climate predictability and preparedness.

In this chapter we discuss the influence of climate variability and large-scale atmospheric circulation, as well as their interactions, on the different mountain cryospheric systems globally. Because the nature of large-scale atmospheric circulation and teleconnections that affect a region's climate is unique to that region, the following sections elaborate on this discussion with a focus on selected high-elevation regions, including (a) the European mountains, (b) the North American Cordillera, (c) the tropical Andes, (d) the Tibetan Plateau and surrounding high-mountain ranges, and (e) Mt. Kilimanjaro in East Africa. For each of

The High-Mountain Cryosphere, ed. Christian Huggel, Mark Carey, John J. Clague and Andreas Kääb. Published by Cambridge University Press. © Cambridge University Press 2015.

these regions the discussion is focused on identifying the important large-scale climate variability and atmospheric circulation that modulates the region's climate. In addition, we briefly discuss the possible influence of anthropogenic climate change on both large-scale and regional processes, as well as possible consequences for the mountain cryosphere.

2.2 European mountains

European climate is controlled by the North Atlantic on the one hand, and the heating/cooling of the European continent on the other. Because of a strong maritime influence due to prevailing onshore westerly winds, coupled with a northward movement of warm water up the eastern side of the Atlantic (the Gulf Stream), winters are particularly mild given the high latitude. Europe has two main mountain systems, the first being the mountains of the Caledonian orogeny, extending from the northwestern UK through Scandinavia. The high latitude means that cryospheric systems are well developed, with the largest glaciers being situated in southern Norway at around 1000–1500 m above sea level in the Jotunheimen National Park, and smaller ice fields farther north [1]. These mountains have a milder climate at lower elevations than may be expected due to the dominant westerly circulation, which strengthens across the North Atlantic as it traverses the warm North Atlantic drift. The western (windward) slopes are subject to an almost constant stream of southwesterly winds, bringing high orographic precipitation totals. However, substantial shelter is provided on the leeward slopes. The eastern slopes in Norway and Sweden have much lower rainfall (as low as 300 mm per year in Abisko, Sweden) and a more continental regime.

The second main mountain system is the 'Alpine' belt, which for the purposes of this chapter will include the Pyrenees (between Spain and France), the Alps, and the Carpathians farther east. In contrast to the Caledonian belt, these mountains run from east to west. Thus they act as a divide, with the storm track either passing to the north (across northern Europe) or through the Mediterranean, more common in the winter. Precipitation in the southern parts of the ranges tends to be lighter in summer, as the Azores High extends its influence. Nevertheless, the combination of steep slopes and high-intensity summer insolation can develop thermal circulations and lead to intense convective rainfall.

The dominant westerly flow in the mid to high latitudes can be measured by the North Atlantic Oscillation (NAO) [2]. The index is based on the normalised pressure difference between Iceland and the Azores. A positive NAO (relatively high pressure near the Azores and low pressure over Iceland) is associated with a stronger westerly flow that is shifted pole-ward, enhancing precipitation in the Caledonian mountains, and also sometimes in the Alpine ranges during winter.

A negative NAO is associated with a suppressed westerly flow that is shifted equator-ward, causing colder and drier conditions in Northern Europe during winter. Relationships between the NAO and mountain climate are generally more predictable in the northern mountains [3] than in the Alpine belt and Mediterranean region [4,5].

The NAO has a strong control on the mountain cryospheric systems; however, the relationship is not straightforward and easily predictable. The high frequency of temperatures near freezing means that winter rain is common even at surprisingly high elevations in all European mountains. Even in the coldest parts of Finnish Lapland (on the lee slopes of the Scandes) it is not unknown for rain to occur in the middle of winter. The result is that the snowpack is extremely irregular at intermediate elevations (500–1000 m) in many European mountains, despite high precipitation totals. A more positive NAO means more snowfall but only above a critical elevation, which is often 2000 m or higher, especially in the more southern mountain ranges. At lower elevations, this increased rainfall will melt any snow that has accumulated. The overall relationship between NAO and snowpack in a mountain basin thus depends on its elevation range. A more negative NAO normally means drier conditions and a higher frequency of anticyclonic weather. Local thermal mountain circulation systems and their attendant phenomena can become well developed. In winter, the radiative loss of heat during the long nights means that katabatic flow can lead to intense temperature inversions. Low temperatures are recorded in the Alpine valleys and in Scandinavia, wherever cold air ponds, and because of high humidity in the maritime climate, fog and low clouds are also common, such that high peaks are exposed to sunlight, while persistent damp cold air stays in the valley.

Relationships between the NAO and local climate are not always strong in all European mountain ranges. In the Caledonian mountains, relationships between surface climate and NAO are most well-defined on the windward slopes. However, in the Alpine belt during summer, the mechanical influence of the jet stream is weak, pressure gradients are often indistinct, and the NAO is of limited relevance.

For the Caledonian belt, most global climate models (GCMs) predict more northerly and stronger storm tracks under climate change [6]. On average, GCMs project warmer (5–7 °C) and wetter (20–30 %) winters over the Fennoscandian continent by 2100 under the intermediate emission scenarios (SRES B2–SRES A2) [7]. While this could mean more snow at higher elevations, the intense warming that is expected will be detrimental for snowpack and glacial systems. Observational studies have suggested that some high-elevation glaciers in the west of Norway were advancing during the late twentieth century because of increased snowfall [8] and a more positive NAO [9]. However, based on climate model projections, we expect a rise in the equilibrium-line altitude (ELA) – the line separating the accumulation from the ablation zone – of the glaciers by 260 ± 50 m, and by 2100 this may cause around 20% of the glaciers in Norway to disappear [1].

In the Alpine belt, most models predict a tendency towards enhanced NAO in winter, and a stronger Mediterranean influence (high pressure) in summer and the transition seasons. Concern over diminishing snow cover conditions in the Alps surfaced in the late 1990s following a series of extremely mild winters [10], and research has also shown that a more positive NAO will preferentially warm higher elevations [11]. Many studies have associated declining snowpack in the Alpine belt with a trend towards a more positive NAO, mainly because of a more amplified warming with elevation and increases in liquid precipitation, as well as a higher snowline associated with increased westerly flow [4], although the influence of circulation on snowpack is more complex [5].

2.3 North American Cordillera

The North American Cordillera covers an extensive area of mountain ranges, inter-montane basins, and plateaus in western North America – from the US state of Alaska to the southern border of Mexico. It includes much of the territory west of the North American Great Plains. Its cryosphere, particularly the seasonal snowpack in the United States and Canada, is of tremendous importance to society and ecosystems. For instance, in most river basins of the western United States, mountain snowpack is the largest component of water storage, and contributes about 75% of annual discharge.

Typically, in the mountains of the western United States and Canada, assessment of the annual snowpack is done by recording the snow water equivalent (SWE) at the observing sites as of April 1. Although there are differences associated with elevation and latitude, most sites attain their climatological maximum SWE at the beginning of April. For any given year, there is a substantial variability in April 1 SWE, which Cayan [12] reported as 25% to 60% of the long-term mean.

Most analyses find that during the cold season (winter and early spring), precipitation has the strongest influence on snowpack variability [12]. Cold-season precipitation variability along the North American Cordillera appears to be stochastic; however, specific spatial patterns of precipitation variability and their relationship with large-scale atmosphere–ocean phenomena do emerge. The multi-year to decadal-scale variability in cold-season precipitation is primarily associated with patterns of variability in the Pacific Ocean – more specifically, the El Niño-Southern Oscillation (ENSO) and the Pacific Decadal Oscillation (PDO) [12–14] These patterns are communicated to the atmosphere through the sea surface temperatures (SSTs). Such air–sea interactions organise the atmospheric circulation – the position of the Northern Hemisphere jet stream and the associated mid-latitude storm tracks – that interacts with the western North American continent. Important aspects of the Pacific climate variability are rooted in the tropics, modulated by the ocean dynamics within the Tropical Pacific thermocline [15,16]. These are most commonly manifested as ENSO, which is the dominant global mode of climate variability on interannual scales [17], but also influences decadal-scale variability [13,15,16]. The influence of variability in the Atlantic Ocean, through the Atlantic Multidecadal Oscillation (AMO), has also been suggested to be relevant for precipitation and temperature variability in this region [18–20].

Snowpack variation in the western United States has a strong interannual and a modest inter-decadal character. It has been suggested that combining PDO and ENSO may enhance the accuracy of North American snowpack forecasts [21]. One study finds that the principal component that explains 45% of the variability in April 1 SWE for the western United States is highly correlated with the PDO [21]. The PDO is a dominant pattern of North Pacific SST anomalies and exhibits variability on both interannual and decadal timescales [15]; however, these fluctuations are most energetic in low-frequency periods: (a) 15–25 years and (b) 50–70 years [22]. More importantly, however, we do not have a good understanding of the dynamical causes for PDO [15,22], although there is evidence that the cause for PDO variability originates in the tropics, and that it is dependent on ENSO on all timescales [16]. Furthermore, random combinations of these and other stochastic processes can give rise to regime shifts in the PDO that markedly affect our ability for long-term (multi-year to decadal) prediction of this phenomenon [15].

Another interesting aspect of the cold-season precipitation variability along the North American Cordillera is the opposite sign of precipitation anomalies primarily as a function of latitude. The variability in the mass balance of northern (Alaska) and southern (British Columbia and Washington) glaciers are out of phase during periods of an enhanced Pacific–North American (PNA) circulation pattern, which is the most prominent mode of low-frequency variability in the Northern Hemisphere extratropics [13]. This pattern, which is associated with intensification of low pressure over the Aleutian Islands and the southeastern United States, and high pressure in the vicinity of Hawaii, and the northwestern United States and British Columbia, facilitates southwesterly flow over the eastern North Pacific, shifts in cyclonic systems to a more northerly storm track, and generally lower levels of storminess. This results in below average snow and ice accumulation over British Columbia and Washington, while the opposite is true for Alaska [13,14]. During the deep Aleutian low phase of the PNA, it is predominantly dry over much of the western United States [12]. The extreme phases of the PNA are also found to be related to the decadal ENSO-like phenomenon [13].

Further south along the North American Cordillera, ENSO has a much stronger and more direct influence on the snowpack variability. Generally, the warm phase of ENSO (El Niño) is associated with a lower snowpack in the Pacific Northwest and northern Rockies, and a higher snowpack in the Sierra Nevada and mountains of Utah, southern Colorado, Arizona and New Mexico. However, the opposite case is more pronounced during the cold phase of ENSO (La Niña) [12].

In the Upper Colorado River Basin (UCRB), which contributes 90% of the water to the Colorado River system, 85% of the water originates in the mountain snowpack. Here, the snowpack is above average during La Niña years, warm phase of the PDO, and the negative (cool) phase of the AMO [18]. Furthermore, the effects of ENSO and PDO are intensified when in phase and suppressed when out of phase [23]. Nowak and co-authors [19] found two significant modes of variability in the annual runoff from the UCRB: a strong decadal (~15 years) and a low frequency (~64 years) mode. The decadal mode was found to be strongly tied to moisture delivery in the region, as well as to SSTs in the equatorial and tropical Pacific. The lower frequency mode, on the other hand, is related to the temperature response in UCRB, and linked to the AMO. Research suggests that the AMO tends to modulate hemispheric temperature – where a warm phase of the AMO (positive SST anomalies in the North Atlantic) will tend to produce increases in Northern Hemispheric temperatures [24]. Furthermore, a model study by Schubert and co-authors [20] shows that the largest precipitation response over the continental United States tends to occur when the two oceans have anomalies of opposite signs – a cold Pacific and warm Atlantic tend to produce the largest precipitation reductions, whereas a warm Pacific and cold Atlantic tend to produce the greatest precipitation enhancements.

2.4 Tibetan Plateau and the surrounding high-mountain ranges

The Tibetan Plateau and the surrounding mountain system covers a very large area of high-elevation terrain between south and central Asia, and is referred to as the third pole because of the vast expanse of snow and ice found there. Some of the major mountain systems include the Himalayas, Karakoram, Hindu Kush, Pamir, and Tien Shan. The moisture flux in these regions is primarily associated with mid-latitude cyclones and the Asian Monsoon. However, the influence of these large-scale circulation processes varies depending on the region's latitudinal positioning, proximity to the ocean and topographical positioning relative to each other.

The dominant large-scale atmospheric processes that affect the climate of the Tibetan Plateau and surrounding regions vary from east to west. On the eastern and southern flanks, monsoon circulations are important contributors to both the thermal and hydrologic regimes. The southeastern part of the Plateau is on the windward slope of the summer monsoon, while the massif on the northern slope of the Himalayas is on its leeward slope [25]. To the west of the Plateau, the westerlies play a more prominent role where the massifs of the central Tien Shan are exposed to western airstreams. In this western region, the Tien Shan and Pamir mountain ranges are the headwaters of rivers that drain into the Aral Sea. Studies of lake water mineralisation suggest that the snow cover in West Central Asia is a winter expression that is controlled by temperature patterns and the moisture-loading capacity over northern Europe and northwest Asia, with more power residing in low-frequency variability during the last 200 years [26]. Isotopic studies of rain, snow, and ice in this same region found that for the central Tien Shan, the moisture originates to the west over the Caspian or Mediterranean seas, while moisture reaching the northern slope of the Himalayas and the southeastern Himalayas is derived from the Indian and Pacific Oceans and, to a lesser extent, from the Atlantic, through both monsoonal circulations as well as extratropical cyclones [25]. In the western Himalayas, one-third of the annual precipitation occurs in the winter owing to the Indian winter monsoon [27].

There has long been evidence of linkages between ENSO and the Indian monsoon [28]. Stronger monsoons are generally associated with La Niña, and El Niño tends to suppress the monsoons [29], although this relationship is not straightforward (Figure 2.1). Kumar and co-authors [30] found that those El Niño events where the warmest SST anomalies are in the central equatorial Pacific have

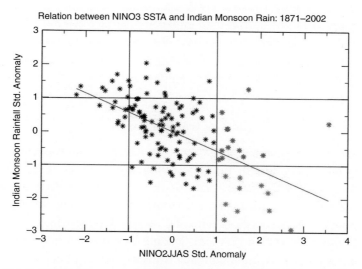

Figure 2.1 Indian monsoon and ENSO. Plot of standardised, all-India summer (June to September (JJAS)) monsoon rainfall and summer NINO3 anomaly index. Severe drought and drought-free years during El Niño events (standardised NINO3 anomalies > 1) are shown in red and green, respectively. Reproduced from [30].

a greater predictability for Indian monsoon suppression. There have been inter-decadal changes in both the variance and coherence between monsoons and ENSO during the last 125 years [28]. However, long-term prediction for the Asian monsoons is very difficult. GCMs reveal major differences between model predictions of monsoon rainfall and the Southern Oscillation Index (SOI). A majority of GCMs simulate enhanced monsoons with La Niña-like conditions although the significance is barely above natural variability. Moreover, GCMs project warming in the eastern Pacific that would change the background state but not necessarily lead to more or fewer ENSO events [31]. The reliability of long-term predictions in this region depends on the ability of global models to improve their representation of tropical circulation, including monsoonal and ENSO features, and more importantly their relationship to each other.

Absorbing aerosols such as dust and black carbon have been suggested to both influence large-scale circulation as well as have a direct effect on mountain cryosphere in the Himalayas and Tibetan Plateau. These aerosols affect the surface radiation budget by absorbing radiation in the troposphere, thus heating the air. Black carbon emissions have been suggested to account for half of the total warming in the Himalayas during the last several decades [32]. Furthermore, the mid to upper tropospheric warming caused by absorbing aerosols over South Asia could intensify the Indian monsoon by creating a warm-core upper-level anticyclonic condition over the Tibetan Plateau in April–May [33]. When deposited on snow, the absorbing aerosols lower the surface albedo, thus increasing the radiation absorbed by the surface and enhancing surface warming. Atmospherically deposited black carbon can increase the absorption of visible radiation by 10–100% in the Tibetan glaciers, and has been a significant contributing factor to the observed rapid glacier retreat in the region [34].

2.5 The tropical Andes

Glaciers can be found along the entire tropical and subtropical Andes from Colombia to northern Chile. In Venezuela only one small ice field, with a size of 0.1 km^2, remains [35]. The largest fraction of ice persists in Peru, in particular in the densely glaciated Cordillera Blanca. South of the Equator the Andes receive their precipitation in association with the seasonal development of the South American Summer Monsoon (SASM), leading to a pronounced wet season that extends from October to April in the inner tropics (southern Ecuador, northern and central Peru) and from December through March in the southern tropical Andes (southern Peru, Bolivia and northernmost Chile). To the north, in northernmost Ecuador and in Colombia, precipitation is more pronounced in the boreal summer. Despite the proximity to the Pacific, the main moisture source for Andean glaciers lies to the east in the Amazon basin, and in the case of Colombia also the Caribbean. Moisture transport occurs mostly at mid and low levels of the atmosphere, but it is regulated by the direction of the zonal wind in the upper troposphere, through downward mixing of momentum over the Andean ridge [36]. Indeed, mass balance records from Bolivia, Peru, and Ecuador all show a statistically significant correlation with the strength of the upper-level zonal wind field, with strong easterlies during the wet season leading to a more positive glacier mass balance [37,38] (Figure 2.2). During the austral winter westerly winds dominate in the upper troposphere, essentially inhibiting any moisture influx from the east. As a result, the tropical Andes south of the Equator experience a pronounced dry season, lasting for 6–8 months.

On interannual timescales this seasonal cycle is significantly affected by the state of the tropical Pacific. Warming in this region during El Niño years leads to enhanced meridional baroclinicity between low and mid latitudes to the east and aloft in the Andes, resulting in an enhanced subtropical jet and strengthened westerly winds over the Andes [39]. As a result, precipitation amounts over the tropical Andes tend to be reduced, following a similar mechanism that causes aridity during austral winter on seasonal timescales. During La Niña events, when equatorial Pacific SST tends to be anomalously cold, this pattern is essentially reversed, and enhanced upper-level easterly winds lead to an enhanced moisture influx and wet conditions with abundant snowfall in the Andes [36]. Temperature in the region is also significantly affected by ENSO, with the entire region experiencing above average temperature during El Niño and significant cooling during La Niña years [40].

All these processes affect the glaciers in the region in manifold ways. Precipitation and temperature are both relevant for glacier mass balance, but their relative importance varies latitudinally. In the inner tropics temperature appears to be more dominant, as it affects the elevation of the snowline, and higher temperature during El Niño implies increased likelihood of rain on the glacier ablation zone [37]. In the subtropical Andes of southern Peru and Bolivia, however, precipitation is generally considered to be more relevant [38] as many glaciers are moisture-starved, especially in the western Cordillera where the ELA of glaciers tends to lie several hundred metres above the atmospheric freezing level. Here a reduced or delayed monsoon season will lead to high ablation rates and very negative mass balance as the low albedo of the dirty glacier surface at the end of the dry season leads to high absorption rates of solar radiation and hence to enhanced melt [41].

While the relationship between the interannual variability in the large-scale circulation and glacier mass balance has been the focus of extensive research, still very little is known about the relevance of atmospheric forcing on decadal to multidecadal timescales. The main reason for this is the lack of long mass balance

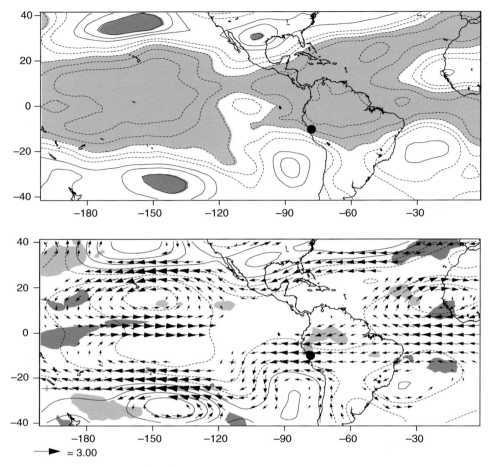

Figure 2.2 Top: correlation between annual (October–September) Cordillera Blanca mass balance time series and austral summer (ONDJFMA) 500 hPa temperature, with significant ($p < 0.05$) positive (negative) correlations shaded in dark (light) grey. Contour interval is 0.1, negative contours are dashed and 0-contour is omitted. Bottom: Same as above but for regression of 250 hPa wind and geopotential height and correlation with outgoing longwave radiation, OLR (1974–1993 only, with 1977–1978 missing). Wind vectors are only plotted where correlation of mass balance with either zonal or meridional component is significant at $p < 0.05$. Scale for wind vector (m/s per std. deviation) is shown in lower left. Contour interval for geopotential height is 4 gpm per std. deviation; negative contours are dashed. Light (dark) grey shading indicates significant ($p < 0.05$) negative (positive) correlation with OLR. Black dots indicate location of Cordillera Blanca. Figure modified from [38].

records from this region, where stake networks were only installed in the early part of the 1990s. Hence, most glacier data sets are too short for a rigid assessment of the influence of modes of variability that operate on decadal or longer timescales. To circumvent this problem Vuille and co-authors [38] used a synthetic 40-year mass balance time series that was reconstructed from hydrologic mass balance modelling. Their results show a clear multi-decadal modulation of the mass balance by ENSO, with a very strong link in the 1950s, 1960s, and early 1970s, but essentially no relationship thereafter. Although they did not attribute this breakdown in the relationship with ENSO to a specific cause, both the AMO and PDO are suspected to significantly affect and modulate the observed ENSO influence on tropical Andean glaciers. This is reasonable, as the AMO is known to influence the intensity of the SASM [42], which in turn is responsible for the moisture delivery to the high Andes. The positive phase (warm North Atlantic SSTs) of the AMO is associated with a northward withdrawal of the Intertropical Convergence Zone (ITCZ), reduced moisture influx to the tropical continent and hence reduced monsoon precipitation [43]. The opposite case occurs during the negative phase of the AMO (cold North Atlantic SSTs). The influence of the PDO on precipitation is rather weak, but it significantly affects temperature on decadal to multi-decadal timescales, with the positive phase of the PDO leading to warmer and the negative phase to cooler conditions in the tropical Andes [44]. Indeed the 1976/1977 Pacific climate shift that accompanied the switch from negative to positive phase of the PDO stands out as a prominent and persistent increase in Andean temperature records [45]. Similarly, the recent shift of the PDO to cooler conditions appears to have slowed down warming rates in the last decade, at least in parts of the Andes.

Superimposed on the interannual to decadal-scale mass balance variability is a long-term negative trend in the cumulative mass balance, which has led to rapid retreat of glaciers throughout the tropical Andes over the past decades. Since there is no clear and coherent decrease in precipitation discernible throughout the entire region, the observed twentieth-century increase in temperature [46] and the associated rise in the freezing level [47] is generally considered to be the main cause for the observed shrinking of the Andean cryosphere [45,47]. Future projections, which imply continued warming in the Andes, irrespective of scenario or concentration pathway, suggest continued retreat and eventual disappearance of at least the lower elevation glaciers [48], as has already happened in a few cases [49].

2.6 Mt. Kilimanjaro: a case study from East Africa

The high mountains of East Africa are volcanic outliers in a broader lowland landscape. The isolated summits reach over 5000 m in Kenya and Tanzania, and this is high enough to support cryospheric systems, albeit on a small scale. Because of the equatorial latitudes, there are no pronounced warm and cold seasons, and seasonality is driven by changes in moisture and the movement of the ITCZ.

The 'melting snows' of Kilimanjaro made famous by Ernest Hemingway have attracted interest ever since their existence was the subject of intense debate by

Figure 2.3 Ice and snow fields at the summit of Mt. Kilimanjaro.
(Source: Nick Pepin)

Rebmann and Cooley in the nineteenth century [50]. The roof of Africa stands at
5895 m above sea level, well above the atmospheric boundary layer, and therefore
has an extremely arid climate (Figure 2.3). The moisture in the tropical trade winds
is concentrated below 3000 m, and this lower elevation heavy moisture distribution
is further accentuated by a temperature inversion at this level which suppresses
convection much of the time. Therefore, as on most tropical mountains, most
rainfall falls on the lower slopes around 2000–3000 m [51]. In the wettest locales
on the southern slopes, the rainforest forms a continuous belt. Below this, the lower
slopes have been cleared for agriculture, with the main crops being banana, maize,
and coffee. Here, the population density is also among the highest in Tanzania.
Above the forest belt, however, the climate grows steadily drier with increasing
elevation, moving from a giant heather belt (3000–4000 m) through alpine moor-
land to a lunar rocky landscape with no vegetation above 5000 m. The reverse
orographic effect on precipitation often comes as a surprise to mid-latitude visitors
who are familiar with the role of orographic enhancement of frontal precipitation.
In the tropics there are limited air–mass contrasts and no fronts, and the isobaric
atmosphere means that the air–mass advection is of limited relevance. Thus a
traditional 'atmospheric circulation' approach has less of a role in explaining
high-mountain climate in the tropics than is the case in mid-latitudes.

Much research has shown that the summit ice fields and glaciers have shrunk considerably over the last century [52], with predictions of their demise now thought to be around 2040 [53]. Tropical glaciers show complex relationships with climate [54]. For Kilimanjaro, it appears that precipitation is more important than temperature in controlling the glacier mass-balance [55]. Any assertion, therefore, that the melting snows on Kilimanjaro are solely a result of atmospheric warming is misleading. It is clear that drying has led to increased sublimation in the intense equatorial sunlight, coupled with a decrease in albedo and a lack of replenishment of the ice through precipitation. The causes of this drying are a focus of debate, especially since they are inconsistent with many model predictions for East Africa [6]. Changes in SSTs in the Indian Ocean (possibly connected with anthropogenic influences on global climate) are thought to be at least partly responsible [56]. ENSO is also known to influence precipitation in East Africa, with El Niño events leading to wetter conditions, particularly in the short rainy season (November–January) [57] and, more recently, a deep persistent snow cover on the summit crater in 2006 [58]. Land-use change (deforestation) may also cause a drying of the local climate, but current research has so far been inconclusive [59,60].

2.7 Conclusions

In this chapter we discussed the influence of interannual to interdecadal climate variability and their interactions with large-scale atmospheric circulation in controlling the cryospheric system in selective mountain regions (Table 2.1). In general, climate variability manifested by specific anomaly patterns in ocean surface temperature and associated atmospheric teleconnections modulates the strength, position, and direction of the relevant large-scale transport of moisture and heat to a specific mountain region. However, at any given time it is usually difficult to predict the regional climate response based on simple associations because in several cases multiple modes of variability are operating at the same time. Furthermore, to obtain a particular regional climate response, some threshold strength associated with a specific mode of variability may be required, e.g. a strong wetting tendency for the southwestern United States under a canonical El Niño. Lastly, random fluctuations in large-scale atmospheric and ocean circulation are quite common, and therefore, have a strong effect on our ability to predict.

Owing to our limited understanding of the mechanisms associated with the different aspects of climate variability, particularly those that exhibit lower frequency modes, as well as our limitations in simulating these processes and teleconnections in our climate models [17], the time limits of current predictive

Table 2.1 *Important components of climate variability associated with different mountain regions covered in this chapter. These relationships should only be used to infer possible climate tendencies based on the current state of large-scale ocean–atmosphere conditions. For any given year, these relationships may have limited to no predictability.*

Regions	Important components of climate variability
European mountains	
Caledonian mountains (Northern Europe)	**+NAO** → Above average snow above a critical elevation, increased liquid precipitation at low elevations
	−NAO → Colder and drier; higher frequency of anti-cyclonic conditions (relationship stronger on the windward slopes)
Southern mountains (Alpine belt)	**+NAO** → Drier and warmer
	−NAO → Increased storm activity and rainfall (relationship between NAO and mountain climate more predictable in northern mountains)
North American Cordillera	
Alaska	**PNA** → Above average snow/ice accumulation (PNA linked to decadal-like ENSO pattern)
	+AO → Above average snow/ice accumulation
Pacific Northwest (BC, WA, OR)	**PNA** → Below average snow/ice accumulation
	La Niña → Above average snowpack (opposite during El Niño)
	+PDO → Below average snowpack
Sierra Nevada	**El Niño** → Above average snowpack (strong relationship)
	+PDO → Below average snowpack at lower elevation and its northern reaches
Upper Colorado River Basin	**La Niña** → Higher winter precipitation anomalies in mountains
	El Niño → Higher precipitation anomalies in most of the region and in all seasons
	+PDO → Above average snowpack
	+AMO → Warmer and drier
Tibetan Plateau and surrounding mountains	
Northwest Tibetan Plateau, Tien Shan, Pamir, and Karakorum	**Mid-latitude cyclones** → Primary moisture source (influenced by the position of the jet stream)
Southeast Tibetan Plateau and Himalayas	**Asian monsoon** → Primary moisture source (influenced by ENSO and Eurasian snow cover variability, among other factors)
Tropical Andes	**El Niño** → Warmer and drier
	La Niña → Colder and wetter
	PDO → Weak effect on precipitation; however +PDO causes warmer conditions and −PDO cooler
	+AMO/−AMO → Reduction/enhancement in SASM precipitation (through ITCZ displacement and reduced/enhanced moisture flux)

Table 2.1 (*cont.*)

Regions	Important components of climate variability
Mt. Kilimanjaro	**Indian Ocean SSTs** → Modulate moisture flux **El Niño** → Wetter in East Africa (precipitation is a more critical driver of the region's cryospheric system than temperature)

PNA = Pacific-North America Pattern, AO = Arctic Oscillation, PDO = Pacific Decadal Oscillation, AMO = Atlantic Multidecadal Oscillation, ENSO = El Niño Southern Oscillation

skills are much shorter than the apparent timescale of oscillations. In the case of ENSO events, for example, which have return times of 2–7 years, we have limited confidence in the forecast beyond a few months. Nonetheless, our understanding of the decadal-scale climate variability and its regional teleconnections continues to improve and has been useful in some areas of prediction. For example, there is some predictive skill with 3–6 months of lead time for the phase of ENSO. By incorporating this understanding and better integrating the different components of global climate variability and their influence on the regional scale, we have the potential to improve the nature of short-term forecasts and preparedness [19].

How will climate change affect the character of the different modes of climate variability? This issue is far from resolved and much research is underway to constrain this better. One analysis of projections from the latest generation of global climate models (CMIP5) suggests that the frequency of the canonical El Niño episodes, analogous to those during 1982/1983 and 1997/1998, will double during the twenty-first century [61], with implications for increases in the frequency of extreme weather events worldwide. However, we need to be cautious when incorporating such information because of the important deficiencies that still exist in the current generation of models to realistically simulate a phenomenon such as ENSO, in addition to other components of climate variability [17]. Furthermore, there are large differences in the nature (amplitude and frequency) of simulated internal variability among these GCMs [62]. Other unresolved issues include (a) changes and shifts in large-scale circulation patterns, such as latitudinal shifts in mid-latitude storm tracks in the Northern Hemisphere, and (b) slowing down of the Northern Hemispheric jet stream because of amplified warming in the Arctic resulting in greater persistence of a particular weather phenomenon, facilitating an increased propensity for extreme weather [63].

The direct influence of atmospheric warming on the mountain cryosphere from increases in anthropogenic greenhouse gas forcing during the twenty-first century will become more and more important, driving the regional climate beyond the envelope of natural variability, and as a consequence significantly changing the nature of the mountain cryosphere. Significant shifts in the atmospheric freezing level are expected from climate change. This would imply a shift from a snow-dominated to rain-dominated regime which would significantly diminish the cyrospheric systems at lower elevations and along the coastal mountains. In higher-elevation and higher-latitude regions, where temperatures remain below freezing and increases in precipitation occur, there would be higher than average snow and ice. For the mid- and high-latitude regions, there would be a strong seasonal response to cryospheric changes in mountains which would include (a) a later start of snow and ice accumulation in autumn, and (b) an earlier melt and transition to rain in spring. These climatic changes will reshape ecosystem function and structure, as well as the different hydrologic parameters in mountains. Nevertheless, even under a severe climate change scenario, climate variability will continue to influence inter-annual to inter-decadal changes and associated climate extremes in the mountain cryosphere globally during the twenty-first century.

Acknowledgements

IR acknowledges insightful discussions and comments from Joe Barsugli (U. Colorado).

References

1. A Nesje, J Bakke, SO Dahl, Ø Lie, JA Matthews. Norwegian mountain glaciers in the past, present and future. *Global and Planetary Change* 2008;**60**(1):10–27.
2. JW Hurrell. Decadal trends in the North Atlantic Oscillation. *Science* 1995;**269**: 676–679.
3. P Imhof, A Nesje, SU Nussbaumer. Climate and glacier fluctuations at Jostedalsbreen and Folgefonna, southwestern Norway and in the western Alps from the 'Little Ice Age' until the present: the influence of the North Atlantic Oscillation. *The Holocene* 2012;**22**(2):235–247.
4. J López-Moreno, S Vicente-Serrano, E Morán-Tejeda, J Lorenzo-Lacruz, A Kenawy, M Beniston. Effects of the North Atlantic Oscillation (NAO) on combined temperature and precipitation winter modes in the Mediterranean mountains: observed relationships and projections for the 21st century. *Global and Planetary Change* 2011;**77**(1): 62–76.
5. SC Scherrer, C Appenzeller. Swiss Alpine snow pack variability: major patterns and links to local climate and large-scale flow. *Climate Research* 2006;**32**(3):187–199.
6. IPCC. Annex I: atlas of global and regional climate projections. In: *Climate Change 2013: The Physical Science Basis. Contribution of Working Group I to the Fifth Assessment Report of the Intergovernmental Panel on Climate Change*, TF Stocker, D Qin, G-K Plattner, M Tignor, SK Allen, J Boschung, A Nauels, Y Xia, V Bex and PM Midgley (eds). Cambridge and New York: Cambridge University Press; 2013.
7. K Jylhä, H Tuomenvirta, K Ruosteenoja. Climate change projections for Finland during the 21st century. *Boreal Environment Research* 2004;**9**(2):127–152.

8. A Nesje, E Jansen, HJB Birks, AE Bjune, J Bakke, C Andersson, *et al.* Holocene climate variability in the northern North Atlantic region: a review of terrestrial and marine evidence. *Geophysical Monograph Series* 2005;**158**:289–322.
9. A Nesje, Ø Lie, SO Dahl. Is the North Atlantic Oscillation reflected in Scandinavian glacier mass balance records? *Journal of Quaternary Science* 2000;**15**(6):587–601.
10. M Laternser, M Schneebeli. Long-term snow climate trends of the Swiss Alps (1931–99). *International Journal of Climatology* 2003;**23**(7):733–750.
11. M Beniston. Variations of snow depth and duration in the Swiss Alps over the last 50 years: links to changes in large-scale climatic forcings. *Climatic Change.* 1997; **36**(3–4):281–300.
12. DR Cayan. Interannual climate variability and snowpack in the western United States. *Journal of Climate* 1996;**9**(5):928–948.
13. C Bitz, D Battisti. Interannual to decadal variability in climate and the glacier mass balance in Washington, western Canada, and Alaska. *Journal of Climate* 1999;**12**(11): 3181–3196.
14. RD Moore, IG McKendry. Spring snowpack anomaly patterns and winter climatic variability, British Columbia, Canada. *Water Resources Research* 1996;**32**(3):623–632.
15. MA Alexander. Extratropical air–sea interaction, SST variability, and the Pacific decadal oscillation (PDO). In: *Climate Dynamics: Why Does Climate Vary*, D Sun, F Bryan (eds). Washington, DC: American Geophysical Union; 2010. pp. 123–148.
16. M Newman, GP Compo, MA Alexander. ENSO-forced variability of the Pacific Decadal Oscillation. *Journal of Climate* 2003;**16**(23):3853–3857.
17. G Flato, J Marotzke, B Abiodun, P Braconnot, S Chou, W Collins, *et al.* Evaluation of climate models. In: *Climate Change 2013: The Physical Science Basis. Contribution of Working Group I to the Fifth Assessment Report of the Intergovernmental Panel on Climate Change*, TF Stocker, D Qin, GK Plattner, M Tignor, SK Allen, J Boschung, A Nauels, Y Xia, V Bex, PM Midgley (eds). Cambridge and New York: Cambridge University Press; 2013.
18. T Hunter, G Tootle, T Piechota. Oceanic–atmospheric variability and western U.S. snowfall. *Geophysical Research Letters* 2006;**33**:L13706.
19. K Nowak, M Hoerling, B Rajagopalan, E Zagona. Colorado river basin hydroclimatic variability. *Journal of Climate* 2012;**25**(12):4389–4403.
20. S Schubert, D Gutzler, H Wang, A Dai, T Delworth, C Deser, *et al.* A US CLIVAR project to assess and compare the responses of global climate models to drought-related SST forcing patterns: overview and results. *Journal of Climate* 2009;**22**(19): 5251–5272.
21. GJ McCabe, MD Dettinger. Primary modes and predictability of year-to-year snow-pack variations in the western United States from teleconnections with Pacific Ocean climate. *Journal of Hydrometeorology* 2002;**3**(1):13–25.
22. NJ Mantua, SR Hare. The Pacific decadal oscillation. *Journal of Oceanography.* 2002;**58**(1):35–44.
23. J Timilsena, T Piechota, G Tootle, A Singh. Associations of interdecadal/interannual climate variability and long-term Colorado River basin streamflow. *Journal of Hydrology* 2009;**365**(3–4):289–301.
24. R Zhang, TL Delworth, IM Held. Can the Atlantic Ocean drive the observed multi-decadal variability in Northern Hemisphere mean temperature? *Geophysical Research Letters* 2007;**34**:L02709.
25. V Aizen, E Aizen, J Melack, T Martma. Isotopic measurements of precipitation on central Asian glaciers (southeastern Tibet, northern Himalayas, central Tien Shan). *Journal of Geophysical Research: Atmospheres (1984–2012)* 1996;**101**(D4):9185–9196.

26. H Oberhänsli, K Novotná, A Píšková, S Chabrillat, DK Nourgaliev, AK Kurbaniyazov, *et al.* Variability in precipitation, temperature and river runoff in W Central Asia during the past ~2000yrs. *Global and Planetary Change* 2011;**76**(1):95–104.

27. A Dimri. Sub-seasonal interannual variability associated with the excess and deficit Indian winter monsoon over the Western Himalayas. *Climate Dynamics* 2013:1–13.

28. C Torrence, PJ Webster. The annual cycle of persistence in the El Nño/Southern Oscillation. *Quarterly Journal of the Royal Meteorological Society.* 1998;**124**(550): 1985–2004.

29. PJ Webster. The annual cycle and the predictability of the tropical coupled ocean–atmosphere system. *Meteorology and Atmospheric Physics* 1995;**56**(1–2):33–55.

30. KK Kumar, B Rajagopalan, M Hoerling, G Bates, M Cane. Unraveling the mystery of Indian monsoon failure during El Niño. *Science* 2006;**314**(5796):115–119.

31. H Paeth, A Scholten, P Friederichs, A Hense. Uncertainties in climate change prediction: El Niño–Southern Oscillation and monsoons. *Global and Planetary Change* 2008;**60**(3):265–288.

32. V Ramanathan, G Carmichael. Global and regional climate changes due to black carbon. *Nature Geoscience* 2008;**1**(4):221–227.

33. K Lau, M Kim, K Kim. Asian summer monsoon anomalies induced by aerosol direct forcing: the role of the Tibetan Plateau. *Climate Dynamics* 2006;**26**(7–8): 855–864.

34. B Xu, J Cao, J Hansen, T Yao, DR Joswia, N Wang, *et al.* Black soot and the survival of Tibetan glaciers. *Proceedings of the National Academy of Sciences* 2009;**106**(52): 22114–22118.

35. C Braun, M Bezada. The history and disappearance of glaciers in Venezuela. *Journal of Latin American Geography* 2013;**12**(2):85–124.

36. R Garreaud, M Vuille, AC Clement. The climate of the Altiplano: observed current conditions and mechanisms of past changes. *Palaeogeography, Palaeoclimatology, Palaeoecology* 2003;**194**(1):5–22.

37. B Francou, M Vuille, V Favier, B Cáceres. New evidence for an ENSO impact on low-latitude glaciers: Antizana 15, Andes of Ecuador, 0 28′ S. *Journal of Geophysical Research: Atmospheres (1984–2012)* 2004;**109**:D18106.

38. M Vuille, G Kaser, I Juen. Glacier mass balance variability in the Cordillera Blanca, Peru and its relationship with climate and the large-scale circulation. *Global and Planetary Change* 2008;**62**(1):14–28.

39. R Garreaud, P Aceituno. Interannual rainfall variability over the South American Altiplano. *Journal of Climate.* 2001;**14**(12):2779–2789.

40. M Vuille, RS Bradley, F Keimig. Interannual climate variability in the Central Andes and its relation to tropical Pacific and Atlantic forcing. *Journal of Geophysical Research: Atmospheres (1984–2012)* 2000;**105**(D10):12447–12460.

41. P Wagnon, P Ribstein, B Francou, J-E Sicart. Anomalous heat and mass budget of Glaciar Zongo, Bolivia, during the 1997/98 El Niño year. *Journal of Glaciology* 2001;**47**(156):21–28.

42. MT Kayano, VB Capistrano. How the Atlantic multidecadal oscillation (AMO) modifies the ENSO influence on the South American rainfall. *International Journal of Climatology* 2014;**34**(1):162–178.

43. CM Chiessi, S Mulitza, J Pätzold, G Wefer, JA Marengo. Possible impact of the Atlantic Multidecadal Oscillation on the South American summer monsoon. *Geophysical Research Letters* 2009;**36**(21). DOI: 10.1029/2009GL039914

44. RD Garreaud, M Vuille, R Compagnucci, J Marengo. Present-day South American climate. *Palaeogeography, Palaeoclimatology, Palaeoecology.* 2009;**281**(3):180–195.

45. M Vuille, B Francou, P Wagnon, I Juen, G Kaser, BG Mark, *et al.* Climate change and tropical Andean glaciers: past, present and future. *Earth-Science Reviews* 2008;**89**(3): 79–96.

46. M Vuille, RS Bradley, M Werner, F Keimig. 20th century climate change in the tropical Andes: Observations and model results. *Climatic Change* 2003;**59**(1–2):75–99.

47. A Rabatel, B Francou, A Soruco, J Gomez, B Cáceres, J Ceballos, *et al.* Current state of glaciers in the tropical Andes: a multi-century perspective on glacier evolution and climate change. *Cryosphere* 2013;**7**(1):81–102.

48. A Soruco, C Vincent, B Francou, JF Gonzalez. Glacier decline between 1963 and 2006 in the Cordillera Real, Bolivia. *Geophysical Research Letters.* 2009;**36**:L03502.

49. E Ramirez, B Francou, P Ribstein, M Descloitres, R Guerin, J Mendoza, *et al.* Small glaciers disappearing in the tropical Andes: a case-study in Bolivia – Glaciar Chacal-taya (16 S). *Journal of Glaciology* 2001;**47**(157):187–194.

50. WD Cooley. Letter to the editor (on Kilimanjaro). *Athenaeum.* **1849**;1125:516–517.

51. PC Røhr, Å Killingtveit. Rainfall distribution on the slopes of Mt. Kilimanjaro. *Hydrological Sciences Journal* 2003;**48**(1):65–77.

52. LG Thompson, E Mosley-Thompson, ME Davis, KA Henderson, HH Brecher, VS Zagorodnov, *et al.* Kilimanjaro ice core records: evidence of Holocene climate change in tropical Africa. *Science* 2002;**298**(5593):589–593.

53. N Cullen, P Sirguey, T Mölg, G Kaser, M Winkler, S Fitzsimons. A century of ice retreat on Kilimanjaro: the mapping reloaded. *Cryosphere* 2013;**7**(2):419–431.

54. J Oerlemans. *Glaciers and Climate Change.* Lisse , PA: A.A. Balkema Publishers; 2001.

55. T Mölg, JC Chiang, A Gohm, NJ Cullen. Temporal precipitation variability versus altitude on a tropical high mountain: observations and mesoscale atmospheric modelling. *Quarterly Journal of the Royal Meteorological Society.* 2009;**135**(643): 1439–1455.

56. T Mölg, M Renold, M Vuille, NJ Cullen, TF Stocker, G Kaser. Indian Ocean zonal mode activity in a multicentury integration of a coupled AOGCM consistent with climate proxy data. *Geophysical Research Letters* 2006;**33**:L18710.

57. G Kaser, T Mölg, NJ Cullen, DR Hardy, M Winkler. Is the decline of ice on Kilimanjaro unprecedented in the Holocene? *The Holocene* 2010;**20**(7):1079–1091.

58. M Latif, D Dommenget, M Dima, A Grötzner. The role of Indian Ocean sea surface temperature in forcing east African rainfall anomalies during December–January 1997/98. *Journal of Climate* 1999;**12**(12):3497–3504.

59. JG Fairman, US Nair, SA Christopher, T Mölg. Land use change impacts on regional climate over Kilimanjaro. *Journal of Geophysical Research: Atmospheres (1984–2012)* 2011;**116**:D03110.

60. N Pepin, W Duane, D Hardy. The montane circulation on Kilimanjaro, Tanzania and its relevance for the summit ice fields: comparison of surface mountain climate with equivalent reanalysis parameters. *Global and Planetary Change* 2010;**74**(2):61–75.

61. W Cai, S Borlace, M Lengaigne, P van Rensch, M Collins, G Vecchi, *et al.* Increasing frequency of extreme El Niño events due to greenhouse warming. *Nature Climate Change* 2014;**4**:111–116.

62. E Hawkins, R Sutton. Time of emergence of climate signals. *Geophysical Research Letters* 2012;**39**(1). DOI: 10.1029/2011GL050087

63. JA Francis, SJ Vavrus. Evidence linking Arctic amplification to extreme weather in mid-latitudes. *Geophysical Research Letters* 2012;**39**:L06801.

3

Temperature, precipitation and related extremes in mountain areas

NADINE SALZMANN, SIMON C. SCHERRER, SIMON ALLEN AND
MARIO ROHRER

3.1 Introduction

The globally observed retreat and vanishing of mountain glaciers, the decrease of snow cover extent and the rise of snow line altitude are among the best visible indicators for ongoing climatic changes. As for snow and ice, the most important atmospheric variables driving climate change impacts on physical and socio-environmental systems are radiation, (near) surface temperature and precipitation. At a multi-decadal global scale, observational trends in near surface temperature show a clear overall warming trend, with spatial and temporal variations only related to its magnitude [1]. For precipitation the overall picture is much less consistent and confidence is generally low [1]. There are regions with an increase and others with a decrease of observed precipitation [1]. At regional to local scales, where impacts are typically felt by society, both variables – air temperature and even more so precipitation – show high spatial and temporal variability, particularly in areas with complex topography. For climate impacts studies in mountain regions, which are typically conducted at regional to local scales, it is thus of great importance to understand the physical processes of spatial and temporal behaviour of air temperature and precipitation and the accuracy, limits and representatives of available climate data products.

Accordingly, the objective of this chapter is to highlight some selected aspects of spatial and temporal characteristics of observed near surface air temperature, precipitation and related extremes in mountain environments. The examples are mostly composed from the Swiss Alps, (1) because it is among the most densely populated and frequently accessed mountain ranges on Earth, and (2) due to the long history of mountain settlements and the relatively small spatial extent, the Swiss Alps have among the densest and longest, and mostly homogenized,

The High-Mountain Cryosphere, ed. Christian Huggel, Mark Carey, John J. Clague and Andreas Kääb. Published by Cambridge University Press. © Cambridge University Press 2015.

observational records available and thus form a perfect basis for fundamental studies. The conclusions drawn from the examples here are in principle valid for any mountain range, but we also point to regionally specific climatic features that can perturb general characteristics.

3.2 Basic characteristics of near surface temperature in mountain topography

Near surface temperature is a key variable for many issues related to the environment, including the cryosphere and high-mountain regions worldwide. A very simple example of a process directly related to temperature is the altitude of the snowfall line (e.g. [2]). Although near surface temperature is strongly influenced by altitude, details of the topography, local phenomena (e.g. local winds, cold air pooling, fog and low stratus, thunderstorms) and the character-istics of the external forcing expressed by large-scale dynamics are key to understanding the details of the local (kilometre-scale) temperature regime. Knowledge of regional and local temperature characteristics is very important to understanding the past, present and future near surface temperature develop-ment and its impacts in mountainous regions [1,3]. In this chapter, gridded as well as station-based Swiss data from 83 carefully homogenized temperature and 60 surface air pressure series (see [4]) recorded by the Federal Office of Meteorology and Climatology MeteoSwiss are analysed.

3.2.1 Altitude dependence of 2m temperature

The complex topography in mountainous regions such as Switzerland leads to temperature altitude relations of monthly and seasonal mean temperature that can be very different from the ISO standard lapse rate of 0.65 °C per 100 m (ISO 1975; see [5,6]). It is crucial to quantify these differences in order to get a better understanding of the local temperature regime.

Figure 3.1 shows monthly mean 1981–2010 normal temperature profiles of all 13 216 grid points of the MeteoSwiss gridded 2 km × 2 km temperature data set [7] for the months January, April, July and October. Note that this data set includes some grid points outside Switzerland. As expected, January is the coldest month, followed by April, October and July, which is the warmest month. In general, temperatures decrease with increasing altitude and geographical latitude. Hence, the temperatures on the southern slopes of the Alps are somewhat higher than those on the northern slopes of the Alps. This creates a certain basic scatter of the data points in the *x*-axis direction. Apart from the north–south gradient, the

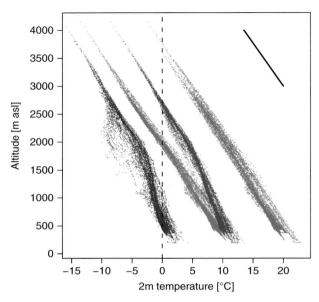

Figure 3.1 Monthly mean 1981–2010 normal 2m temperature (°C) on the *x*-axis vs altitude (m asl) on the *y*-axis of all 2 km × 2 km grid points of the MeteoSwiss gridded temperature data set for the months January (blue), April (green), July (red) and October (brown). The solid black line in the upper right corner shows the temperature lapse rate (0.65 °C/100 m) of the standard atmosphere (ISO, 1975).

lapse rate and the scatter substantially depend on the month. In July and April, the temperature decreases with altitude are almost linear, with a lapse rate very close to the 0.65 °C per 100 m of the standard atmosphere. In addition, the scatter in the *x*-axis direction is relatively small. The Pearson correlation coefficient *r* of temperature and altitude is very high, with 0.995 for April and 0.992 for July. For January and October, more scatter (r_{Jan} = 0.963; r_{Oct} = 0.988) and a reduced temperature lapse rate is found, especially for grid points below approximately 2000 m asl. The behaviour is similar for January and October but with a much larger scatter in January than in October, especially for altitudes between 1000 and 2000 m asl. The scatter is mainly the imprint of strong cold air pools in some of the higher-altitude valleys such as the Goms (Valais) and the Engadina (Grisons) and therefore of strong temperature inversions (up to more than 5 °C temperature depression in the monthly mean, see Figure 3.1). The smaller lapse rate for stations below 1500 m asl in the winter half-year shows that cold air pooling is also found often on the Swiss Plateau. Scherrer and Appenzeller [8] found on average 30 days with strong inversions (range: 10–55 days) between October and March in the 1901–2012 period.

3.2.2 Altitude dependence of daily temperature anomalies

The amplitude of daily 2m temperature anomalies with respect to the normal value is greatly determined by surface properties and exposition of the measuring location, but also depends a lot on local weather phenomena. For example, snow pack is able to reduce 2m temperature and can potentially dampen the daily cycle; vanishing snow due to a warming climate can lead to stronger temperature trends compared to places where snow is still present. Scherrer [9] showed that in the 1961–2012 period the daily mean 2m temperature of a spring day without snow cover is, on average, 0.4 °C warmer than one with snow cover at the same location. The temperature trend increases caused by this effect were in the order of 3–7% of the total observed trend. This is only one example of local surface properties potentially being important in this respect. Others, such as the vegetation type and soil moisture, might also influence the daily cycle but are not discussed here in detail. Local weather phenomena like Föhn winds also play an important role here (see comments below). The largest day-to-day temperature variations in Switzerland are caused in conjunction with air mass changes and can reach up to 33 °C temperature change in 12 hours and 38 °C temperature change in 24 hours at the most prominent Swiss cold pool station, La Brévine [9].

Figure 3.2 shows the standard deviation of the daily mean 2m temperature anomalies with respect to station altitude in January (black points) and July (grey points). The variability of daily mean 2m temperature is clearly larger in January, i.e. winter with ~3–6 °C, than in July, i.e. summer with 2.5–4 °C. For low-lying stations on the southern slopes of the Alps, the difference between winter and summer is small and can be almost 0 °C, e.g. for the station Lugano, which is very close to Lake Lugano. For typical cold air pooling stations, the seasonal difference is large (~3 °C for the cold pool station La Brévine, see [9]). There is a small tendency for larger variability at high-altitude stations, but there are also low-altitude stations that show similar variability to the high-mountain stations. The low-altitude stations with high variability are often influenced by Föhn winds (e.g. Vaduz station). The main conclusion of the altitude dependence of daily temperature anomalies in complex terrain is that due to local exposition and weather processes that can be of similar importance as the station altitude, there is no universally valid altitude variability relation.

3.2.3 The relation between surface air pressure and 2m temperature

Another interesting question to tackle is how monthly and seasonal surface air pressure and 2m temperature means are related and how this relation depends on station altitude. Figure 3.3 shows seasonal anomalies of 1961–2013 homogeneous 2m temperature vs seasonal anomalies of homogeneous surface air pressure at the high Alpine station Jungfraujoch (3580 m asl) and at the Plateau station Zürich/

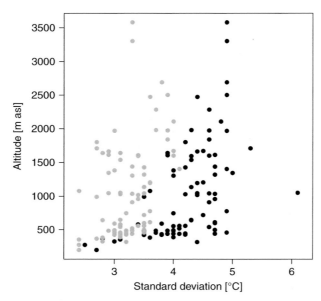

Figure 3.2 The standard deviation of the daily mean 2m temperature anomalies with respect to station altitude in January (black dots) and July (grey dots) for 83 Swiss meteorological stations in mid-January and mid-July for the normal period 1981–2010.

Fluntern (557 m asl) for winter (DJF) and summer (JJA). At Jungfraujoch the seasonal temperature and air pressure anomalies are almost linearly related, with a very high Pearson correlation coefficient r of 0.81 in summer and 0.86 in winter. At Zürich/Fluntern the relation is not even statistically significant on the 5% level (r = 0.27 in summer and r = 0.29 in winter). How can these differences be explained? It is very well known that in general high-mountain peak stations are much more directly influenced by large-scale flow than low-lying stations, which are part of and heavily influenced by processes of the planetary boundary layer (e.g. [10]). Another very important effect is that air pressure at mountain stations has to react on temperature of the atmosphere below the station, i.e. even more with increasing altitude (because 'more atmosphere' is below the station). These two effects together lead to these large differences of the correlation coefficients between air pressure and 2m temperature means at different station altitudes.

Next we want to illustrate how different the relation can be for different stations and regions in Switzerland. Figure 3.4 shows the summer (grey points) and winter (black points) correlation coefficients for 60 Swiss stations of different altitude and exposition. The range of correlation coefficients is between about 0 and 0.86. As expected, the correlation coefficients are reasonably well related to station altitude for most stations and are not too different between summer and winter

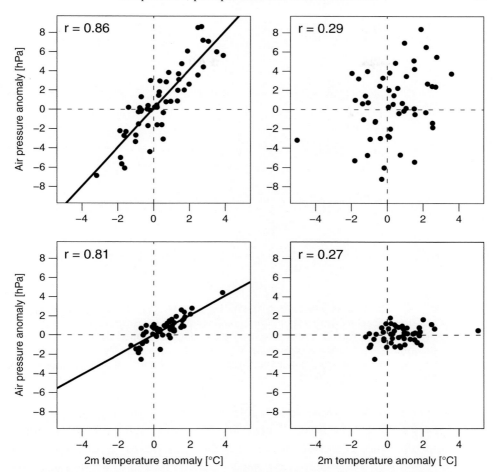

Figure 3.3 Seasonal anomalies of 1961–2013 homogeneous 2m temperature (°C) vs seasonal anomalies of homogeneous surface air pressure (hPa) for the high-altitude station Jungfraujoch (3580 m asl, left panels) and the Swiss Plateau station Zürich/Fluntern (557 m asl, right panels). The top panels show the winter season (DJF), the bottom panels the summer season (JJA). The Pearson correlation coefficient is shown in the top left corner. For the station Jungfraujoch, also a least-square linear fit is shown (solid line).

seasons (r is 0.79 in summer and 0.80 in winter using all stations). Very low correlation coefficients are again found for stations that show a climate, which is influenced by very local effects such as Föhn winds ($r_{DJF, Vaduz} < 0$) or cold air pool stations, e.g. Samedan with r = 0.32 in winter compared to r = 0.55 in summer where cold air pooling is a much less important effect. Stations south of the Alpine main divide show considerably higher correlations with surface pressure than their peers north of the Alpine main crest on the Plateau (r = ~0.6 on the southern slopes

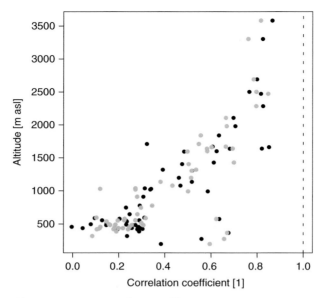

Figure 3.4 The Pearson correlation coefficient (units: [1]; shown on *x*-axis) between homogeneous 2m temperature anomalies and homogeneous surface air pressure anomalies for 60 Swiss meteorological stations plotted against station altitude (units: [m asl]; shown on *y*-axis). Winter (DJF) values are shown in black, summer (JJA) values in grey. The overall correlation coefficient of all DJF values is 0.80; the one for JJA is 0.79.

vs 0.1–0.3 on the Plateau in winter). A potential reason for this large difference in winter is certainly that the Swiss Plateau, in contrast to southern Switzerland, is much more often subject to cold air pooling and long phases of fog and low stratus (see above and [8]). These phases are rather cold and related to positive pressure anomalies and are thus counteracting the normally positive correlation of temperature anomalies and surface pressure.

3.3 Temperature extremes

While long-term changes in the high-mountain cryosphere are primarily linked to warming of mean temperature, some of the most rapid and dramatic short-term responses can be associated with changes in the extremes of temperature. Responses to unusually warm temperatures on the order of days to weeks include rapid glacier melt [11,12], thawing of permafrost and triggering of high-mountain rockfall [13,14], and changes in runoff from the cryosphere [15].

Definitions of what determines an extreme weather or climate event vary, but commonly daily, monthly, seasonal or annual values that are observed with a frequency of less than 10% during a longer term (normally 30-year) climatological reference period are considered extreme. Therefore typical extreme temperature indices consider the number or proportion of days where maximum temperature (T_{max}) or minimum temperature (T_{min}) are below the 1st, 5th or 10th percentiles (cold extremes), or above the 90th, 95th or 99th percentiles (warm extremes) with respect to the reference period (for full discussion see [16]).

At the global scale, some of the most robust statements concerning observed and future climate relate to the widespread increase in extremes of temperature (e.g. [17,18]). There is high confidence that the frequency of warm days and nights have increased, while cold days and nights have decreased since 1950 [1]. The role of humans in contributing to these observed changes in extremes though greenhouse gas emissions and other anthropogenic influences is considered very likely (likelihood of 90–100%; [1]). In Europe, for instance, the frequency of extremely warm days has almost tripled during the period 1880–2005 [19], with the most rapid changes in temperature extremes observed over the past two decades [20]. There is generally less confidence concerning observed changes in prolonged warm or cold spells.

Based on homogeneous long-term records from the network of climate stations across Switzerland, significant warming has been confirmed in both maximum and minimum daily temperatures during the twentieth century, with seasonal and altitudinal variations. At Säntis (2503 m asl), increases in the extremes of temperature have been shown to exceed corresponding changes in mean temperatures [21]. However, differences in analytical methods, reference periods and statistical approaches led to some apparently contrasting findings, particularly concerning observed changes in winter. Using fixed temperature thresholds, Appenzeller *et al.* [22] considered changes in the annual frequency of days across Switzerland where minimum temperature exceeded 0 °C (so-called 'thawing days'). Strongest trends were observed in winter at elevations between 1000 and 1600 m asl, with an increase of more than 260% recorded in the ski resort village of Davos (1590 m asl) for the period 1961–2005.

Based on the well-established TX90p and TN90p indices, which define the percentage of time when daily maximum or minimum temperature exceeds the 90th percentile (after [16]), a clear seasonal pattern is seen in the changing frequency of warm days and nights, particularly for the highest elevation sites in Switzerland (Figure 3.5). Increases in both warm days and nights are largest in early summer and are smallest (or even become negative) during autumn and early winter, consistent with the pattern observed across lower elevation sites in Switzerland over the time period 1965–2011. What is most notable is the exaggerated increase in the frequency of warm days at the highest elevation site of Jungfraujoch (3580 m asl) over the summer months, and particularly in June. This was previously noted in the results from Allen and Huggel [13], where warming of up to 0.68 °C per decade was observed in the 95th quantile of summer daily

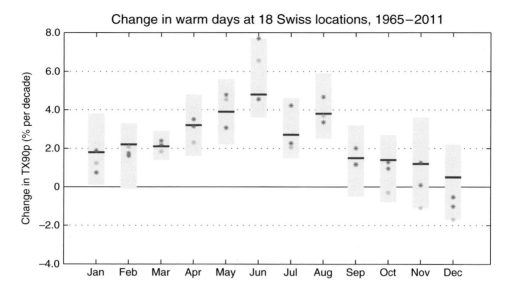

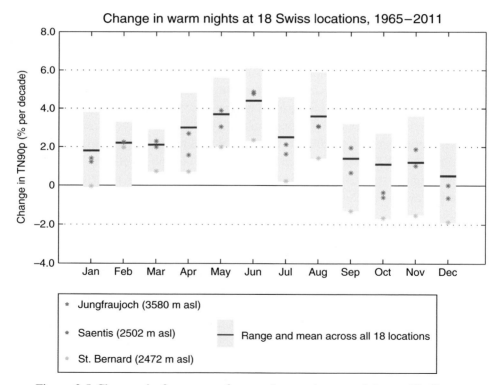

Figure 3.5 Changes in frequency of warm days and warm nights at 18 climate stations across Switzerland where long-term homogenized data are available. Changes are calculated as linear trends over the time period 1965–2011. The extreme indices are calculated relative to the reference period of 1971–2000.

Table 3.1 *Change in frequency in annual extremes of temperature at the three highest-elevation, long-term climate stations in the Swiss Alps. Changes (in % per decade) are calculated as linear trends for the time period 1965–2011. Bolded values indicate significance at the 95% level of confidence.*

	Trend in frequency (%/decade)			
	Warm days (TX90p)	Warm nights (TN90p)	Cold days (TX10p)	Cold nights (TN10p)
Jungfraujoch (3580 m)	**2.74**	**1.75**	**−1.21**	**−1.75**
Saentis (2503 m)	**2.01**	**2.06**	**−1.48**	**−1.67**
Col du Grand St. Bernard (2472 m)	**1.87**	0.27	**−1.58**	−0.49

maximum temperatures at Jungfraujoch over the period 1961–2011. The lack of significant autumn warming at higher elevations in Switzerland over this time period is not specific to the extremes of temperature [22], and over most of Europe mean temperature trends have generally been weakest in autumn [23].

On an annual basis, warm days and nights, and cold days and nights, have all shown statistically significant trends in frequency at two of the three highest-elevation long-term stations in the Swiss Alps (Table 3.1). However, at Col du Grand St. Bernard, located south of the main Alpine divide, there has been no significant trend in the extremes of minimum temperature over the period 1965–2011. The use of single high-elevation stations to represent and characterize changes in extremes across the Alps is therefore not advised. Further studies are needed to clarify whether features such as the anomalous increase in warm summer days at Jungfraujoch, or absence of trends in extremes of minimum temperature at Col du Grand St. Bernard, are local-scale phenomena or indicative of larger spatial patterns across the Alps.

Regardless of which extreme temperature index is used, summer 2003 distinguishes itself in virtually all station data across Switzerland and much of central Europe. In fact, summer 2003 was by far the hottest summer in Europe since 1500 [24]. At Jungfraujoch around 50% of the days and nights in summer 2003 can be considered extremely warm, while there were no cold nights recorded (Figure 3.6). Therefore, the unusual feature of the 2003 heatwave was not the absolute value of the extreme temperatures, but rather the persistence of very warm days and lack of cool temperatures. At high elevations this prolonged warming had significant impacts on the sensitive cryosphere, rapidly melting the snow cover and lowering the reflectivity of glacial ablation areas, further exacerbating melting [12]. Mass losses (measured in water equivalent per year) from glaciers measured in the European Alps were up to four times larger in 2003 than average annual Alpine

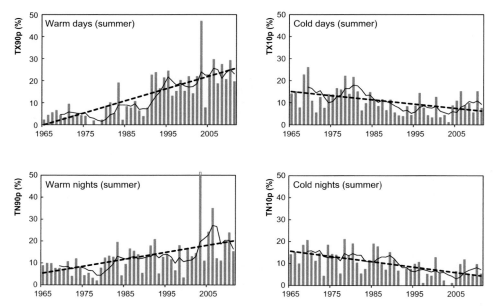

Figure 3.6 Annual frequency of warm days, cold days, warm nights and cold nights, as calculated for the summer months (JJA) at Jungfraujoch (3580 m asl), Switzerland. The extreme indices are calculated relative to the reference period of 1971–2000. The linear trend (dashed line) and five-year moving average (solid line) are also indicated.

values over the previous 20-year period [25]. In permafrost areas the seasonal summer thaw penetrated deeper into bedrock slopes than in previous years, resulting in exceptional rockfall activity [14].

3.4 Precipitation patterns in mountain areas

The precipitation pattern in mountain regions is characterized by large spatial variability. The relationship between elevation and precipitation is complex but generally, at monthly to inter-annual temporal scales, precipitation increases with elevation [26]. However, when the relief rises above the height of an atmospheric moist boundary layer, precipitation only increases with elevation on the lower slopes, while the upper slopes are rapidly drying and precipitation decreases. The elevation at which the maximum precipitation is reached in such situations is variable, and depends on several factors such as the thickness of the moist boundary layer or wind speed and direction [27]. In many mountain areas, 'rain shadow' is another common effect, usually near major moisture sources, such as oceans or great lakes (e.g. [28]). It leads to dry areas on the lee side of a mountain range. Prominent examples here are the Tibetan Plateau, the Atacama Desert

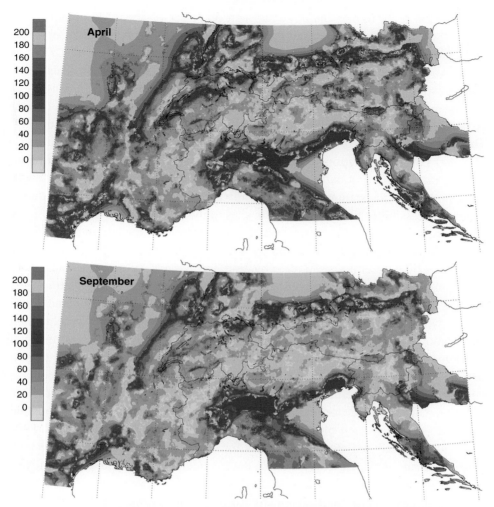

Figure 3.7 Relative vertical precipitation gradients [%/km height] for mean April (upper panel) and mean September (lower panel) for the period 1971–1990. Derived using the PRISM method [50], taken from [46].

(Andes) or the Colorado Front Range (Rocky Mountains). An example of mean monthly precipitation gradients (see Figure 3.7) and spatial patterns of annual, seasonal and monthly distribution for the European Alps and neighbouring mountain chains like the Black Forest or the Vosges, can be found in [29].

3.4.1 Measuring and monitoring precipitation

The scales at which elevation-depending properties are effective range typically over around 5–10 km horizontal distance [26]. In fact, spatial patterns of precipitation seem to be generally persistent on scales of tens of kilometres (e.g. Nepal [30], the Alps [27] or the Olympic Mountains [31]). An optimal observational network should thus have an inter-station distance of about 10–15 km to reveal variability in precipitation over spatial-temporal scales. Rain-gauge networks, especially in mountainous areas, typically do not reach such a density. In Europe networks exhibit an inter-station distance of about 10–50 km (even coarser in remote areas) and a temporal resolution between 10 min and one month [32]. The network maintained by MeteoSwiss is close to such an optimum, with about 450–500 rain-gauge stations (corresponding to an inter-station distance of about 30 km) that measure in daily resolution or less [33]. However, also in the Swiss Alps, high-elevation stations are significantly underrepresented [27], mainly due to difficult accessibility and technical issues, as outlined in the following.

Accuracy and reliability of precipitation measurements for long-term monitoring is critical, particularly in mountain environments (e.g. [34]). While newer techniques like satellite-based precipitation estimations or weather radars are increasingly important for various applications (e.g. flood warning), rain-gauge stations as a direct physical record of precipitation at a specific location and in use for many decades, are considered the reference data for long-term precipitation observations. Limitations of rain-gauge accuracy are mostly due to instrumental problems and intrinsic error sources, making calibration, correction and station maintenance an important part of precipitation monitoring. Important systematic errors are due to wind effects, particularly for light rain or snow, and evaporation or heavy rainfall can distort readings. In cold climates such as in high-mountain areas, snowfall can even block or cover the instruments, leading to record gaps. There are international recommendations for best practices in order to cope with these challenges [35]. However, despite these guidelines and improved techniques (e.g. by adding smart electrical heating to weighing gauges, optimized wind shields), in particular in high-mountain areas several technical challenges remain for accurate and reliable observations. Moreover, in these regions accessibility to the stations for maintenance can be very difficult and is often not possible year-round, and/or is highly time and resource intensive. As a consequence, first, precipitation observations in mountain regions are generally associated with significant uncertainties and require intense post-processing; and second, the density of observational networks in mountain areas is generally low, leading to insufficient spatial coverage of reliable long-term observations.

Consequently, as there is not a clear and confident global trend for precipitation [1], general statements about trends are even more difficult and lacking confidence for mountain regions.

Nevertheless, knowledge on the local precipitation regime is badly needed in order to reduce local people's vulnerability to risks from climatic changes and related extreme events. In mountain areas this includes precipitation as a key component of water resources (immediate or stored in the form of snow and ice) for irrigation or fresh water, or as an important trigger of floods, droughts, debris flows, etc. Therefore, where no local observations are available, alternative data sets such as gridded observations are increasingly used instead, for impact assessments, and also for the evaluation of global and regional climate models [36]; however, this is often done without knowing their quality at the local scale (see Section 3.5).

3.5 Precipitation extremes

Precipitation extremes are often local in scale, and in high mountain regions negative impacts can be multiplied by gravitational processes when debris flows or GLOFs (glacier outburst floods) are triggered.

In general, the number and intensity of heavy precipitation events has likely increased in more regions of the world than decreased, but the changes are not uniform [1]. This is probably even more so for mountain regions, but the data basis here is weak, as outlined above. Regions affected by the Indian monsoon or ENSO, which apply to large parts of the Himalayas and the Andes, respectively, will likely encounter more intensive precipitation events, while confidence in any specific projected change remains low mainly due to natural variations [1]. An example of the disastrous impacts of heavy monsoon floods combined with triggered landslides and flash floods was the 2014 event in northern India and Pakistan, or in August 2005 the heavy (often referred to as '100-year floods') precipitation event in the Swiss Alps, which resulted in severe damage caused by floods and the triggered landslides [37]. In the Swiss case the mountain orography, including its small-scale features, played a key role in both the control of the spatial distribution and persistence of the precipitation pattern, and on the landslide processes. On August 21–22 the rainfall reached its peak due to a combination of approaching moist air from the Mediterranean Sea and convective processes, which led to an exceptionally heavy rainfall intensity that has rarely been seen in the past [38].

As station observations are typically rare in mountain areas and gridded data sets are increasingly used for impact assessments, in the following we discuss the potential and limitations of selected gridded precipitation products for the analysis of extreme events, using the example of the August 2005 event in the Swiss Alps.

3.5.1 *Selected gridded data products*

3.5.1.1 *Reanalyses*

Reanalyses are retrospective analyses of the atmosphere using data assimilation methods and a numerical model. Most reanalyses are global in scale and provide spatial and temporal homogeneous and physically consistent data going back several decades at a spatial resolution in the order of 50–250 km. Since accuracy of the data at a specific site depends significantly on the available and assimilated observational data, areas with good observational coverage tend to have higher reanalysis accuracy then those in data-scarce areas. Accuracy furthermore depends on the variable. While some observations like mean sea level pressure depend strongly on real observations (diagnostic variable), other variables, including precipitation, are generated as prognostic variables by the model. For the validation of reanalysis, only observations that have not previously been included in the assimilation process should thus be used. And although precipitation is not nominally assimilated in reanalyses, validation with observation includes the risk of self-reference due to the fact that, for instance, clear sky radiance observations are a major input into reanalysis. This implies that information on what constitutes a rainy grid point is being included in the reanalysis [34].

Here, we use two different reanalysis products: *NCEP/NCAR R1 Reanalysis* (Figure 3.8a), one of the first reanalyses (first generation) that became available [39] and *MERRA (Modern Era Retrospective-Analysis for Research and Applications*; [40]) (Figure 3.8c), among the newest reanalysis types (third generation), with a significant overall improvement in precipitation and water vapour climatology. NCEP/NCAR R1 provides data from 1948 to the present day at a grid resolution of about 200 km, and at six-hourly, daily and monthly temporal resolution. MERRA provides hourly, daily and monthly values at a grid resolution of about 50 km since 1979.

3.5.1.2 Combined observations

The Global Precipitation Climatology Project (GPCP) established under the World Climate Research Programme (WCRP) provides several products that are intended to approach Climate Data Record standards, by combining multi-satellite field with rain-gauge analyses (over land). The precipitation products are generated by optimally merging estimates computed from microwave, infrared and sounder data observed by the international constellation of precipitation-related satellites, and – for the monthly data sets – precipitation-gauge analyses [41]. Here, we use 1DD (Figure 3.8b), a global precipitation estimate at 1° and daily resolution since 1996 [42]. Verification over contiguous areas of the United States shows generally a good agreement between 1DD and ground truth. However, for individual cells, differences can be significant [43]. Rubel *et al.* [44] state that daily precipitation estimations of this product – even when averaged over the entire region of the European Alps – may differ significantly from observations.

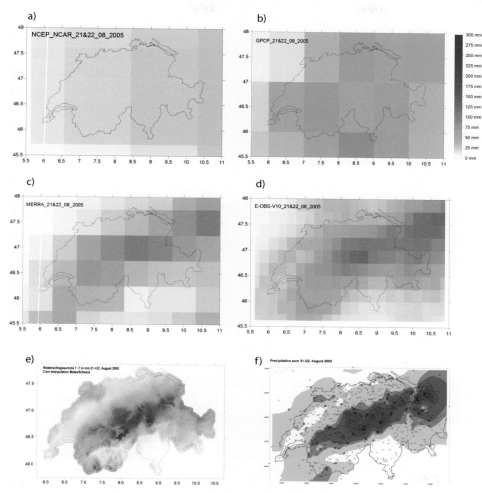

Figure 3.8 Gridded data products showing the two-day sums of the observed precipitation on August 21 and 22, 2005 in Switzerland. Further details of the single data products are provided in the text.

3.5.1.3 Interpolated rain-gauge station data

E-OBS (Figure 3.8d) is a European land-only daily gridded data set based on meteorological station observations across Europe for 1950–2006 (currently updated to 2013) using adopted standard interpolation techniques [45]. It was generated to provide an observational basis for the analysis of RCM simulations. E-OBS data are available for two grid resolutions and two grid types. Here we use the 0.25° resolution product of Version 10.0.

Meteoswiss 2 km-grid (Figure 3.8e) is a spatial analysis of daily precipitation at 2 km horizontal resolution, covering the whole of Switzerland and extending over

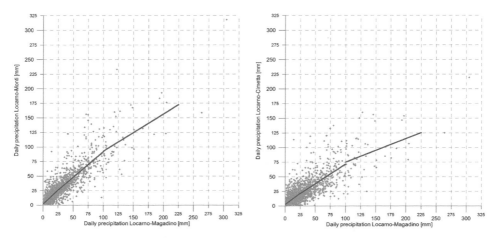

Figure 3.9 Correlation between altitude and precipitation is not significant for daily single extreme precipitation events. Here, daily precipitation values above 1 mm for Locarno-Magadino (204 m asl) vs Locarno-Monti (374 m asl) and Locarno-Monti vs Locarno-Cimetta (1662 m asl) and linear correlation from 1 to 25 mm, from 25 to 100 mm and from 100 mm to 225 mm precipitation in Locarno-Magadino for 1982–2013 are shown as an example.

a multi-decadal period (1961–present; daily updates as preliminary results available). The product is generated in several interpolation steps and target use includes environmental planning and monitoring.

Meteodat interpolation (Figure 3.8f), unlike the other products above, is not continuously available, but is generated for specific analyses only. It is a conservative interpolation using universal Kriging, not incorporating any drift by the orography or other variables. As outlined by Schwarb [46] for climatological precipitation fields of the European Alps, there is a significant correlation between altitude and precipitation depending on the respective month and on the geographic location (see Figure 3.7). However, for the analysis of single extreme precipitation events, this correlation has no significance (see Figure 3.9). Thus, a method providing a best linear unbiased estimation was chosen here.

3.5.2 Comparison and discussion of the gridded data products

Figure 3.8a–f show the two-day sums of the observed precipitation on August 21 and 22, 2005 in Switzerland for the six data products introduced above. The grid resolutions range from about 200 km down to 2 km. It is clear that the two products with the lowest resolution (NCAR/NCEP Reanalysis and GPCP) are not able to represent the heavy rainfall event independent of the base data and techniques used to generate the data product. However, not only for extreme precipitation events, but also for 'average days' the spatial variability of precipitation cannot be captured realistically at grid resolutions above about 50 km × 50 km (about the resolution of MERRA), particularly not in areas with significant topography. Such coarse grid resolutions should thus be used to conduct large-scale climate studies only.

Products at 20–50 km spatial resolutions, here represented by a reanalysis (MERRA) and an interpolated station data product (E-OBS), show much more realistic patterns. The heavy rainfall event is localizable in both products, yet underestimated in its magnitude due to spatial smoothing of features smaller than the grid resolution. Although the result for MERRA is remarkable for the case here, it must be added that in a data-scarce area, where the number of assimilated observations does not reach the level as over central Europe, a comparable rainfall event would probably not be similarly well represented. That is, the use of a product for one region cannot simply be transferred to other regions.

Finally, for scales at 2 km and larger (Figure 3.8e,f), and based on a high inter-station distance (15–20 km), the rainfall event is clearly identifiable, as are many local-scale topographic features. The 2 km grid of MeteoSwiss is probably among the highest-quality, long-term precipitation products currently available for a mountain range. Nevertheless, it should be stressed that despite the high accuracy assumed from an overall visual estimation, direct use of single (or very few) grid points – for example, for local impact studies – must still be done with greatest care as outlined by the MeteoSwiss product descriptions.

In summary, gridded data sets provide a great opportunity for many application-oriented uses, but can have significant limitations, for instance for daily precipitation extremes (see also [47]). Critical evaluation of a product in line with the purpose and the geographic region prior to their use is thus of great importance [48]. New and upcoming techniques and methods will continuously improve available climate products, while rain-gauge stations remain a major pillar for any observational product. Detailed spatial analysis of single events (e.g. by [49] for the August 29, 2003 flash flood in the Eastern Italian Alps) will be easier and opens new possibilities for high-resolution, merged gauge–radar products. At the global scale, the Global Precipitation Measurement Programme was launched on 27 February 2014 and continues the success of the Tropical Rainfall Measuring Mission (TRMM). It relies on an international network of satellites and provides the next generation of global observations of rain and snow.

3.6 Conclusions

The cases and examples presented and discussed in this chapter are mostly taken from the Swiss Alps, for reasons outlined in the introduction. Despite the specific geographic focus, the main processes described are valid for any mountain range, although in their effective occurrence may be influenced by local/regional features. However, to understand local characteristics and changes of key climate variables such as temperature, precipitation and related extremes, adequate long-term local observations are a fundamental prerequisite for any impact assessment.

Acknowledgments

We acknowledge the E-OBS data set from the EU-FP6 project ENSEMBLES (http://ensembles-eu.metoffice.com) and the data providers in the ECA&D project (www.ecad.eu), the gridded 2 km × 2 km air temperature and precipitation data set of MeteoSwiss, MERRA Reanalysis and GPCP-1DD data from NASA, NCEP-NCAR Reanalysis data from NOAA and gauge data from MeteoSwiss.

References

1. IPCC. *Climate Change 2013: The Physical Science Basis. Contribution of Working Group I to the Fifth Assessment Report of the Intergovernmental Panel on Climate Change*. TF Stocker, D Qin, G-K Plattner, M Tignor, SK Allen, J Boschung, A Nauels, Y Xia, V Bex and PM Midgley (eds). Cambridge and New York: Cambridge University Press; 2013

2. G Serquet, C Marty, M Rebetez, Monthly trends and the corresponding altitudinal shift in the snowfall/precipitation day ratio. *Theoretical and Applied Climatology*, **114**:3–4 (2013), 437–444. doi: 10.1007/s00704-013-0847-7.

3. P Ceppi, SC Scherrer, AM Fischer, C Appenzeller, Revisiting Swiss temperature trends 1959–2008. *International Journal of Climatology*, **32** (2012), 203–213. doi:10.1002/joc.2260

4. M Begert, T Schlegel, W Kirchhofer, Homogeneous temperature and precipitation series of Switzerland from 1864 to 2000. *International Journal of Climatology*, **25** (2005), 65–80. doi:10.1002/joc.1118

5. C Rolland, Spatial and seasonal variations of air temperature lapse rates in Alpine regions. *Journal of Climate* **16** (2003), 1032–1046.

6. H Kunz, SC Scherrer, MA Liniger, C Appenzeller, The evolution of era-40 surface temperatures and total ozone compared to observed Swiss time series. *Meteorologische Zeitschrift*, **16**:2 (2007), 171–181. doi:10.1127/0941-2948/2007/0183.

7. C Frei, Interpolation of temperature in a mountainous region using nonlinear profiles and non-Euclidean distances. *International Journal of Climatology*, **34** (2013), 1585–1605. doi: 10.1002/joc.3786.

8. SC Scherrer and C Appenzeller, Fog and low stratus over the Swiss Plateau: a climatological study. *International Journal of Climatology*, **34** (2013), 678–686. doi: 10.1002/joc.3714.

9. SC Scherrer, Die grössten Temperatursprünge im automatischen Messnetz der MeteoSchweiz, *Fachbericht MeteoSchweiz*, **248** (2014), 1–42 (in German).

10. TR Oke, *Boundary Layer Climates*, 2nd edn. London: Methuen; 1987.

11. PD Hughes, Response of a Montenegro glacier to extreme summer heatwaves in 2003 and 2007. *Geografiska Annaler*, **90** :4 (2008), 259–267.

12. F Paul, H Machguth, A Kääb, On the impact of glacier albedo under conditions of extreme glacier melt: the summer of 2003 in the Alps. *EARSeL eProceedings*, **4** (2005), 139–149.

13. SK Allen, C Huggel, Extremely warm temperatures as a potential cause of recent high mountain rockfall. *Global and Planetary Change*, **107** (2013), 59–69.
14. S Gruber, M Hoelzle, W Haeberli, Permafrost thaw and destabilization of Alpine rock walls in the hot summer of 2003. *Geophysical Research Letters*, **31** (2004), L13504.
15. M Zappa, C Kan, Extreme heat and runoff extremes in the Swiss Alps. *Natural Hazards and Earth System Sciences*, **7** (2007), 375–389.
16. X Zhang, L Alexander, GC Hegerl, P Jones, A Klein Tank, TC Peterson, B Trewin, FW Zwiers, Indices for monitoring changes in extremes based on daily temperature and precipitation data. *Wiley Interdisplinary Reviews: Climate Change*, **2** (6) (2011), 851–870. doi: 10.1002/wcc.147
17. B Orlowsky, SI Seneviratne, Global changes in extremes events: regional and seasonal dimension. *Climatic Change*, **110** (3–4) (2012), 669–696.
18. SE Perkins, LV Alexander, JR Nairn, Increasing frequency, intensity and duration of observed global heatwaves and warm spells. *Geophysical Research Letters*, **39** (2012), L20714.
19. PM Della-Marta, MR Haylock, J Luterbacher, H Wanner, Doubled length of western European summer heat waves since 1880. *Journal of Geophysical Research: Atmospheres*, **112** (2007), D15103.
20. M Beniston, Decadal-scale changes in the tails of probability distribution functions of climate variables in Switzerland. *International Journal of Climatology*, **29** (10) (2009), 1362–1368.
21. M Beniston and DB Stephenson, Extreme climatic events and their evolution under changing climatic conditions. *Global and Planetary Change*, **44**:1–4 (2004), 1–9.
22. C Appenzeller, M Begert, E Zenklusen, SC Scherrer, Monitoring climate at Jungfaujoch in the high Swiss Alpine region. *Science of the Total Environment*, **391** (2008), 262–268.
23. AMG Klein Tank, GP Können, FM Selten, Signals of anthropogenic influence on European warming as seen in the trend patterns of daily temperature variance. *International Journal of Climatology*, **25** (2005), 1–16.
24. J Luterbacher, T Dietrich, E Xoplaki, M Grosjean, H Wanner, European seasonal and annual temperature variability, trends, and extremes since 1500. *Science*, **303** (2004), 1499–1503.
25. M Zemp, R Frauenfelder, W Haeberli, M Hoelzle, Worldwide glacier mass balance measurements: general trends and first results of the extraordinary year 2003 in Central Europe. In: Russian Academy of Sciences (ed.), *XIII Glaciological Symposium, Shrinkage of the Glaciosphere: Facts and Analysis. Data of Glaciological Studies [Materialy Glyatsiologicheskikh Issledovaniy]*, vol. **99**, pp. 3–12. St. Petersburg; 2005.
26. C Daly, Guidelines for assessing the suitability of spatial climate data sets. *International Journal of Climatology*, **26** (2006), 707–721. doi:10.1002/joc.1322.
27. C Frei, C Schär, A precipitation climatology of the Alps from high-resolution rain-gauge observations. *International Journal of Climatology*, **18** (1998), 873–900.
28. RG Barry, *Mountain Weather and Climate*. 3rd edition, Cambridge: Cambridge University Press; 2008.
29. M Schwarb, C Daly, C Frei, C Schär, Mean annual/seasonal precipitation throughout the European Alps, 1971–1990. In *Hydrological Atlas of Switzerland*, Plate 2.6 & 2.7. Switzerland: University of Berne; 2001. www.hades.unibe.ch/en
30. AP Barros, M Joshi, J Putkonen, DW Burbank, A study of the 1999 monsoon rainfall in a mountainous region in central Nepal using TRMM products and rain gauge observations. *Geophysical Research Letters*, **27** (2000), 3683–3686. doi: 10.1029/2000GL011827.

31. AM Anders, GH Roe, DR Durran, DR, Montgomery, B Hallet, Precipitation and the form of mountain ranges. *Bulletin of the American Meteorological Society*, **85** (2004), 498–499.

32. M New, M Hulme, P Jones, Representing twentieth-century space-time climate variability. Part I: development of a 1961–1990 mean monthly terrestrial climatology. *Journal of Climate*, **12** (1999), 829–856.

33. M Wüest, C Frei, A Altenhoff, M Hagen, M Litschi, C Schär, A gridded hourly precipitation dataset for Switzerland using rain-gauge analysis and radar-based disaggregation. *International Journal of Climatology*, **30** (2010), 1764–1775. doi: 10.1002/joc.2025

34. FJ Tapiador, FJ Turk, W Petersen, *et al.*, Global precipitation measurement: methods, datasets and applications. *Atmospheric Research*, **104–105** (2012), 70–97.

35. B Sevruk, M Ondrás, B Chvílac, The WMO precipitation measurement intercomparison. *Atmospheric Research*, **92** (3) (2009), 376–380.

36. S Kotlarski, K Keuler, OB Christensen, A Colette, M Déqué, A Gobiet, V Wulfmeyer, Regional climate modeling on European scales: a joint standard evaluation of the EURO-CORDEX RCM ensemble. *Geoscientific Model Development Discussions*, **7** (1) (2014), 217–293. doi:10.5194/gmdd-7-217-2014.

37. M Rotach, C Appenzeller, PE Albisser, Starkniederschlagsereignis August 2005, Arbeitsberichte 211, MeteoSchweiz (2006), Zürich, Switzerland (in German).

38. M Beniston, August 2005 intense rainfall event in Switzerland: not necessarily an analog for strong convective events in a greenhouse climate. *Geophysical Research Letters*, **33** (2006), L05701. doi:10.1029/2005GL025573.

39. E Kalnay, M Kanamitsu, R Kistler, *et al.*, The NCEP/NCAR 40-year reanalysis project. *Bulletin of the American Meteorological Society*, **77** (1996), 437–471.

40. MM Rienecker, MJ Suarez, R Gelaro, *et al.*, MERRA: NASA's modern-era retrospective analysis for research and applications. *Journal of Climate*, **24** (2011), 3624–3648.

41. RF Adler, GJ Huffman, A Chang, *et al.*, The Version 2 Global Precipitation Climatology Project (GPCP) monthly precipitation analysis (1979–present). *Journal of Hydrometeorology*, **4** (2003), 1147–1167.

42. GJ Huffman, RF Adler, M Morrissey, *et al.*, Global precipitation at one-degree daily resolution from multi-satellite observations. *Journal of Hydrometeorology*, **2** (2001), 36–50.

43. J McPhee, S Margulis, Validation and error characterization of the GPCP-1DD precipitation product over the contiguous United States. *Journal of Hydrometeorology*, **6** (2005), 441–459. doi: http://dx.doi.org/10.1175/JHM429.1

44. F Rubel, P Skomorowski, B Rudolf, Verification scores for the operational GPCP-1DD product over the European Alps. *Meteorologische Zeitschrift*, **11** (5) (2002), 367–370. doi:10.1127/0941-2948/2002/0011-0367.

45. MR Haylock, N Hofstra, AMG Klein Tank, EJ Klok, PD Jones, M New, A European daily high-resolution gridded dataset of surface temperature and precipitation. *Journal of Geophysical Research: Atmospheres*, **113**, (2008), D20119. doi:10.1029/2008JD 10201.

46. M Schwarb, The Alpine precipitation climate, PhD thesis, ETH-Zürich, no. 13911 (2000), http://e-collection.library.ethz.ch/view/eth:23937?lang=en.

47. M Turco, AL Zollo, C Ronchi, C De Luigi, P Mercogliano, Assessing gridded observations for daily precipitation extremes in the Alps with a focus on northwest Italy. *Natural Hazards and Earth System Science*, **13** (6) (2013), 1457–1468. doi:10.5194/nhess-13-1457-2013.

48. N Salzmann, OL Mearns, Assessing the performance of multiple regional climate model simulations for seasonal mountain snow in the Upper Colorado River Basin. *Journal of Hydrometeorology*, **13** (2012), 539–556. doi:http://dx.doi.org/10.1175/2011JHM1371.1

49. M Borga, P Boscolo, F Zanon, M Sangati, Hydrometeorological analysis of the 29 August 2003 flash flood in the Eastern Italian Alps. *Journal of Hydrometeorology*, **8** (2007), 1049.

50. C Daly, RP Neilson, DL Phillips, A statistical-topographic model for mapping climatological precipitation over mountainous terrain. *Journal of Applied Meteorology*, **33** (1994), 140–158.

4

Snow and avalanches

SVEN FUCHS, MARGRETH KEILER AND SERGEY SOKRATOV

4.1 Introduction

4.1.1 Snow cover

After seasonally frozen ground, seasonal snow cover has the second largest extent of any component of the cryosphere, with a mean annual area of approximately 26 million km^2, most of it located in the Northern Hemisphere [1]. In many mountain ranges snow and ice are key components of the hydrological cycle with the duration and depth of the seasonal snow cover being key climatic factors of the alpine ecosystem [2]. In mountain regions snow cover plays an important role as an economic factor (e.g. tourism, hydro-power, agriculture, etc.; [3]). Also, snow cover is a determinant of potential snow avalanches and other hazards in mountain areas [4,5].

Spatial and temporal variability of snow cover and snow depth are strongly related to regional and local precipitation patterns and temperature regimes, both parameters interacting with the terrain [6]. Changing snowpack affects subsurface temperatures and permafrost distribution, accumulation as well as ablation of glaciers and vegetation growth in the high-mountain area. The generation of runoff in the high mountains is primarily determined by snowmelt and thus by spring temperature [7], and during summer also by ice melt of the glaciated areas. For the society, especially in arid mountain regions such as the Southwestern United States or Central Asia, freshwaters from high-mountain areas are the most important perennial water resource [8]. On the other hand, a fast and early-season onset of snowmelt may lead to snowmelt-generated floods in the mountains and the lowland [9].

In a very broad and long-term perspective, snow cover is also influencing the climate through albedo. Considering the multiple interactions of snow with other

The High-Mountain Cryosphere, ed. Christian Huggel, Mark Carey, John J. Clague and Andreas Kääb. Published by Cambridge University Press. © Cambridge University Press 2015.

phenomena of high-mountain areas, and their local and regional effects, it is essential to obtain a better understanding of the current and future snow cover dynamic in space and time [10]. Yet, measured and observed data regarding snow (solid precipitation, snow water equivalent, snow cover duration, snow depth) in high-mountain areas are limited. Ground-based observations of different snow parameters such as snow depth are very sparse, thus the analysis of satellite snow cover data is considered as an efficient alternative assessment method [11].

4.1.2 Snow avalanche hazard and risk

An avalanche is defined as the sudden release of snow masses and ice on slopes, and may contain a certain portion of rocks, soil and vegetation; the dislocation on the trajectory is more than 50 m downhill. Due to the speed of the moving mass, snow avalanches can be distinguished from creeping and gliding movements of snow. A number of classifications of snow avalanches exists, developed in different countries and based on different classification principles. De Quervain *et al.* [12] suggested a scheme to classify avalanches according to their release type, the shape of the trajectory and the type of movement, which is still used by the majority of scientists and practitioners in the field (Table 4.1). The evolution of the snowpack from the start of accumulation of solid precipitation until the snow cover melt is crucial regarding the release of snow avalanches. The conditions that lead

Table 4.1 *International snow avalanche classification [12].*

Zone	Criterion	Characteristic and denomination	
Origin	Manner of starting	From a point	From a line
		Loose snow avalanche	*Slab avalanche*
	Position of failure layer	Within the snowpack	On the ground
		Surface-layer avalanche	*Full-depth avalanche*
	Liquid water in snow	Absent	Present
		Dry-snow avalanche	*Wet-snow avalanche*
Transition	Form of path	Open slope	Gully or channel
		Unconfined avalanche	*Channelled avalanche*
	Form of movement	Snowdust cloud	Flowing along ground
		Powder-snow avalanche	*Flowing snow avalanche*
Deposition	Surface roughness of deposit	Coarse	Fine
		Coarse deposit	*Fine deposit*
	Liquid water in deposit	Absent	Present
		Dry deposit	*Wet deposit*
	Contamination of deposit	No apparent contamination	Rock debris, soil, branches, trees
		Clean deposit	*Contaminated deposit*

to the release of avalanches, and also a possible increase in avalanche hazard, are often quite widespread, but the prediction of individual avalanches is extremely difficult due to the high spatial variability and transient/dynamic nature of the snowpack [13]. As a result, however, whole valleys may be endangered by snow avalanches during a winter season.

Different mechanisms of snow avalanche formation correspond to different volumes, repeatability and dynamic characteristics of the events [4]. Loose snow avalanches are released from a more or less definable point in a relatively cohesionless surface layer of either dry or wet snow. Slab avalanches, in contrast, involve the release of a cohesive slab over an extended plane of weakness. Typically, natural slab avalanche activity is highest soon after snowstorms because of the additional load of the deposited snow [13]. The existence of a weak layer below a cohesive slab layer is a prerequisite for the development of dry snow slab avalanches. This weak layer is either buried surface hoar or a result of the metamorphism in the snowpack; during this metamorphism the properties of the snowpack are changing. Crystals formed by kinetic grain growth such as surface hoar or depth hoar [14], together with changes in response to temperature and variability in water vapour gradients, can also be accompanied by formation of solid and icy layers on top of the snowpack. Such surfaces restrict the connection of new-fallen snow with the older snow below the solid layer, and often forms the horizon at which the snow masses start to move downhill. Slab thickness is usually less than 1 m, typically about 0.5 m, but can reach several metres in the case of large, disastrous avalanches [15]. In general, snow avalanches start from terrain that favours snow accumulation and is steeper than about 30–45°. On terrain of less than about 15° snow avalanches start to decelerate and finally stop. Differently to the causes of snow avalanches release, the mechanism of avalanche movement and corresponding distances and forces are rather well described and can be modelled (e.g. [16]).

Avalanche flow velocities vary between 50 and 200 km/h for large dry snow avalanches, whereas wet avalanches are considerably denser and slower (20–100 km/h, [4]). If the avalanche path is steep, dry snow avalanches may generate a powder cloud. Depending on the type of avalanche the moved amount of snow is variable, but in combination with the high velocities the induced damage may vary significantly. In general, slab avalanches and dry snow avalanches with a powder cloud are most disastrous.

Besides natural triggering by overloading or internal weakening of the snow-pack, snow slab avalanches can also be triggered artificially – unlike most other rapid mass movements – through localised, rapid, near-surface loading by, for example, people (usually unintentionally) or by explosives (intentionally) used as part of avalanche control programmes or industrial activities [17]. The industrial development, especially in previously non-exploited regions, is often associated with the increasing degree of hazard occurrence, including snow avalanches [18]. In addition, the artificial change in the vegetation and slope morphology, for example by mining activity [19,20] or during a new ski resort construction [21], can change the position of avalanche-endangered areas at a modified territory or change the run-out distances at the existing avalanche tracks [22]. Occasionally, snow avalanches have been triggered by large earthquakes [23]. In general, naturally released avalanches mainly threaten residents and infrastructure, whereas human-triggered avalanches are the main threat to recreationists.

The threat of avalanches on the anthroposphere can be quantified by the concept of risk. It has been introduced in disaster management since experiences from past years suggested that elements at risk and vulnerability should be increasingly considered within the framework of hazard management in order to reduce losses (e.g. [24]). Starting with the 1990s as the United Nations International Decade for Natural Disaster Reduction, the primary focus was shifted from hazards and their physical consequences to the processes involved in the physical and socio-economic dimensions of risk and a wider understanding, assessment and management of natural hazards. This highlighted the integration of approaches to risk reduction into a broader context between sciences and humanities [25].

Taking the perspective of the sciences, the risk concept is given by a quantifying function of the probability of occurrence of a hazard scenario (p_{Si}) and the related consequences on objects exposed. The consequences can be further quantified by the elements at risk and their extent of damage, and specified by the individual value of elements j at risk (A_{Oj}), the related vulnerability in dependence on scenario i ($v_{Oj,\,Si}$) and the probability of exposure ($p_{Oj,\,Si}$) of elements j exposed to scenario i (Eq. 4.1).

$$R_{i,j} = f(p_{Si}, A_{Oj}, v_{Oj,Si}, p_{Oj,Si}) \qquad (4.1)$$

If snow avalanche risk is considered by the potential loss to an exposed system, resulting from the convolution of hazard and consequences at a certain site and during a certain period of time, it becomes obvious that dynamics in risk has different sources. These will be discussed in the following sections separately. The main challenge of risk assessment is rooted in the system dynamics driven by both geophysical and social forces, stressing the need for an integrative risk-management approach based on a multidisciplinary concept that takes into account different theories, methods and conceptualisations, including environmental and socio-economic change.

Embedded in the overall concept of risk management, mitigating snow avalanches is pillared by technical mitigation, land use regulations, risk transfer, organisational measures and information. Conventional mitigation concepts – which influence both the magnitude and the frequency of avalanches – mainly consider technical structures within the catchment, along the channel system or track and in the run-out area. Throughout many mountain regions, conventional mitigation of snow avalanche hazards can be traced back to the late nineteenth century [26]. According to the approach of disposition management (reducing the probability of occurrence of avalanches) and event management (interfering with the transport process of the hazard itself), a wide range of technical measures is applicable. These measures were supplemented by efforts to afforest high altitudes. Conventional technical measures against avalanche hazards, such as deflection and retention walls, as well as snow rakes in the avalanche starting zones, are not only very costly in construction, but, because of a limited lifetime and therefore an increasing complexity of maintenance, the feasibility of technical structures is restricted due to a scarceness of financial resources provided. Since conventional technical measures neither guarantee reliability nor complete safety, a residual risk of damage remains, which may be reduced by local structural protection [27]. Experiences from past decades suggested that the reduction of exposure should be increasingly considered within the framework of avalanche hazard risk reduction by land use regulations [28,29], risk transfer and organisational measures [15] and information [26].

4.2 Environmental change

4.2.1 Climate change and mountain snow cover

Information on changes of snowfall is limited and mostly restricted to Northern Hemisphere areas (North America and Eurasia), and has to be discussed on a region-by-region basis [30]. Regions with increases of snowfall are located in Canada and Northern Europe; however, a high number of areas show a decline in snowfall events. A decrease in snowfall events can be caused by a variety of reasons: (1) decrease of winter precipitation (e.g. Japan); (2) increase of temperature in winter (more precipitation as rain rather than snow); and (3) earlier onset of spring [30].

Based on analysis of satellite records, the extent of snow cover is significantly decreasing in the Northern Hemisphere during spring time [30]. This trend is confirmed by most station observations of snow, though the results depend on considered snow variables, station elevation and period of record. Correlated to the changes in snow cover duration are the trends in earlier timing in snowmelt-driven streamflows (e.g. in Northwestern America, [31]) and for the earlier spring floods in snow-dominated regions [32]. Nevertheless, the distribution of snow cover in mountains is highly influenced by both variable meteorological conditions and local topography. For two case studies in the European Alps and one in Central Asia, Dedieu *et al.* [11] indicated that elevation is the dominant topographic parameter and changes in snow cover duration can be compared to the changes in temperature and precipitation. Yet, the highest variation between the comparison of snow cover duration and the meteorological parameters are attributed to winters with a scarcity of snow. Stewart [7] concluded that the response to temperature and precipitation change has to be interpreted in the context of physical characteristics for a particular location. At low and mid elevations of mountains (near freezing temperature in winter season) a decrease of the snowpack and the snow cover duration could be observed. At elevations that remain well below freezing during winter, increasing temperatures have had little or no effect on snowpack accumulation and melt; in areas with increasing precipitation a high variable response was detected. The duration of snow cover is also variable, as illustrated in Figure 4.1 for the region of Sochi (Krasnaya Polyana) in the Russian Federation. From 1960 to 1985 the amount of days with snow cover, as well as the amount of days with 'reliable' (deposition for more than 30 days without disappearance for more than three days) snow cover, was increasing; for the last 25 years it has been decreasing.

Projected changes for snow consider mainly the decrease of snow cover extent and are related to both precipitation and temperature changes [30]. Projected

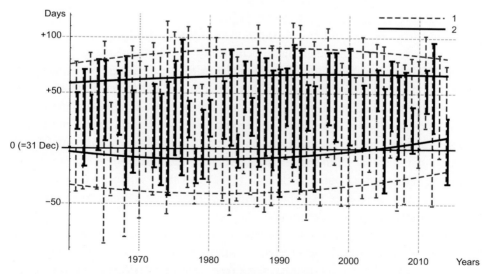

Figure 4.1 Duration of snow cover (1) and duration of 'reliable' snow cover (2) since 1960 in Sochi (Krasnaya Polyana), Russian Federation. Note: the value of 0 corresponds to December 31 in each year; negative values indicate the days before, and positive values the days after the turn of the year.

changes for the next century may be inconsistent: Scandinavia can expect an increase in snow-related floods for the next 50–80 years, but this trend might be reversed within decades by a substantial change from solid precipitation to rain, given that temperatures continue to increase [33]. Diffenbaugh *et al.* [9] evaluated the response of snow-dependent regions under global warming and stressed that extreme change in snow accumulation and melt remains a key unknown for assessing climate change impacts. Their results indicate that many snow-dependent regions of the Northern Hemisphere are likely to experience increasing stress from low-snow years within the next three decades. For mountain areas, a diverse response is expected: due to warmer temperatures in the next decades, the snow volume may respond with reduction at mid-elevation sites by 90% (1000 m) to 50% (2000 m) and at high-elevation sites by 35% in the European Alps [3,5]. Regarding the different projections of climate change, important caveats are that the general circulation models (GCMs) do not resolve the complex topography of the snow-dominated mountain regions and the shifts in liquid/solid threshold of the precipitation [5,9].

4.2.2 Effects on snow avalanches

While it seems to be evident that climate change will affect temperatures and precipitation responsible for avalanche activity (see Figure 4.2; e.g. [30,34,35]), it is not as evident that avalanche events will increase in the near future.

The number of studies focusing on the effect of future environmental change on the occurrence and magnitude of snow avalanches is limited. However, a few papers provide insight into the climatic control of snow avalanches (e.g. [36]) but do not address recent changes in avalanche activity. Changes of temperature, precipitation (amount and solid–liquid thresholds) and wind characteristics influence the structure and stratigraphy of the snowpack and consequently the release and properties of snow avalanches. In general, a classification to different snow climates (e.g. two basic types of snow climate are maritime and continental, based on dominant weather and snow characteristics; and transitional snow climate exhibiting features associated with both types) and their influence on snow avalanche activity is necessary [4,37,38]. Shifts from one snow climate to another may lead to changing avalanche activities.

Germain *et al.* [39] analysed climatic conditions that account for avalanche activity in a Canadian case study with a maritime influence and a mean annual temperature of 0 °C. Five climatic categories were identified: (1) above-average total snowfall, (2) high-frequency of snowstorms, (3) major rain events and facet–crust development, (4) sequences of freezing rain and strong winds and (5) early-season weak layers of faceted crystals and depth hoar. Categories (1) and (2)

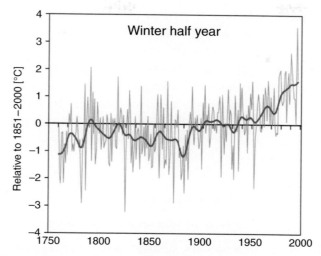

Figure 4.2 Annual mean winter temperatures in the greater Alpine area, 1760–2007. Annual means (grey bars) and 21-year low-pass filtered data (bold line) are shown as deviation from the average 1851–2000. It is shown that in the European Alps the average winter temperatures – adjusted to measurement uncertainties – increased about 2.5 °C during the last 250 years.
Data based on 32 LSS monthly series of the HISTALP database at ZAMG [35,85].

are crucial for dry snow avalanches triggered by the load of new snowfall. These categories should be considered for mountain areas with increasing snow precipitation or a high frequency of snowstorms. Categories (3) and (4) may indicate a shift in the solid–liquid precipitation thresholds, and category (5) accounts for warm periods and unfrozen ground during the first snowfall; all three influence the characteristics of the snowpack.

In the European Alps the long-term natural avalanche activity seems to be constant [40,41], although it is pointed out that the variability of events makes an exact statement difficult. Baggi and Schweizer [42] investigated the occurrence of dry and wet snow avalanches in a small study area in Switzerland over a period of 20 years. The results indicate that loose snow avalanches occurred when air temperature was high and/or after a (liquid) precipitation period. Slab avalanche occurrence was primarily related to warm air temperatures and snowpack properties. Regarding a transitional snow climate, they concluded that wet snow avalanches are also often related to rain events (overloading), but wet slab instability strongly depends on snowpack properties in relation to warming of the snowpack (weakening) and meltwater production (infiltration and storage). Changing climate conditions will supposedly affect the wet snow avalanche activity as far as time and elevation of occurrence are concerned [42]. According to modelling results for the Aspen Mountains, wet snow avalanches will likely occur 2–19 days earlier in the season compared to historical records [43]. Eckert *et al.* [36] focused on changing annual avalanche run-out and correlated it to climate variability using an advanced statistical framework. The results indicate no change in the mean avalanche run-out altitude during the last 60 years in the French Alps, despite the increase in temperature (Figure 4.2). Corresponding to the high variability of snow depth and snow cover in mountain areas, possible effects on snow avalanche activity will cover a wide range, from decreasing or increasing occurrence to a shift from dry snow avalanches to wet snow avalanches.

4.3 Socio-economic change

4.3.1 Drivers of socio-economic change

Socio-economic change in mountains includes land use changes, such as deforestation and urban development in mountain regions, but also population growth, migration and the associated changes such as the development of traffic infrastructure and tourism facilities. Starting in the 1990s, these issues were increasingly addressed on the scientific but also political level [44] – the United Nations International Decade for Natural Disaster Reduction or the implementation of Agenda 21 at the 1992 United Nations Conference on Environment and Development (UNCED) in Rio de Janeiro are prominent examples. Agenda 21 acknowledges the importance of mountain regions and promotes generating and strengthening knowledge about the ecology and sustainable development of mountain ecosystems and providing the public with knowledge concerning mountain-related global change issues, including natural hazard risk management. The importance of mountains in the global ecosystem, as well as their provision of livelihood for considerable parts of the world population, has been further expressed by the UN declaration of the year 2002 as the International Year of the Mountains.

Population density and land use are direct drivers for socio-economic change in mountain regions. Apart from the overall population number, it is also the population distribution and composition, such as the level of urbanisation and household size, as well as the increasing effects of counterurbanisation [45,46] which defines the level of exposure to mountain hazards [47]. If population density is taken as a proxy for the intensity of human activities in mountain areas, considerable parts of the high-mountain areas are potentially at risk (Table 4.2).

Therefore, a sustainable use of mountain areas must include the analysis, assessment and management of socio-economic change due to the relative scarceness of living space. Taking countries in the European Alps as an example, only 38.7% of the territory is suitable for land development purposes in the Republic of Austria, while in the western part of the country (Federal State of Tyrol) it is only

Table 4.2 *Mountain area, population number and density for selected mountain areas* [47].

Article I.	Mountain area (1000s of km^2)	Population in the mountains (millions)	Population density in mountains (persons/km^2)
Svalbard	48	0.001	0.02
Japan	185	15	81
Ethiopia	471	35.2	77
Tajikistan	131	2.9	22
Ecuador	108	5.3	49
Austria	55	3.3	60
British Columbia	750	0.5	0.7

11.9%. In Switzerland, 26% of the territory is classified as non-productive and approximately 68% of the territory is classified as an area for agriculture and forestry; as a result only around 7% is suitable for the development of settlements and infrastructure. In the Russian Federation, approximately 10% of the Russian territory with an average population density of 8.3 persons per km^2 is prone to mountain hazards. The historical shift of a traditionally agricultural society to a service industry- and leisure-oriented society is reflected by an increasing pressure on alpine areas for human settlement, industry and recreation. Accordingly, a conflict between human requirements and naturally determined conditions such as steep terrain leads to an increasing concentration of tangible assets and population in certain regions, in particular with respect to agglomerations along the larger valley bottoms (e.g. [48]).

One of the major industries in the mountain regions is winter tourism [49]. It is crucial for the local economies, and for many mountain-region communities it is unfavourably affected by climate change [50]. Climate change increasingly threatens winter tourism, starting with lower mountain ranges and extending towards high-mountain areas [51]. In particular, communities below 2500 m asl will be affected from this regionally differing trend [49]. Apart from abandonment in low-mountain regions, the adaptation technique involves artificial snow production. The latter affects the whole ecosystem, hydrological and biological cycles, and often has negative impacts on anything other than skiing-related businesses [52], such as an increased demand for water and a higher energy consumption [53].

An increased population density in mountain areas is accompanied by a development of infrastructure, such as sanitation and power lines, but also traffic infrastructure. As a result, an increasing amount of network infrastructure is exposed to snow avalanches [54–59].

Unlike previous booms in mining, cattle or energy, the development wave in land use changes is driven by growth in the secondary and tertiary economies such

as services, recreation and information businesses, instead of commodity production. The result is sprawling land use conversion, mostly from agricultural to
residential, in even the most rural areas. Such changes have been investigated for
mountain areas world-wide, but not explicitly and solely directed towards the
exposure to snow avalanches [60–64].

Closely related to these challenges are the continuous spatio-temporal changes
of landscape processes and of society that are subject to dynamical but also
interactive changes [65]. Moreover, research questions relating to these changes
on the interlinkage between individual landscape processes (e.g. coupled and
multi-hazards, [66]) as well as between landscape systems and human systems,
have not been sufficiently studied so far.

4.3.2 *Effects on snow avalanche risk*

Socio-economic change is a major driver for the dynamics of avalanche risk since
the concept of risk is rooted in the connected system dynamics driven by both
geophysical and social forces. The social system (and therefore land use), elements
at risk exposed and vulnerability are hence not constant over time and space
[29,67,68]. Socio-economic change is also an important part of avalanche risk
management in order to plan and implement tailored management solutions and
adaptation strategies [27].

For the European Alps, a statistically significant trend with regard to an increase
in the annual cost of snow avalanche loss could not be proven. While the large
avalanche events in 1951, 1954, 1968, 1975 and 1984 can clearly be traced, a data
set for the Swiss Alps had not shown any trend [69]. Due to the construction of
mitigation measures, the number of devastating avalanches [70], as well as the
corresponding losses, has declined over the last 50 years in Switzerland [71].
Within the period 1946–1992, 295 individuals were buried inside buildings,
135 of which (46%) died and 56 (19%) were injured. A detailed study within the
canton of Grisons in the eastern part of Switzerland concluded that there was a
reduction of annual damage costs between 1950 and 2000, in particular with
respect to the years with above-average avalanche activity. The total sum of
avalanche losses due to direct building damage in eastern Switzerland amounted
to €63.3 million. This is 40% of the sum paid by the mandatory building insurer for
natural hazards losses in the canton, but avalanches make up only 15% of the
number of all incidents. This means an average loss of €1.25 million per year,
compared with €1.77 million per year for losses due to other natural hazards, such
as debris flows, rockfall events and floods. Damage resulting from avalanches
amounted to an average of €17 500 per event, while losses caused by other types of
natural hazard processes cost an average of €6000 per event.

In the Eastern European Alps, information related to destructive snow avalanches is rather sparse. Between 1967 and 1992 a total of 5135 avalanches had been reported [72], 4032 of which caused damage to settlements and infrastructure. The data did not show any trend; however, large events were reported from 1969, 1974, 1980, 1981 and 1983 during the period under investigation. An analysis of destructive avalanches between 1950 and 2008 from the written reports, which were compiled in the course of the implementation of hazard maps by the Austrian Torrent and Avalanche Control Service, shows a decreasing trend related to the overall number [73].

4.3.2.1 Temporal dynamics of socio-economic changes

The temporal variability of exposure has an important influence on the assessment of avalanche risk since socio-economic developments in the human-made environment have led to an asset concentration and a shift in urban and suburban population in many mountain regions. Long-term changes are related to a significant increase in numbers and values of buildings endangered by snow avalanches [59,74–78]. Short-term fluctuations in exposure supplemented the underlying long-term trend, in particular with respect to temporary variations of migrating or commuting citizens in settlements and of vehicles on the infrastructure network [28,55,79], as well as with respect to different management strategies [80]. By implementing a quantifying fluctuation model it was shown that strong variations could be observed for mountain resorts during the winter season as well as throughout the day [59,79].

If the exposure on traffic corridors is considered, short-term variability becomes obvious [59]: the number of persons or the freight traffic potentially affected by snow avalanches is subject to high fluctuations on different temporary scales. As a consequence, risk (resulting from the daily traffic during the period of investigation, the mean number of passengers and the mean value of good being transported, the speed of the vehicles crossing the endangered sections of the traffic corridor, etc.; [54,56,81]) is variable with a high temporal resolution.

If exposure of settlements is considered, the long-term variability becomes evident: Based on a model to quantify the long-term evolution of the built environment, Fuchs and Keiler [68] reported a significant increase in the number and value of elements at risk exposed for many alpine regions, while other regions show an opposing trend [71]. In many rural and urban settlements of the European Alps the total number of buildings exposed to snow avalanches had almost tripled since the 1950s, and the total value increased by a factor of almost four. The proportional increase in the number of buildings was significantly lower than the proportional increase in the value of buildings. Buildings inside hazard-prone areas showed a lower average value than buildings outside those areas [82]. A major part of this increase was found within the category of residential buildings: in 1950, the proportion of residential buildings was less than 15% of the total amount of endangered buildings. By 2000 this ratio had changed to almost 50%. The number of endangered persons has increased substantially since 1950. The increase in residential population was about 60%, while the increase in temporal population and tourists was a factor of ten [29,83].

To conclude, both long-term and short-term temporal changes of exposure contribute considerably to the risk level, and should therefore be included in operational risk analyses. The vast majority of avalanche fatalities in the Western world nowadays, however, are recreationists exploring the uncontrolled backcountry, making their own decisions. The societal impact of avalanches in Europe and North America has transitioned from an issue affecting settlements and infrastructure (often named involuntary exposure) to more of a recreational issue (voluntary exposure).

4.3.2.2 Spatial dynamics of socio-economic changes

The analysis of spatial dynamics of societal changes is equally crucial for risk assessment, and provides an important factor for tailored mitigation concepts. Spatial dynamics influence settlement patterns; risk management, including its analytical tools and policy recommendations, is inherently geospatial in nature, affecting the location, type and density of development [84]. The concept of space refers to the location of exposure, including distribution and regional patterns.

Until now, there were only a few approaches targeting at a small-scale spatial analysis of exposure in mountain regions; therefore, such information was only accessible through a time-consuming, and therefore costly, detailed on-site analysis [67,71,79]. With respect to the exposure to snow avalanches, a recent study has shown considerable spatial variation throughout mountain communities in the eastern European Alps [73]. Around 2.45 million buildings exist in Austria, 123 040 of which are exposed (for the definition of exposure, see [26,73]) to mountain hazards (torrents: 113 876; snow avalanches: 9164). Subtracting those buildings which are exposed to both torrents and snow avalanches (= corrected sum), approximately 120 400 buildings remain (around 5% of the building stock), with an overall value of €67.25 billion (torrents: €61.14 billion; snow avalanches: €6.11 billion). In sum, around 430 000 people are exposed in these buildings (torrents: 399 253; snow avalanches: 30 158). Taking an overall population of 8.44 million this equals around 5% of the residents.

The results were further analysed according to the construction period, and it was shown that the increase in the building density is significantly higher in potentially endangered areas than outside these areas (Figure 4.3), which in turn requires adaptation and risk mitigation.

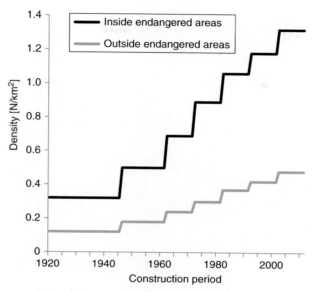

Figure 4.3 Analysis of exposure for buildings in Austria according to the density. After [82]

If queried spatially on a municipal level, considerable differences become evident throughout the country (Figure 4.4). To give an example, in the Federal State of Salzburg around 17% and in Tyrol 15% of all buildings are exposed to torrents and snow avalanches, whereas in Vienna, Burgenland and Lower Austria this value is considerably lower – which is not only a result of different socio-economic development, but also related to the drivers of socio-economic change. In the Federal State of Salzburg, moreover, the number of communities with a clearly above-average exposure is evident.

Identifying and analysing socio-economic dynamics is still a challenge in avalanche risk management, even if it is undoubted that these (1) influence the level of risk a society is exposed to on different temporal and spatial scales; and (2) provide the fundamentals for a sustainable and tailored management concept.

4.4 Conclusions

Due to the effects of climate change, snow avalanche hazards are a dynamic risk component in high mountains. Climate change affects the global temperature and precipitation patterns – the two primary driving factors for the development of the seasonal snowpack and avalanche hazard – but very little is currently known about the effect of climate change on avalanche hazard. Given the increasing knowledge of local-scale changes in temperature and precipitation – being either observed or

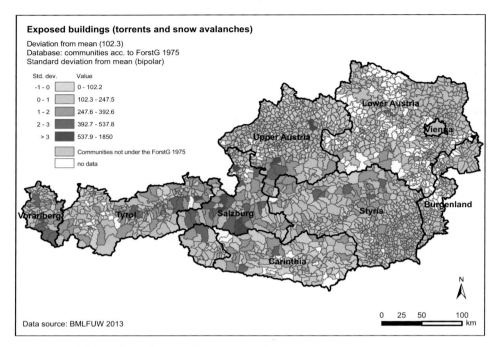

Figure 4.4 Number of exposed buildings, shown as the deviation from the mean (102.3 buildings per municipality).
After [82]

the result of modelling – the understanding of these dynamics is growing. So far, however, studies on changing avalanche frequencies and magnitudes are focusing on individual case studies in high-mountain areas, and remain therefore fragmentary for an area-wide regional hazard analysis. So far, no significant long-term trends in natural avalanche activity have been identified.

Despite the significant increase in population density and exposure of infrastructure and settlements to avalanche hazards over the last 50 years, which have been observed in many mountain regions, there are only a few studies available on the local-scale dynamics of elements at risk, which makes a regional-scale or even national risk assessment challenging. Major losses in high-mountain regions were repeatedly associated with such an increase in land use and economic activities; in contrast, a decrease in annual cost of snow avalanche loss has been reported. Currently, the vast majority of avalanche fatalities are recreationists voluntarily exposing themselves to avalanche hazards.

The concept of risk is increasingly used to track these challenges with respect to economically efficient and societally desirable management options, such as technical mitigation, spatial planning or evacuation. In practice, however, risk assessment and subsequent risk management are regularly undertaken by taking a static viewpoint, while losses are the predictable result of interactions among three major dynamic systems: (1) the specific physical environment of high mountains, which includes snow cover and snow avalanches; (2) the social and demographic characteristics of the communities that experience them; and (3) the elements at risk such as buildings, infrastructure and other components of the built environment.

Focusing on climate and global change in high-mountain areas, risk management strategies have to acknowledge the underlying dynamics in order to be prepared for adaptation and mitigation. Long-term changes are superimposed by short-term fluctuations, and both have to be considered when evaluating risk resulting from mountain hazards. Moreover, the uncertainties of global change underlying the hazard scenarios, but also the lack of knowledge with respect to socio-economic changes, have to be communicated to the stakeholders and the general public; the activities of the Intergovernmental Panel on Climate Change (IPCC) on the global scale, but also of other international organisations such as the UN/ISDR or the World Bank, focusing more on regional adaptation are indispensable. Moreover, stakeholders and the administration in charge on the local level should be aware of the drivers beyond these dynamics, and include them in their local management strategies.

References

1. R Barry, YG Thian (2011) *The Global Cryosphere*. Cambridge University Press, Cambridge.
2. M Beniston, F Keller, S Goyette (2003) Snow pack in the Swiss Alps under changing climatic conditions: an empirical approach for climate impacts studies. *Theoretical and Applied Climatology* **74** (1–2):19–31
3. TV Callaghan, M Johansson, RD Brown, *et al.* (2011) Changing snow cover and its impacts. In: AMAP (ed.) *Snow, Water, Ice and Permafrost in the Arctic (SWIPA): Climate Change and the Cryosphere*. Arctic Monitoring and Assessment Programme, Oslo, pp. 4.1–4.58.
4. D McClung, P Schaerer (2006) *The Avalanche Handbook*. The Mountaineers, Seattle.
5. M Keiler, J Knight, S Harrison (2010) Climate change and geomorphological hazards in the eastern European Alps. *Philosophical Transactions of the Royal Society of London Series A: Mathematical, Physical and Engineering Sciences* **368**:2461–2479
6. T Grünewald, J Stötter, JW Pomeroy, *et al.* (2013) Statistical modelling of the snow depth distribution in open alpine terrain. *Hydrology and Earth System Sciences* **17** (8):3005–3021
7. IT Stewart (2009) Changes in snowpack and snowmelt runoff for key mountain regions. *Hydrological Processes* **23** (1):78–94
8. D Viviroli, HH Dürr, B Messerli, M Meybeck, R Weingartner (2007) Mountains of the world, water towers for humanity: typology, mapping, and global significance. *Water Resources Research* **43** (7):W07447
9. NS Diffenbaugh, M Scherer, M Ashfaq (2013) Response of snow-dependent hydrologic extremes to continued global warming. *Nature Climate Change* **3** (4):379–384

10. T Grünewald, M Schirmer, R Mott, M Lehning (2010) Spatial and temporal variability of snow depth and ablation rates in a small mountain catchment. *The Cryosphere* **4** (2): 215–225

11. JP Dedieu, A Lessard-Fontaine, G Ravazzani, E Cremonese, G Shalpykova, M Beniston (2014) Shifting mountain snow patterns in a changing climate from remote sensing retrieval. *Science of the Total Environment* **493**:1267–1279

12. MR de Quervain, L de Crécy, ER LaChapelle, K Lossev, M Shoda, T Nakamura (1981) *Avalanche Atlas. Illustrated International Avalanche Classification*. UNESCO, Paris.

13. J Schweizer, B Jamieson, M Schneebeli (2003) Snow avalanche formation. *Review of Geophysics* **41** (4):1016

14. C Fierz, R Armstrong, Y Durand, *et al.* (2009) *The International Classification for Seasonal Snow on the Ground*. UNESCO, Paris.

15. M Bründl, P Bartelt, J Schweizer, M Keiler, T Glade (2010) Review and future challenges in snow avalanche risk analysis. In: I Alcántara-Ayala, A Goudie (eds) *Geomorphological Hazards and Disaster Prevention*. Cambridge University Press, Cambridge, pp. 49–61.

16. M Christen, Y Bühler, P Bartelt, *et al.* (2012) Integral hazard management using a unified software environment: numerical simulation tool 'RAMMS' for gravitational natural hazards. In: G Koboltschnig, J Hübl, J Braun (eds) *Internationales Symposion Interpraevent. Proceedings* Vol. **1**. International Research Society Interpraevent, Klagenfurt, pp. 77–86.

17. E Mokrov, P Chernouss, Y Fedorenko, E Husebye (2000) The influence of seismic effect on avalanche release. In: *Proceedings of the 2000 International Snow Science Workshop*, October 1–6 , Big Sky, Montana, pp. 338–341.

18. J Qiu (2014) Avalanche hotspot revealed. *Nature* **509** (7499):142–143

19. Y Fedorenko, P Chernouss, E Mokrov, E Husebye, E Beketova (2002) Dynamic avalanche modelling including seismic loading in the Khibiny mountains. In: International Research Society Interpraevent (ed.) *Interpraevent 2002 in the Pacific Rim, Matsumoto, 14–18 October 2002*. International Research Society Interpraevent, Tokyo, pp. 705–714.

20. S Fuchs, M Keiler (2013) Space and time: coupling dimensions in natural hazard risk management? In: D Müller-Mahn (ed.) *The Spatial Dimension of Risk: How Geography Shapes the Emergence of Riskscapes*. Earthscan, London, pp. 189–201.

21. K Scharr, E Steinicke, A Borsdorf (2012) Sochi/Сочи 2014: Olympic Winter Games between high mountains and seaside. *Revue de Géographie Alpine* **100** (4):1–14.

22. SA Sokratov, YG Seliverstov, AL Shnyparkov, KP Koltermann (2013) Antropogennoe vliyanie na lavinnuyu i selevuyu aktivnist' [Anthropogenic effect on avalanche and debris flow activity]. *Lyed i sneg [Ice and Snow]* **122** (2):121–128.

23. C Stethem, B Jamieson, P Schaerer, D Liverman, D Germain, S Walker (2003) Snow avalanche hazard in Canada: a review. *Natural Hazards* **28** (2–3):487–515.

24. C Aubrecht, S Fuchs, C Neuhold (2013) Spatio-temporal aspects and dimensions in integrated disaster risk management. *Natural Hazards* **68** (3):1205–1216.

25. S Fuchs, C Kuhlicke, V Meyer (2011) Editorial for the special issue: vulnerability to natural hazards – the challenge of integration. *Natural Hazards* **58** (2):609–619.

26. M Holub, S Fuchs (2009) Mitigating mountain hazards in Austria: legislation, risk transfer, and awareness building. *Natural Hazards and Earth System Sciences* **9** (2): 523–537.

27. M Holub, J Suda, S Fuchs (2012) Mountain hazards: reducing vulnerability by adapted building design. *Environmental Earth Sciences* **66** (7):1853–1870.

28. A Zischg, S Fuchs, M Keiler, J Stötter (2005) Temporal variability of damage potential on roads as a conceptual contribution towards a short-term avalanche risk simulation. *Natural Hazards and Earth System Sciences* **5** (2):235–242.

29. M Keiler, R Sailer, P Jörg, *et al.* (2006) Avalanche risk assessment: a multi-temporal approach, results from Galtür, Austria. *Natural Hazards and Earth System Sciences* **6** (4):637–651.

30. TF Stocker, D Qin, G-K Plattner, *et al.* (eds) (2013) *Climate Change 2013: The Physical Science Basis. Contribution of Working Group I to the Fifth Assessment Report of the Intergovernmental Panel on Climate Change.* Cambridge University Press, Cambridge.

31. HG Hidalgo, T Das, MD Dettinger, *et al.* (2009) Detection and attribution of stream-flow timing changes to climate change in the Western United States. *Journal of Climate* **22** (13):3838–3855.

32. SI Seneviratne, N Nicholls, D Easterling, *et al.* (2012) Changes in climate extremes and their impacts on the natural physical environment. In: CB Field, V Barros, TF Stocker, *et al.* (eds) *Managing the Risks of Extreme Events and Disasters to Advance Climate Change Adaptation. Special Report of the Intergovernmental Panel on Climate Change.* Cambridge University Press, Cambridge, pp. 109–230.

33. EN Støren, Ø Paasche (2014) Scandinavian floods: from past observations to future trends. *Global and Planetary Change* **113**:34–43.

34. TG Glazovskaya (1998) Global distribution of snow avalanches and changing activity in the Northern Hemisphere due to climate change. *Annals of Glaciology* **26**:337–342.

35. I Auer, R Böhm, A Jurkovic, *et al.* (2007) HISTALP: historical instrumental climatological surface time series of the Greater Alpine Region. *International Journal of Climatology* **27** (1):17–46.

36. N Eckert, H Baya, M Deschatres (2010) Assessing the response of snow avalanche runout altitudes to climate fluctuations using hierarchical modeling: application to 61 winters of data in France. *Journal of Climate* **23** (12):3157–3180.

37. SS Sharma, A Ganju (2000) Complexities of avalanche forecasting in Western Himalaya: an overview. *Cold Regions Science and Technology* **31** (2):95–102.

38. P Haegeli, DM McClung (2007) Expanding the snow-climate classification with avalanche-relevant information: initial description of avalanche winter regimes for southwestern Canada. *Journal of Glaciology* **53** (181):266–276.

39. D Germain, L Filion, B Hétu (2009) Snow avalanche regime and climatic conditions in the Chic-Choc Range, eastern Canada. *Climatic Change* **92** (1–2):141–167

40. M Laternser, C Pfister (1997) Avalanches in Switzerland 1500–1990. In: JA Matthews, D Brunsden, B Frenzel, B Gläser, MM Weiß (eds) *Rapid Mass Movements as a Source of Climate Evidence for the Holocene.* Gustav Fischer Verlag, Stuttgart, pp. 241–266.

41. M Laternser, M Schneebeli (2002) Temporal trend and spatial distribution of avalanche activity during the last 50 years in Switzerland. *Natural Hazards* **27** (3): 201–230

42. S Baggi, J Schweizer (2009) Characteristics of wet-snow avalanche activity: 20 years of observations from a high alpine valley (Dischma, Switzerland). *Natural Hazards* **50** (1):97–108.

43. B Lazar, M Williams (2008) Climate change in western ski areas: potential changes in the timing of wet avalanches and snow quality for the Aspen ski area in the years 2030 and 2100. *Cold Regions Science and Technology* **51** (2–3):219–228.

44. B Messerli (2012) Global change and the world's mountains. *Mountain Research and Development* **32** (S1):S55–S63.

45. R Löffler, E Steinicke (2006) Counterurbanization and its socioeconomic effects in high mountain areas of the Sierra Nevada (California/Nevada). *Mountain Research and Development* **26** (1):64–71.
46. BP Kaltenborn, O Andersen, C Nellemann (2009) Amenity development in the Norwegian mountains: effects of second home owner environmental attitudes on preferences for alternative development options. *Landscape and Urban Planning* **91** (4):195–201.
47. O Slaymaker, C Embleton-Hamann (2009) Mountains. In: O Slaymaker, T Spencer, C Embleton-Hamann (eds) *Geomorphology and Global Environmental Change.* Cambridge University Press, Cambridge, pp. 37–70.
48. W Bätzing (2002) *Die aktuellen Veränderungen von Umwelt, Wirtschaft, Gesellschaft und Bevölkerung in den Alpen. Im Auftrag des Umweltbundesamtes, gefördert durch das Bundesministerium für Umwelt,Naturschutz und Reaktorsicherheit*, vol. P26. Umweltbundesamt, Berlin.
49. R Steiger (2012) Scenarios for skiing tourism in Austria: integrating demographics with an analysis of climate change. *Journal of Sustainable Tourism* **20** (6):867–882
50. C Gonseth (2013) Impact of snow variability on the Swiss winter tourism sector: implications in an era of climate change. *Climatic Change* **119** (2):307–320.
51. S Agrawala (ed.) (2007) *Climate Change in the European Alps: Adapting Winter Tourism and Natural Hazards Management.* OECD, Paris.
52. C de Jong (2012) Zum Management der Biodiversität von Tourismus-und Wintersportgebieten in einer Ära des globalen Wandels. *Jahrbuch des Vereins zum Schutz der Bergwelt* **2011/2012** (76/77):131–168.
53. M Olefs, A Fischer, J Lang (2010) Boundary conditions for artificial snow production in the Austrian Alps. *Journal of Applied Meteorology and Climatology* **49** (6): 1096–1113.
54. K Kristensen, C Habritz, A Harbitz (2003) Road traffic and avalanches: methods for risk evaluation and risk management. *Surveys in Geophysics* **24** (5–6):603–616.
55. J Hendrikx, I Owens (2008) Modified avalanche risk equations to account for waiting traffic on avalanche prone roads. *Cold Regions Science and Technology* **51** (2–3): 214–218.
56. S Margreth, L Stoffel, C Wilhelm (2003) Winter opening of high alpine pass roads: analysis and case studies from the Swiss Alps. *Cold Regions Science and Technology* **37** (3):467–482.
57. C Rheinberger, M Bründl, J Rhyner (2009) Dealing with the White Death: avalanche risk management for traffic routes. *Risk Analysis* **29** (1):76–94.
58. M Bründl, H Etter-J, M Steiniger, C Klingler, J Rhyner, W Ammann (2004) IFKIS: a basis for managing avalanche risk in settlements and on roads in Switzerland. *Natural Hazards and Earth System Sciences* **4** (2):257–262.
59. S Fuchs, M Keiler, SA Sokratov, A Shnyparkov (2013) Spatiotemporal dynamics: the need for an innovative approach in mountain hazard risk management. *Natural Hazards* **68** (3):1217–1241.
60. J-C Castella, PH Verburg (2007) Combination of process-oriented and pattern-oriented models of land-use change in a mountain area of Vietnam. *Ecological Modelling* **202** (3–4):410–420.
61. B Martin, F Giacona (2009) Analyse géohistorique du risque d'avalanche dans le massif des Vosges. *Houille Blanche* **2009** (2):94–101
62. H Cammerer, AH Thieken, PH Verburg (2013) Spatio-temporal dynamics in the flood exposure due to land use changes in the Alpine Lech Valley in Tyrol (Austria). *Natural Hazards* **68** (3):1243–1270.

63. K Culbertson, D Turner, J Kolberg (1993) Toward a definition of sustainable development in the Yampa Valley of Colorado. *Mountain Research and Development* **13** (4):359–369.

64. WE Riebsame, H Gosnell, DM Theobald (1996) Land use and landscape change in the Colorado mountains I: theory, scale, and pattern. *Mountain Research and Development* **16** (4):395–405.

65. G Hufschmidt, M Crozier, T Glade (2005) Evolution of natural risk: research framework and perspectives. *Natural Hazards and Earth System Sciences* **5** (3):375–387.

66. M Kappes, M Keiler, K von Elverfeldt, T Glade (2012) Challenges of analyzing multi-hazard risk: a review. *Natural Hazards* **64** (2):1925–1958.

67. M Keiler (2004) Development of the damage potential resulting from avalanche risk in the period 1950–2000, case study Galtür. *Natural Hazards and Earth System Sciences* **4** (2):249–256.

68. S Fuchs, M Keiler (2008) Variability of natural hazard risk in the European Alps: evidence from damage potential exposed to snow avalanches. In: J Pinkowski (ed.) *Disaster Management Handbook.* CRC Press and Taylor & Francis, Boca Raton, FL and London, pp. 267–279.

69. Schneebeli M, Laternser M, Ammann W (1997) Destructive snow avalanches and climate change in the Swiss Alps. *Eclogae Geologicae Helvetiae* **90** (3):457–461

70. M Schneebeli, M Laternser, P Föhn, W Ammann (1998) *Wechselwirkungen zwischen Klima, Lawinen und technischen Massnahmen.* vdf Hochschulverlag an der ETH, Zürich

71. S Fuchs, M Bründl (2005) Damage potential and losses resulting from snow avalanches in settlements of the canton of Grisons, Switzerland. *Natural Hazards* **34** (1):53–69.

72. R Luzian (2002) *Die österreichische Schadenslawinen-Datenbank. Forschungsanliegen – Aufbau – erste Ergebnisse. Mitteilungen der forstlichen Bundesversuchsanstalt Wien 175.* Forstliche Bundesversuchsanstalt, Wien.

73. S Fuchs (2013) Vulnerability landscape Austria. *Wildbach-und Lawinenverbau* **172**:154–165.

74. M Keiler, A Kellerer-Pirklbauer, J-C Otto (2012) Concepts and implications of environmental change and human impact: studies from Austrian geomorphological research. *Geografiska Annaler Series A, Physical Geography* **94** (1):1–5

75. C Campbell, L Bakermans, B Jamieson, C Stethem (eds) (2007) *Current and Future Snow Avalanche Threats and Mitigation Measures in Canada.* Canadian Avalanche Centre, Revelstoke, BC.

76. J Gardner, J Dekens (2007) Mountain hazards and the resilience of social–ecological systems: lessons learned in India and Canada. *Natural Hazards* **41** (2):317–336.

77. U Sharma, A Scolobig, A Patt (2012) The effects of decentralization on the production and use of risk assessment: insights from landslide management in India and Italy. *Natural Hazards* **64** (2):1357–1371.

78. AL Shnyparkov, S Fuchs, SA Sokratov, KP Koltermann, YG Seliverstov, MA Vikulina (2012) Theory and practice of individual snow avalanche risk assessment in the Russian arctic. *Geography, Environment, Sustainability* **5** (3):64–81.

79. M Keiler, A Zischg, S Fuchs, M Hama, J Stötter (2005) Avalanche related damage potential: changes of persons and mobile values since the mid-twentieth century, case study Galtür. *Natural Hazards and Earth System Sciences* **5** (1):49–58.

80. S Fuchs, M Thöni, MC McAlpin, U Gruber, M Bründl (2007) Avalanche hazard mitigation strategies assessed by cost effectiveness analyses and cost benefit analyses: evidence from Davos, Switzerland. *Natural Hazards* **41** (1):113–129.

81. MA Vikulina, AL Shnyparkov (2006) K voprosu o terminologii i pokazatelyakh lavinnoi deyatel'nosti [To the question on terminology and characteristics of the avalanche actions]. In *Proceedings of the III international conference 'Avalanches and related subjects', Kirovsk, Russia, September 4–8, 2006 [Trudy III Mezhdunarodnaya konferentsiya "Laviny i smezhnye voprosy", Kirovsk, 4–8 sentyabrya 2006]*. Apatit-media, Kirovsk.

82. S Fuchs, A Zischg (2013) *Vulnerabilitätslandkarte Österreich*. Universität für Bodenkultur, Institut für alpine Naturgefahren, Wien.

83. S Fuchs, M Keiler, A Zischg, M Bründl (2005) The long-term development of avalanche risk in settlements considering the temporal variability of damage potential. *Natural Hazards and Earth System Sciences* **5** (6):893–901.

84. P Berke, G Smith (2009) Hazard mitigation, planning, and disaster resiliency: challenges and strategic choices for the 21st century. In: U Fra Paleo (ed.) *Building Safer Communities: Risk Governance, Spatial Planning and Responses to Natural Hazards*. IOS Press, Amsterdam, pp. 1–20.

85. R Böhm (2009) Klimarekonstruktion der instrumentellen Periode – Probleme und Lösungen für den Großraum Alpen. In: R Schmidt, C Matulla, R Psenner (eds) *Klimawandel in Österreich*. Innsbruck Univerity Press, Innsbruck, pp. 145–164.

5

The frozen frontier

The extractives super cycle in a time of glacier recession

JEFFREY BURY

5.1 Introduction

Over the course of the past several decades, accelerating climate change has begun to profoundly transform the cryosphere. As noted in this volume and elsewhere, global mean temperature has been increasing since the late nineteenth century and average maximum and minimum temperatures over land have also increased over the past 50 years [1]. Because the cryosphere is one of the most sensitive barometers of these shifting climatic conditions, sea ice, mountain glaciers, and ice sheets across the planet have been receding at a rapidly increasing pace [2–4]. During this same time period, the confluence of rapid economic growth, deepening integration of global markets, and burgeoning demand for natural resources have initiated what has often been referred to as a "super cycle" of growth in mineral and energy extraction across the planet [5,6]. These rapidly expanding frontiers of social and economic change have transformed the pace, scale, and extent of global natural resource extraction activities, particularly along the edges of the cryosphere [7–13].

The complex conjunction of these shifting social and natural processes of change is the focus of this chapter. The objectives of the chapter are to provide a combined review of current research addressing these themes, which are often the domain of inquiry in very disparate research fields. In addition, the chapter seeks to integrate the findings of recent research addressing these themes in order to examine the interaction of these processes of change and the ways in which new natural and social relationships are being created. The ultimate goal is to provide a more comprehensive survey of the environmental changes underway and the risks and opportunities they pose for societies, as well as to highlight some of the key questions these transformations pose for high-mountain glacial environments,

The High-Mountain Cryosphere, ed. Christian Huggel, Mark Carey, John J. Clague and Andreas Kääb. Published by Cambridge University Press. © Cambridge University Press 2015.

downstream systems, and high-latitude mountain glacial areas for future investigation across the social and natural sciences.

This chapter begins with a brief review of recent research focusing on the rapid recession of glaciers and ice cover across the cryosphere and the recent extractives super cycle. It then focuses on the emerging opportunities that these changes are providing for new mining and energy exploration and extraction activities, highlights ways in which these activities are also linked to a rapidly expanding global complex of physical infrastructure and the global economy and the cryosphere, and then examines the "liquid relations" that link extractive industries to the water resources that the cryosphere provides for mining and energy operations. The chapter concludes with a discussion of the possible future trajectories for these dynamics in frozen environments as mining and energy operations continue to explore the frigid fringes of the cryosphere and beyond.

5.2 The icy edge of climate change

One of the most significant drivers of change currently affecting the cryosphere is global climate change. The Fifth Assessment Report (AR5) of the Intergovernmental Panel on Climate Change (IPCC) argues that it is *certain* that global mean surface temperature has been increasing for the past century and that climate changes have caused impacts on natural and human systems on all continents and across the oceans [1,14]. As the report states, annual sea ice in the Arctic has declined precipitously over the past several decades, permafrost and snow cover have decreased over the Northern Hemisphere, the Greenland and Antarctic ice sheets are decreasing at an accelerating rate, and mountain glaciers world-wide have continued to recede at an accelerating pace [15].

While there has been a growing consensus that glacier recession is one of the early harbingers of planetary warming, a host of recent studies conducted over the past decade analyzing glacier length, area, and volume, have further strengthened our understanding of the significance of the changes underway. As the AR5 report suggests, and while there are some exceptions, glaciers are receding across the planet at a rapidly increasing pace. Overall, it is estimated that the total mass of ice lost from glaciers since 1971 has been approximately 802 gigatons per year, which has contributed to approximately 2.21 millimeters of sea level increase on an annual basis [15]. Moreover, these new studies demonstrate that over the past 60 years glacier recession has uncovered vast new expanses of land. As Table 5.1 illustrates, between 1952 and 2011 glaciers have receded from approximately 20 242 km^2 of land across the planet [4,9,16–21]. This amount is likely significantly less than the actual amount as much work has yet to be completed for the Caucus Mountains in Central Asia, the Russian Arctic, and much of Antarctica [21,22].

Table 5.1 *Changes in glacial coverage 1952–2011.*

Region*	Measurement period	Glacial coverage (km²)**	Area of recession (km²)	% change
Antarctica (King George and Keguelen Islands)	1963–2008	1953	235	−12.0
Canada	1952–2007	280 241	12 826	−4.6
China/Central Asia	1956–2009	20 454	999	−4.9
Europe	1969–2006	2278	505	−22.2
Greenland	1990–2000	n/a***	1368	n/a
Himalayas	1962–2010	10 737	1568	−14.6
Iceland/Svalbard	1990–2011	6209	316	−5.1
Indonesia	1942–1972	10	3	−30.3
New Zealand	1978–2002	513	85	−16.6
Norway	1965–2003	1473	99	−6.8
Russia	1952–2010	1108	145	−13.1
South America – north (Colombia, Peru)	1955–2010	1585	483	−30.5
South America – south (Chile, Argentina)	1942–2011	28 479	1284	−4.5
Tanzania	1962–2011	7	6	−76.0
United States (including Alaska)	1952–2007	1449	321	−22.2

Sources: [18,21]

  * Measurements for Antarctic continent, Caucus Mountains and Russian Arctic n/a.

  ** Area covered when measurements began.

*** 39 glaciers, width: 420 km; from [18]

While a significant amount of land has been uncovered over the past century due to glacier recession, the rate of these changes is expected to accelerate over the next few decades in response to global warming. This is largely due to the fact that continued increases in temperature are likely to continue into the future due to the continued growth of global greenhouse gas emissions and because it will take many decades for the warming that has already taken place to be fully incorporated into widely varying individual glacier dynamic change regimes. Overall, given the non-linear nature of accelerating change affecting the cryosphere, even if no further warming were to take place, new land will continue to be uncovered across the planet at an accelerating rate as the glaciers continue to recede.

5.3 The icy edge of the global extractives super cycle

One key social and economic driver of change along the icy edge of the planet's rapidly receding glaciers is the increasing demand for mineral resources and hydrocarbons that has grown rapidly over the past several decades. A host of social

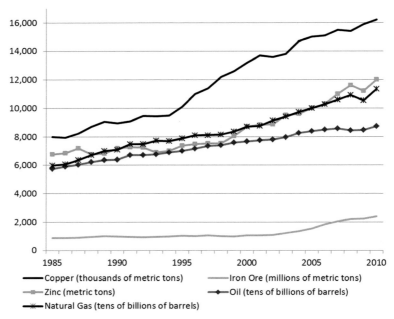

Figure 5.1 World hydrocarbons consumption and mineral production.
Source: [92]

and economic shifts such as population growth, the intensification of energy use, and the rapid development of countries such as China and India have, in combination, stimulated a "super cycle" of growth in energy and mineral exploration, production, and consumption. This recent global "boom" in natural resource prices and extractive activities has been extensively treated by scholars examining the implications of these shifts in regional contexts such as Africa and Latin America, as well as across individual relations of production and consumption for resources such as gold, oil, and natural gas [8,23–30]. As Figure 5.1 illustrates, global consumption of petroleum and natural gas and global production of key metals have increased significantly since 1985.

The growth of the extractives super cycle since the late 1980s has put significant pressure on mining and hydrocarbons corporations to continually discover and extract ever-increasing amounts of natural resources to meet global demand. While the processes through which global investment flows and capital accumulation produce extractive commodities vary significantly, one commonality most resources share is that they are becoming harder to find and more costly to produce. Most mining and hydrocarbon operations today are limited to diffusely concentrated elements because the highly concentrated deposits have already been discovered. Locating new deposits to meet increasing demand has therefore become more costly, frequently takes place in the most remote areas of the planet, and

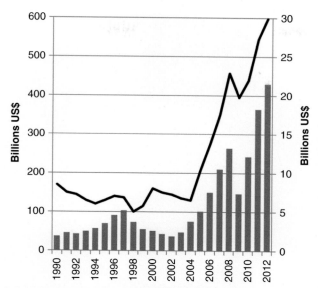

Figure 5.2 Global hydrocarbons exploration (black, left axis) and non-ferrous mineral exploration (gray, right axis) expenditures 1990–2012.
Sources: [31,32]

requires complex and often unproven new extractive technologies that significantly increase their production costs. Figure 5.2 illustrates the rapidly increasing hydrocarbon and non-ferrous mineral exploration expenditures by large mining corporations between 1990 and 2000. Between 1996 and 2012, large mining corporations deployed approximately US$125 billion in exploration expenditures [31]. In addition, hydrocarbons exploration expenditures have increased nearly 500% since the late 1990s and mineral exploration expenditures have increased by nearly 800%. Figure 5.3 illustrates the average percentage of exploration expenditures by region. The largest expenditures were directed towards remote areas in Latin America, northern Canada, Australia, and Africa [32].

While debates among scholars about the continued pace and scale of the global extractives super cycle remain inconclusive due to recent global economic crises and the growth of new demand from newly industrializing countries such as China [5,6,28,33], the vast expanses of land that have or will soon be uncovered by receding glaciers across the planet have already become the frozen frontier for the global extractives sector.

One of the key factors influencing new interest in these post-glacial landscapes is their unknown mineral and hydrocarbon potential. While many new deposits of oil, gas, and minerals have recently been discovered along the edges of glaciers or beneath newly navigable waters near ice sheets or sea ice, current exploration technologies are incapable of surveying subterranean deposits located entirely

J. Bury

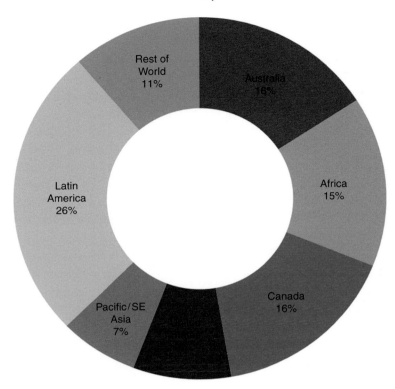

Figure 5.3 Average percentage of global mining exploration expenditures by region 1996–2012.
Source: [32]

beneath glaciers or underneath rapidly moving ice due to lack of surface access and equipment limitations. Consequently, many of the rock, sediment, and mineral or hydrocarbon deposits that are being uncovered by rapidly receding glaciers have been buried in ice for most, if not all, of modern human history. The new lands spread across this *terra glaciem* of the planet are thus providing vast new impetus for the speculative intrigue that often defines resource booms and short-term economic growth cycles. The dynamics of this emerging boom of exploration and discovery are similar to a wide array of historical mining and oil booms that have long been the focus of scholars. These sudden expansions of the resource frontier began as the newly discovered resources were integrated into the global economy. A recent example that illustrates the dynamics of resource booms is the wave of hydraulic fracturing hydrocarbon activities that have rapidly expanded across the United States over the past decade [25,26,29,34–39].

Another factor influencing the expansion of extractive industries onto post-glacial landscapes and into the cryosphere is the unique geographic location of these new lands and the economics of large-scale extractive operations. Most glaciers at lower latitudes are located at high elevations and in mountainous environments and are therefore concentrated at or near the crests of mountain ranges. This is also the same zone where mining activities are the most cost-effective because diffuse mineral deposits that are located close to the surface require less removal of overburden. In addition, because glaciers are one of the most powerful erosive forces on the planet, the inexorable force of glacial ice over long periods of time has created a vast array of sediment deposits and landscape features such as deep canyons and proglacial lakes that enhance placer mining, mineral discovery activities, or facilitate the use of water-intensive mining technologies such as placer, hydraulic, and cyanide heap leach mining.

5.4 New mountains of mines and the frozen north

As glacier recessions have unveiled vast new lands, or decreased the logistical challenges for operations, new exploration activities for mineral and petroleum deposits have expanded rapidly. While many small-scale extractive opportunities for minerals and hydrocarbons have likely been emerging from the ice for a long time, a number of recent hydrocarbons and mineral discoveries indicate that there are significant new resources beneath the ice.

In the mining sector, recent mineral discoveries demonstrate the magnitude of the mineral deposits that are being uncovered by glacial recession. In Kyrgystan, for example, new gold deposits were discovered at the Kumtor Mine both near and underneath several glaciers in the Tien Shan Mountains (see Figure 5.4). In order to extract the gold, the mining company Cameco removed nearly 39 million cubic meters of glacial ice. Between 1997 and 2009, 7.2 million ounces of gold were extracted [11].

One of the most widely discussed intersections between glaciers and mining is in the Central Chilean Andes. Barrick Gold Corporation, one of the largest mining companies in the world, discovered extensive gold deposits underneath glaciers on the border between Chile and Argentina [11,40–42]. While construction of the Pascua Lama project was halted in mid-2013, the project is emblematic of the ways in which glaciers and mining are increasingly coming into conflict in glaciated and high-mountain environments (see Figure 5.5). Similar examples of these new dynamics are taking place across Latin America in the Central Andes of Chile, the Cordillera Blanca in Peru, and in Southern Argentina [40,43,44]. New mineral and gem discoveries have also taken place across northern Canada and Central Asia [45,46].

The most significant new discoveries of mineral deposits along the icy edges of receding glaciers are currently taking place in Greenland. Sustained dynamic ice thinning and rapidly accelerating glacial retreat have been occurring along the maritime boundaries of the continent [18,47]. These changes are rapidly revealing a diverse array of new resources with extraordinary extractive potential. Between 2000 and 2013, the amount of land under license for mineral exploration increased from 11 289 km^2 to 48 684 km^2. According to the new Greenland Bureau of

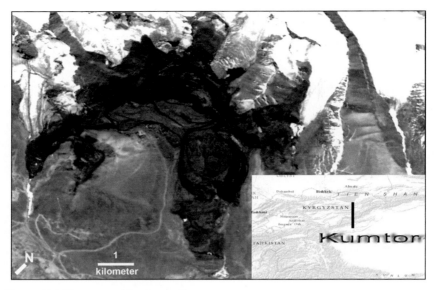

Figure 5.4 Kumtor gold mine in Kyrgyzstan.
(Source: USGS/NASA Landsat, accessed on July 1 2014)

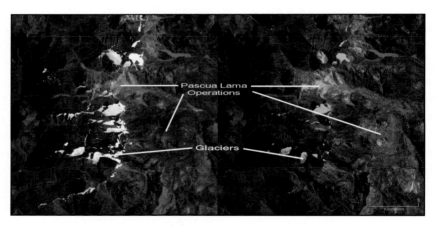

Figure 5.5 Glacier recession and mining at the Pascua Lama mine, Chile (left 1994; right 2014).
(Source: USGS, Landsat 5, 1994; Landsat 8, 2014)

Minerals and Petroleum, more than 100 new prospecting licenses have been granted, and several large-scale mining operations are expected to begin in the next few years. This includes iron ore, polymetallic zinc and lead, rare earths, gold, molybdenum, and ruby and sapphire mining operations [48].

   In the hydrocarbons sector, recent discoveries of significant petroleum and gas reserves have largely taken place in areas of the Arctic underneath the diminishing seasonal ice-covered waters of the Arctic Ocean and near receding glaciers around the Arctic Circle. According to a recent Circum-Arctic assessment, it is estimated that the region contains approximately 90 billion barrels of oil, 1669 trillion cubic feet of gas, and 44 billion barrels of natural gas liquids [49]. While the majority of these hydrocarbons occur offshore, significant portions of the gas and oil fields lie beneath the coastal margins of Greenland and Canada, including the East Greenland Rift Basins that contain an estimated 31 billion barrels of oil and oil-equivalent natural gas (BOE), the Amerasia Basin that contains 19 billion BOE, the West Greenland–East Canada Basin that contains 17 billion BOE, and the Sverdrup Basin that contains 2.4 BOE [49,50].

Overall, as these examples demonstrate, while new exploration activities have only recently begun, the potential for new mineral and hydrocarbon activities across the post-glacial landscapes and waterways is profound. While recent research has begun to document the size and extent of new hydrocarbon and mineral discoveries in these areas, more is needed that examines the rapidly expanding frozen frontiers of the extractives super cycle and the ways in which it is increasingly imbricated with the cryosphere.

5.5 Extending the extractives complex into the cryosphere

As new mining and hydrocarbon activities are extended into the receding cryosphere, extractive industries will not only be producing new resources for the global economy, they will also be linking glaciers and glacial landscapes to the global economy and commodity production networks [5,30,40]. The massive machinery and equipment that are necessary to extract hydrocarbons and minerals from the earth to create commodities that can then be circulated through global networks of production and consumption are but one component of an extractives "complex" that includes a vast assemblage of humans, animals, plants, highways, railroads, power lines, energy infrastructure, communications networks, exploration equipment, mineral processing facilities, refineries, pipelines, storage facilities, and ports that are necessary to support activities that locate, extract, process, and transport extracted materials across the planet.

One of the most important issues related to the extension of the extractives complex into post-glacial landscapes is that new extractive operations pose unusual environmental risks in these environments. The construction of infrastructure, the use of novel exploration and extraction technologies, building refining and processing facilities, and siting extraction by-products and waste are critical features of any extractive operation, but are unproven in the extreme environments near the cryosphere in many cases. Debates have thus emerged focusing on the acceptability of risks associated with technologies such as oil platforms in the Arctic Ocean, the consistent failure of hydrocarbon pipelines in extreme environments, high-elevation hydraulic fracturing and mining, and the impacts of roads, wells, open-pit mines, and infrastructure such as roads and power lines on fragile plant and animal species [51]. Furthermore, recent debates have also highlighted the importance of understanding not only the potential ecological impacts of extractive operations on specific zones where extraction takes place, but on the base array of networks and nodes that link the operation to the global extractives complex. This includes features such as shipping lanes, air transport corridors, under-sea telecommunications, and hydrocarbon supply chains.

One example that usefully illustrates the unusual environmental risks of new extractive industries in the cryosphere is oil exploration and production in the Arctic. Because both glacial and sea ice coverage are rapidly diminishing at high latitudes in the Northern Hemisphere, oil drilling operations have recently expanded into these new areas. In 2012, after nearly a decade of planning and US\$4.5 billion in expenditure, one large oil company experienced a string of weather- and cold-related difficulties culminating in the grounding of an oil platform during a severe storm. Subsequent analyses of the event raised questions about the suitability of current technologies for operations in such extreme environments and the significant social and environmental risks of such accidents [52].

The extension of the extractives complex into post-glacial landscapes also raises several important theoretical and empirical research questions for further consideration. First, because the extension of the extractives complex requires the siting and creation of a vast array of new infrastructure, the political, social, and economic processes that shape these new networks provide unique research opportunities to examine the role of the global economy, nation-states, and other key actors in shaping these systems. These new research avenues differ significantly from the bulk of historical research focusing on these questions as it has often been limited to examining the evolution of such relationships in a post-hoc fashion. This is particularly important because all of the new nodes and networks of the complex, at least in the initial stage, will likely be created to exclusively support extractive operations. Addressing these questions would enable scholarship to broaden and expand on recent commodities chain research and extractive industries debates [24,26,53–57].

Finally, new examinations of the ways in which the extractives complex is being extended into the cryosphere can also contribute to recent debates focused on mining in extra-glacial environments that examine natural resource management under varying governance and institutional configurations, as well as the social, economic, and political implications of these forms of natural resource governance [55,58–64]. Furthermore, new research focusing on these questions might also build upon recent debates focusing on the relationships between natural resources extractive operations and social networks and movements [8,30,61,62,65]. This could also include examinations of a variety of issues such as geopolitical struggles over access to and control of mineral and hydrocarbon resources in the Arctic, disputes over emerging terrain and boundaries, and access to and control of coastal margins and deep-sea mineral and hydrocarbon discoveries [66].

5.6 Liquid relations and stratified societies

One of the most important linkages between the cryosphere and society is through the water that is released by glaciers. These liquid relations are particularly important in the semi-arid tropics and subtropics because over 80% of freshwater for downstream populations originates in mountain regions [67–70]. Discussions of these dynamic relations are largely centered on water quantity and quality. While there are currently very few examples that illustrate these relationships in post-glacial environments, all of the new mining operations previously discussed in this chapter have been engaged in legal and social disputes over their impacts on nearby water resources. In addition, because new extractive activities in post-glacial landscapes will be operating in even more extreme environments with a greater degree of ecological fragility than current extractive zones, recent research related to water resources can inform these discussions.

In terms of water quantity, rapidly receding glaciers affect water resources in a non-linear fashion as they melt [9,68,71]. During the earliest phases of recession, glaciers often release more water and therefore increase water supply. During later phases, particularly in the tropics, water releases decline precipitously and therefore provide less water for downstream purposes such as extractive industries. Another factor that affects water supplies is the use of water resources during extractive operations. Most large-scale mining operations use large amounts of water for mine operations such as cyanide heap leaching, to control sedimentation and dust, or for mine personnel. This use of water can put significant pressure on relatively scarce water resources in semi-arid environments and can lead to struggles over access to water, the institutions that manage it, or among major water users or sectors of society [8,63,66,72–74].

The other component of the liquid relations between glaciers and extractive industries is through the impacts of extractive activities on water quality. The impacts of large-scale mining and hydrocarbon operations are often very large due to the enormous size of open-pit mines, the extent of drilling and pumping stations and hydrocarbon transportation infrastructure, and the volume of earth excavated over the course of the operation. Large-scale mining activities generate long-term and persistent negative environmental impacts on surficial water resources and subterranean aquifers through the generation of toxic acid mine drainage and releases of heavy metals and sediments. In addition, tailings failures, landslides, and atmospheric releases of wind-borne contaminants also pose significant downstream risks. Even small-scale artisanal mining can cause extensive damage to water sources due to unregulated storage and use of heavy metals such as cyanide, arsenic, and mercury. Historically, most mining operations have left behind environmental degradation and ghost towns in the familiar "boom–bust" cycle that has long typified the industry. The overall magnitude of extractive operations' impacts on water resources is very significant. For example, in the United States, metal mining operations accounted for approximately 40% of all of the toxic materials released into the environment. In 2012 this amounted to more than 657 tons of toxic materials that will persist in the natural environment for millennia into the future [75].

The negative impacts of extractive industries on water resources leads to a series of cascading impacts on other systems. Though extractive activities expanding into new post-glacial landscapes are likely to affect smaller populations, the long-standing debates related to these questions will likely still be highly relevant. The negative economic, environmental, and social effects of extractive industries have often been described as a "resource curse," which has been the subject of extensive debates in academic and development discussions [61,62,76–78]. These social impacts can be extensive due to the scale of most extraction operations and have recently been the focus of extensive research. For example, open-pit mines require large amounts of land, water for ore processing, and extensive energy and transportation infrastructure to support mine operations. These requirements frequently lead to the displacement of large numbers of people. When mining companies seek access to the necessary water resources and land and water rights for their operations from governments, this frequently leads to increasing friction with local communities and is often the source of social mobilizations and protests. In addition, while mineral extraction companies do create new employment opportunities in local communities, most physical labor has been replaced by massive machinery in large-scale operations and therefore the growth of employment is much less than during previous periods of mineral expansion. Consequently, social tensions are often exacerbated when it is discovered that there are fewer employment opportunities than local communities expect. Overall, there has been a significant increase in social conflicts related to the expansion of mining operations in mountainous areas over the past decade [72,79,80]. In Peru, for example, conflicts related to mining operations have increased more than 300% over the past decade [81].

Other social impacts of mining operations include risks to human health through exposure to environmental contaminants and mine safety incidents. In terms of human health, the risks from exposure to toxic heavy metals and other contaminants can be severe and very long term. For example, a recent World Health Organization study conducted in Peru suggests that approximately 1.6 million people in the country have been exposed to lead contamination from several centuries of mining operations [82]. In terms of mine safety and accidents, the drama of mine rescues, from Chile to China, in recent years emphasizes the continuing dangers to local miners, for whom mining is often the only job available.

5.7 The frigid fringe: extractive bio-futures and the freezing depths

While the previous sections explored the host of new relationships that are emerging between the expanding global extractives super cycle and the receding cryosphere, this concluding section seeks to broaden its focus by examining new frontiers of change related to new mining technologies that are being developed for use, and new locations for extractive operations in extreme environments that are frequently very cold but are not considered to be part of the cryosphere. The rationale for broadening the focus of the chapter to include extra-cryospheric locations is to highlight the ways in which extractive technologies and "frontiers" are rapidly evolving and to suggest that these new technologies will likely significantly influence future extractive operations in the cryosphere through their use in post-glacial operations or as components of new technologies that are developed specifically for operations in the cryosphere. However, the nature of these influences on future extractive operations is still unclear, but as the following section illustrates, a number of remarkable transformations are underway.

One future frontier of change is related to the introduction of a new set of extractive technologies that are called "biomining," which are already reconfiguring the global extractive complex. Biomining refers to new techniques that allow for the processing of metal-containing ores with biological technologies. In the early 1940s mining engineers discovered a new bacterium (*Acidithiobacillus ferroxidans*) that was partially responsible for chemical reactions involved in acid mine drainage. The bacterium was soon patented and utilized in new biohydrometallurgical processes to leach copper from depleted ores. Over time, a variety of new "extremophiles," as these new microorganisms are now known, were discovered in extremely acidic and highly toxic heavy metal environments. Because many of these new organisms actually engage in novel forms of chemosynthesis, as opposed to carbon-based life-forms that use photosynthesis, they facilitate the dissolution and release of concentrated minerals in ores and as such can significantly increase operational efficiencies for heap leach mining operations [83–85]. Biomining technologies are particularly well-suited for new operations near or within the cryosphere due to the fact that one of the most significant by-products of the chemosynthesis process is heat, which is well-suited to cold environments as it dissipates more easily. One additional benefit of generating heat through the use of new biomining technologies is that this might provide some degree of protection for mineral infrastructure and equipment. Biomining processes have already been widely introduced across the globe and are currently involved in the production of roughly 25% of all copper on the planet, and recent innovations suggest that there is widespread potential for the use of bio-leaching techniques in other heavy metal mining operations, as well as in new "bio-remediation" efforts [86–88].

Another important frontier for extractive activities in the near future is in the freezing depths of the oceans along the fracture zones of major crustal plates. Recent advances in pressurization as well as remote and semi-autonomous vehicle technologies have created opportunities for deep-ocean mineral and hydrocarbon exploration. Recent exploration activities have identified highly concentrated polymetallic nodules in many locations adjacent to hydrothermal sea-floor vent structures [89]. One company, Nautilus Minerals, has already developed a suite of sea-floor production tools, risers and lifting systems, and production support vessels to successfully extract these resources and is developing production plans for new operations in Papua New Guinea [90]. These new mining technologies are likely to accelerate rapidly in the next few years in response to potentially very significant discoveries of rare earth elements in nodules distributed fairly widely across the ocean floor [91].

Overall, the frontiers of extractive technologies in the near future share several common elements. First, they are located in very remote locations, including the bottom of the oceans, the edges of the cryosphere, and in the next few decades among asteroids in outer space. The other element they all share is that all of these technologies will require unprecedented amounts of capital expenditure and the development of new technologies. Finally, they will share a cold future that links them together along a new frigid and frozen extractives frontier.

5.8 Conclusion

Rapidly receding glaciers across the planet are transforming the cryosphere and have become the frozen frontier for new extractive industries. While historical research seeking to examine these social and natural processes of change has often been disparate, new research needs to evaluate their combined and interconnected nature in order to understand the future of these frontiers. New integrated assessments of these processes will strengthen our ability to examine the ways in which natural and social processes interact and shift over time, particularly as the extractives super cycle of the past decade continues to shift in the near future. The expansion of extractive industries into the cryosphere has thus far received relatively little research attention in either the social or natural sciences. However, given the remarkable magnitude of these discoveries, and the fact that they are motivating a host of new exploration activities in similar areas, new research is needed that focuses on the dynamics of extractive industries operating in the cryosphere, the ways in which the extractives complex is being extended into the cryosphere, the impacts of extractive industries on water resources social relationships, and the ways in which new technologies and operations in other frigid environments affect the evolution of extractive industries operating in the cryosphere.

References

1. IPCC, *Climate Change 2013: The Physical Science Basis. Contribution of Working Group I to the Fifth Assessment Report of the Intergovernmental Panel on Climate Change*. TF Stocker, D Qin, G-K Plattner, M Tignor, SK Allen, J Boschung, A Nauels, Y Xia, V Bex, PM Midgley (eds) (New York, NY: Cambridge University Press, 2013).
2. J Yin, JT Overpeck, SM Griffies, A Hu, JL Russell, RJ Stouffer, Different magnitudes of projected subsurface ocean warming around Greenland and Antarctica. *Nature Geosciences*, **4**: 8 (2011): 524–528.
3. Y Nie, Q Liu, S Liu, Glacial lake expansion in the central Himalayas by Landsat images, 1990–2010. *PLoS ONE*, **8**: 12 (2013): e83973.
4. MJ Willis, AK Melkonian, ME Pritchard, A Rivera, Ice loss from the Southern Patagonian Ice Field, South America, between 2000 and 2012. *Geophysical Research Letters*, **39**: 17 (2012): L17501.
5. J Bury, A Bebbington, *New geographies of extractive industries in Latin America*. In A Bebbington and J Bury (eds.) *Subterranean Struggles: New Dynamics of Mining, Oil, and Gas in Latin America* (Austin, TX: University of Texas Press, 2013), pp. 27–66.
6. DB Silver, Super cycle: past, present and future. *Mining Engineering*, **60**: 6 (2008): 72–77.
7. J Bury, BG Mark, M Carey, *et al.*, New geographies of water and climate change in Peru: coupled natural and social transformations in the Santa River watershed. *Annals of the Association of American Geographers*, **103**: 2 (2013): 363–374.
8. A Bebbington, J Bury, (eds.) *Subterranean Struggles: New Dynamics of Mining, Oil, and Gas in Latin America* (Austin, TX: University of Texas Press, 2013).
9. M Baraer, BG Mark, JM McKenzie, *et al.*, Glacier recession and water resources in Peru's Cordillera Blanca. *Journal of Glaciology*, **58**: 207 (2012): 134–150.
10. JT Bury, BG Mark, JM McKenzie, *et al.*, Glacier recession and human vulnerability in the Yanamarey watershed of the Cordillera Blanca, Peru. *Climatic Change*, **105**: 1–2 (2011): 179–206.
11. J Kronenberg, Linking ecological economics and political ecology to study mining, glaciers and global warming. *Environmental Policy and Governance*, **23**: 2 (2013): 75–90.
12. BS Orlove, E Wiegandt, BH Luckman, *Darkening Peaks: Glacier Retreat, Science, and Society* (Oakland, CA: University of California Press, 2008).
13. M Carey, *In the Shadow of Melting Glaciers: Climate Change and Andean Society* (New York: Oxford University Press, 2010).
14. IPCC, *Climate Change 2014: Impacts, Adaptation, and Vulnerability* (New York: Cambridge University Press, 2014).
15. DG Vaughan, JC Comisol, I Allison, *et al.*, *Observations: cryosphere*. In TF Stocker, D Qin, G-K Plattner, M Tignor, SK Allen, J Boschung, A Nauels, Y Xia, V Bex, PM Midgley (eds.) *Climate Change 2013: The Physical Science Basis. Contribution of*

Working Group I to the Fifth Assessment Report of the Intergovernmental Panel on Climate Change (New York: Cambridge University, 2013).

16. A Arendt, T Bolch, J Cogley, *et al.*, *Randolph Glacier Inventory: A Dataset of Global Glacier Outlines. Global Land Ice Measurements from Space* (Boulder, CO: Institute of Arctic and Alpine Research, 2012).

17. T Bolch, A Kulkarni, A Kääb, *et al.*, The state and fate of Himalayan glaciers. *Science*, **336**: 6079 (2012): 310–314.

18. JE Box, DT Decker, Greenland marine-terminating glacier area changes: 2000–2010. *Annals of Glaciology*, **52**: 59 (2011): 91–98.

19. JS Kargel, GJ Leonard, MP Bishop, A Kaab, BH Raup, *Global Land Ice Measurements from Space: Satellite Multispectral Imaging of Glaciers* (New York: Springer Praxis, 2014).

20. A Rabatel, B Francou, A Soruco, *et al.*, Review article of the current state of glaciers in the tropical Andes: a multi-century perspective on glacier evolution and climate change. *The Cryosphere Discuss*, **6**: 4 (2012): 2477–2536.

21. DG Vaughan, JC ComisoI, I Allison, *et al.*, *Observations: Cryosphere Supplementary Material*. In TF Stocker, D Qin, G-K Plattner, M Tignor, SK Allen, J Boschung, A Nauels, Y Xia, V Bex, PM Midgley (eds.) *Climate Change 2013: The Physical Science Basis. Contribution of Working Group I to the Fifth Assessment Report of the Intergovernmental Panel on Climate Change* (New York: Cambridge University, 2013).

22. E Rignot, J Mouginot, M Morlighem, H Seroussi, B Scheuchl, Widespread, rapid grounding line retreat of Pine Island, Thwaites, Smith and Kohler glaciers, West Antarctica from 1992 to 2011. *Geophysical Research Letters*, **41**: 10 (2014): 3502–3509.

23. G Bridge, Mapping the bonanza: geographies of mining investment in an era of neoliberal reform. *The Professional Geographer*, **56**: 3 (2004): 406–421.

24. G Bridge, P Le Billon, *Oil* (Malden, MA: Polity Press, 2013).

25. J Ghazvinian, *Untapped: The Scramble for Africa's Oil* (Orlando, FL: Harvest Books, 2008).

26. MT Huber, *Lifeblood: Oil, Freedom, and the Forces of Capital* (Minneapolis , MN: University of Minnesota Press, 2013).

27. D Humphreys, The great metals boom: a retrospective. *Resources Policy*, **35**: 1 (2010): 1–13.

28. M Radetzki, RG Eggert, G Lagos, M Lima, JE Tilton, The boom in mineral markets: how long might it last? *Resources Policy*, **33**: 3 (2008): 125–128.

29. S Sawyer, *Crude Chronicles: Indigenous Politics, Multinational Oil, and Neoliberalism in Ecuador* (Durham, NC: Duke University Press, 2004).

30. M Watts, *Crude Politics: Life and Death on the Nigerian Oil Fields* (Berkeley, CA: University of California, 2009).

31. Barclays, *Energy and Power Spending Outlook* (London: Barclays Capital, 2012).

32. MEG, *World Exploration Trends* (Halifax: Megals Economics Group, 2013).

33. N Salidjanova, *Going Out: An Overview of China's Outward Foreign Direct Investment* (Washington, DC: US–China Economic & Security Review Commission, 2011).

34. D Brading, C Harry, *Colonial Silver Mining : Mexico and Peru* (Berkeley, CA: Center for Latin American Studies, Institute of International Studies, University of California, 1972).

35. EH Galeano, *Open Veins of Latin America* (New York: Monthly Review Press, 1973).

36. JC Nash, *We Eat the Mines and the Mines Eat Us: Dependency and Exploitation in Bolivian Tin Mines* (New York: Columbia University Press, 1982).

37. A Willow, S Wylie, Politics, ecology, and the new anthropology of energy: exploring the emerging frontiers of hydraulic fracking. *Journal of Political Ecology*, **21**: (2014): 222–236.

38. A Hudgins, A Poole, Framing fracking: private property, common resources, and regimes of governance. *Ecology*, **21**: (2014): 222–348.

39. BC Black, *Crude Reality: Petroleum in World History* (Lanham, MD: Rowman & Littlefield Publishers, 2012).

40. A Brenning, The impact of mining on rock glaciers and glaciers. In BS Orlove, E Wiegandt, BH Luckman (eds.), *Darkening Peaks: Glacier Retreat, Science, and Society* (Oakland, CA: University of California Press, 2008) pp. 196–205.

41. S Fields, The price of gold in Chile. *Environmental Health Perspectives*, **114**: 9 (2006): A536.

42. L Urkidi, A glocal environmental movement against gold mining: Pascua-Lama in Chile. *Ecological Economics*, **70**: 2 (2010): 219–227.

43. AM Cervantes, H Charahua, *Identification of Environmental Impacts in Quebrada Honda, Huaraz, in Environmental Sciences* (Huaraz: National University of Ancash Santiago Antunez de Mayolo, 1999).

44. *Peruano* (March 14, 2013). Illegal mining threatens glaciers in the Cordillera Blanca. El Peruano, 1.

45. O.B. Group, *The Report: Mongolia 2012* (Oxford: Oxford Business Group, 2012).

46. USGS, *2011 Minerals Yearbook Canada* (Washington, DC: United States Geological Survey, 2011).

47. SA Khan, KH Kjær, M Bevis, *et al.*, Sustained mass loss of the northeast Greenland ice sheet triggered by regional warming. *Nature Climate Change*, **4** (2014): 292–299.

48. GPMP, *Report to Inastsisartut, the Parliament of Greenland, Concerning Mineral Resource Activities in Greenland* (Kallallit Nunaat: Greenland Bureau of Minerals and Petroleum, 2013).

49. USGS, *Circum-Arctic Resource Appraisal: Estimates of Undiscovered Oil and Gas North of the Arctic Circle* (Menlo Park, CA: USGS, 2008).

50. USGS, *Assessment of Undiscovered Oil and Gas Resources of the East Greenland Rift Basins Province* (Menlo Park, CA: USGS, 2007).

51. D O'Rourke, S Connolly, Just oil? The distribution of environmental and social impacts of oil production and consumption. *Annual Review of Environment and Resources*, **28**: 1 (2003): 587–617.

52. JJ Efstathiou (July 28, 2013). Rig grounding revives debate over Shell's Arctic drilling. *Bloomberg*, 28.

53. J Bair, *Frontiers of Commodity Chain Research* (Palo Alto, CA: Stanford University Press, 2009).

54. G Gereffi, M Korzeniewicz, *Commodity Chains and Global Capitalism* (Westport, CT: Greenwood Press, 1994).

55. JS Rolston, The politics of pits and the materiality of mine labor: making natural resources in the American West. *American Anthropologist*, **115**: 4 (2013): 582–594.

56. RA Schroeder, Tanzanite as conflict gem: certifying a secure commodity chain in Tanzania. *Geoforum*, **41**: 1 (2010): 56–65.

57. MT Huber, Energizing historical materialism: fossil fuels, space and the capitalist mode of production. *Geoforum*, **40**: 1 (2009): 105–115.

58. G Bridge, Mapping the bonanza: geographies of mining investment in an era of neoliberal reform. *The Professional Geographer*, **56**: 3 (2004): 406–421.

59. G Bridge, Past peak oil: political economy of energy crises. In R Peet, P Robbins, M Watts (eds.) *Global Political Ecology* (London: Routledge, 2011), pp. 307–324.

60. T Dunning, *Crude Democracy: Natural Resource Wealth and Political Regimes* (Cambridge: Cambridge University Press, 2008).

61. A Bebbington, *Social Conflict, Economic Development and the Extractive Industry: Evidence from South America* (London: Routledge, 2011).

62. ML Ross, *The Oil Curse: How Petroleum Wealth Shapes the Development of Nations* (Princeton, NJ: Princeton University Press, 2012).

63. M Himley, Geographies of environmental governance: the nexus of nature and neoliberalism. *Geography Compass*, **2**: 2 (2008): 433–451.

64. K Eaton, Backlash in Bolivia: regional autonomy as a reaction against indigenous mobilization. *Politics & Society*, **35**: 1 (2007): 71–102.

65. T Perreault, *Extracting justice: natural gas, indigenous mobilization, and the Bolivian state*. In S Sawyer, DT Gomez (eds.) *The Politics of Resource Extraction: Indigenous Peoples, Multinational Corporations and the State* (London: Palgrave, 2012), pp. 75–102.

66. A Bebbington, J Bury, Institutional challenges for mining and sustainability in Peru. *Proceedings of the National Academy of Sciences*, **106**: 41 (2009): 17296–17301.

67. LO Fresco, Challenges for food system adaptation today and tomorrow. *Environmental Science & Policy*, **12**: 4 (2009): 378–385.

68. AM Milner, LE Brown, DM Hannah, Hydroecological response of river systems to shrinking glaciers. *Hydrological Processes*, **23**: 1 (2009): 62–77.

69. D Viviroli, H Dürr, B Messerli, M Meybeck, R Weingartner, Mountains of the world, water towers for humanity: typology, mapping, and global significance. *Water Resources Research*, **43**: 7 (2007): 7447.

70. J Xu, R Grumbine, A Shrestha, *et al.*, The melting Himalayas: cascading effects of climate change on water, biodiversity, and livelihoods. *Conservation Biology*, **23**: 3 (2009): 520–530.

71. I Juen, G Kaser, C Georges, Modelling observed and future runoff from a glacierized tropical catchment (Cordillera Blanca, Perú). *Global and Planetary Change*, **59**: 1–4 (2007): 37–48.

72. J Bury, Livelihoods in transition: transnational gold mining operations and local change in Cajamarca, Peru. *The Geographical Journal*, **170**: 1 (2004): 78–91.

73. J Bury, B Mark, JM McKenzie, A French, M Baraer, Glacier recession and human vulnerability in the Yanamarey watershed of the Cordillera Blanca, Peru. *Climatic Change*, **105**: (2011): 179–206.

74. BG Mark, J Bury, A French, J McKenzie, M Baraer, K Huh, Climate change and tropical Andean glacier recession: evaluating hydrologic changes and livelihood vulnerability in the Cordillera Blanca, Peru. *Annals of the Association of American Geographers*, **100**: 4 (2010): 1–12.

75. EPA, *Toxics Release Inventory* (Washington, DC: Environmental Protection Agency, 2012).

76. A Buxton, *MMSD+10: Reflecting on a Decade of Mining and Sustainable Development* (London: International Institute for Environment and Development, 2012).

77. M Humphreys, J Sachs, J Stiglitz, *Escaping the Resource Curse* (New York: Columbia University Press, 2007).

78. JP Richards, (ed.) *Mining, Society, and a Sustainable World* (New York: Springer, 2009).

79. A Behrends, S Reyna, G Schlee, *Crude Domination: An Anthropology of Oil* (New York: Berghahn Books, 2011).

80. M Himley, Regularizing extraction in Andean Peru: mining and social mobilization in an age of corporate social responsibility. *Antipode*, **45**: 2 (2013): 394–416.
81. A Bebbington, The new extraction: rewriting the political ecology of the Andes. *NACLA Report on the Americas*, **42**: 5 (2009): 12–20.
82. A van Geen, C Bravo, V Gil, S Sherpa, D Jack, Lead exposure from soil in Peruvian mining towns: a national assessment supported by two contrasting examples. *Bulletin of the World Health Organization*, **90**: 12 (2012): 878–886.
83. DE Rawlings, BD Johnson, *Biomining* (New York: Springer Verlag, 2007).
84. T Rohwerder, T Gehrke, K Kinzler, W Sand, Bioleaching review part A: progress in bioleaching – fundamentals and mechanisms of bacterial metal sulfide oxidation. *Applied Microbiology and Biotechnology*, **63**: 3 (2003): 239–248.
85. M Labban, Deterritorializing extraction: bioaccumulation and the planetary mine. *Annals of the Association of American Geographers*, **104**: 3 (2014): 560–576.
86. CL Brierley, How will biomining be applied in future? *Transactions of Nonferrous Metals Society of China*, **18**: 6 (2008): 1302–1310.
87. GJ Olson, JA Brierley, CL Brierley, Bioleaching review part B: progress in bioleaching – applications of microbial processes by the minerals industries. *Applied Microbiology and Biotechnology*, **63**: 3 (2003): 249–257.
88. H Watling, The bioleaching of sulphide minerals with emphasis on copper sulphides: a review. *Hydrometallurgy*, **84**: 1–2 (2006): 81–108.
89. M Nimmo, *NI 43–101 Technical Report: Clarian-Clipperton Zone Project, Pacific Ocean* (Boulder, CO: Golder Associates, 2012).
90. Nautilus. Solwara 1 Project-High Grade Copper and Gold (2014), accessed May 1, 2014 at www.nautilusminerals.com/s/Projects-Solwara.asp.
91. D Mohwinkel, C Kleint, A Koschinsky, Phase associations and potential selective extraction methods for selected high-tech metals from ferromanganese nodules and crusts with siderophores. *Applied Geochemistry*, **43**: (2014): 13–21.
92. USGS, *Mineral Commodity Summaries* (Washington, DC: USGS, 2012).

6

Cultural values of glaciers

CHRISTINE JURT, JULIE BRUGGER, KATHERINE W. DUNBAR,
KERRY MILCH, AND BEN ORLOVE

6.1 Introduction

The previous chapters have illustrated the many ways glaciers are linked to society. As important components of hydrological systems, they provide water for irrigation, energy production, and consumption. They are associated with natural hazards, particularly floods and debris flow. Because glaciers interact with natural resources that underpin economic activities, the consequences of glacier retreat can be valued in economic terms. But many people – those who live near glaciers and those who live farther away – express a deep appreciation and concern for glaciers beyond issues of water and hazards. In this chapter we examine cultural values of glaciers that derive from culturally specific standards, rather than from universal market-based monetary systems.

An examination of cultural values can lead to a deeper understanding of climate change responses and risks and the limits of adaptation – a discussion that is gaining traction in climate change debates [1], particularly concerning the concepts of loss and damage [2,3] – and is crucial for future responses to climate change as well as to glacier retreat in particular.

6.2 Three cases in the Alps, the Andes, and the North Cascades

For a consideration of cultural values, we draw material from case studies of communities located in regions with glaciated mountain ranges: Stilfs in the Italian Alps, Siete Imperios in the Cordillera Blanca in Peru, and Concrete and Glacier in the North Cascades, United States. We include two cases where local populations consider glacier retreat to be a serious threat to their ways of life and another case where local populations do not see glacier retreat as having a major impact.

The High-Mountain Cryosphere, ed. Christian Huggel, Mark Carey, John J. Clague and Andreas Kääb. Published by Cambridge University Press. © Cambridge University Press 2015.

The comparison of the three cases – each the subject of several years' worth of ethnographic work – shows the ways in which context shapes the cultural values of glaciers.

This research relies on a number of complementary methods. As anthropologists, we conducted extensive field work, residing in or near the communities for months, interacting regularly with residents in their daily lives. We conducted interviews and organized focus groups. We also collected unpublished historical documents written by local people and archival material from local meetings, governments, and community organizations. Taken together, these methods allowed us to get a full view of the lives of the inhabitants and the meaning of glaciers for their lives and their places.

It is beyond the scope of this chapter to provide a complete picture of the residents of each region, not least because of their heterogeneous character. In the North Cascades we drew on a group of 48 white American residents (male = 25, female = 23) who are likely different from Native American residents in their perspectives of glaciers. For local tribes, glaciers are crucial to their identity because they provide the habitat in which salmon – important for food and ceremonies – have been able to thrive. In Stilfs, the sample (male = 38 and female = 16) was drawn from the German-speaking population of the municipality of Stilfs and a small number of residents of nearby villages working in institutions in the municipality who have direct contact with the glaciers. In Siete Imperios we mainly included Quechua-speaking farmers (male = 24, female = 27) who are representative for the site. See Figures 6.1 and 6.2 for additional information.

6.3 Understanding the cultural values of glaciers

The term "value" refers to two related but distinct concepts. It denotes general standards of assessment, by which the relative worth or goodness of different entities may be judged; these standards may be widely shared, or held by specific groups of communities. In this sense, someone may speak of economic, moral, or aesthetic values, or other sorts of standards. The term also denotes the application of one or more of these standards to a specific entity. For example, one can say that Rembrandt's painting "The Night Watch" has aesthetic value and historic value; though it is inconceivable that it would ever be sold, it also has an economic value, which can be estimated from the prices of other Rembrandt paintings that have sold and from insurance premiums of other paintings in museum collections [4–6]. We use the term "cultural value" in both senses. The first indicates that a specific group of people who share the frameworks and symbols that we call culture can have specific standards of

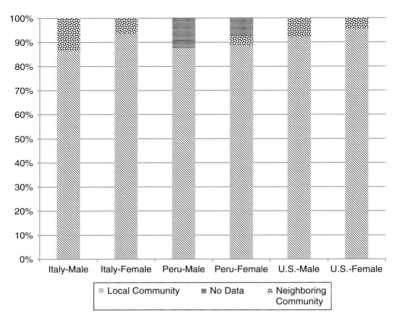

Figure 6.1 Current residence.

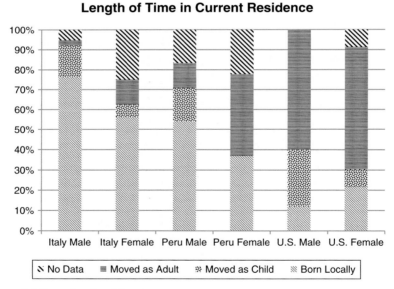

Figure 6.2 Length of residence.

worth. The second can be used to refer to the value that a culture places on specific entities, whether these are culturally specific (e.g., traditional artifacts, ritual performances) or not [7].

In this chapter we look extensively at cultural values in the first sense – the standards of assessment in mountain communities that live near glaciers – in order to arrive at cultural values in the second sense – the application by these communities of their standards to glaciers. The concerns that people have about glaciers reflect the standards of their place and time – what can be termed the cultural framing of glaciers [8]. Carey examined the historical roots of glacier narratives [9]. He specified that different groups, such as climbers, scientists, and artists, "value glaciers on many levels and consume them in diverse ways." Indeed, glaciers can serve to show the differences in value systems between local people and outsiders. In her work on the relations between native populations and scientists in Yukon, Canada, Cruikshank showed that differences between the indigenous religious values of the former and the Enlightenment values of the latter were so great that they impeded communication and collaboration [10]. In a similar vein, current residents of mountain communities, whether indigenous or not, may hold values quite different from outsiders such as scientists and representatives of national and international agencies.

We discuss three aspects of cultural values related to glaciers which emerged from our research: community, identity, and self-reliance. In the early development of anthropology, a *community* was understood as a group that shared common interests and a common social structure and was often based in a single locality; the German sociologist Toennies' notion of *Gemeinschaft* – a German word that refers to "community" – held great influence [11]. We retain his meaning of the word community as a set of people who interact with one another and carry out activities in common. Recent work has recognized that communities are often far from homogeneous, and that communities are not simple objective social realities, but are subjectively perceived by actors [12]. In this sense, a community is a "symbolic construct and a contrastive one; it derives from the situational perception of a boundary which marks off one social group from another" [13]. The differences that mark communities can be between not only other sets of people, but also places [14,15].

The concept of *identity* has gained prominence somewhat more recently, reflecting the emergence of racial, ethnic, gender, and other forms of identity, in politics, replacing earlier forms of identity based on nationality and class. Other work has associated identity with classification systems of individuals and with the institutions which impose these classifications [16]. Here, we refer to identity that arises from membership in a category of individuals who share important characteristics and a sense of commonality. We note that identity, like community, can often be associated with places as well as with social collectivities – though, as in the cases of religious and ethnic identities, it can be distinct from place as well [17,18].

The third value considered here, *self-reliance*, can also be considered to be interconnected with identity and community. We view self-reliance as inward-looking: it suggests that it is possible to do what is needed by oneself. It is closely linked to the notion of autonomy that we consider more outward-looking: it suggests that it is possible to do what is needed without others. Especially for communities with histories of political marginalization, self-reliance and autonomy are important for community continuity.

6.4 Case study 1: Stilfs, South Tirol, Italian Alps

The municipality of Stilfs is situated in the Autonomous Province of Bolzano, in northern Italy on the border of Switzerland. It has some of the highest mountains in the Eastern Alps, like the Ortles (3905 m), attracting mountaineers from all over the world [19]. The Ortles' total glaciated area shrank by about 25% between 1987 and 2009. Twenty-one ice bodies became extinct [20].

The study area is composed of three villages (Trafoi, Stilfs, and Sulden) in the Trafoi and Sulden valleys, located between 1033 and 3905 meters, which together have about 1200 inhabitants. In 2011, 98.46% of the inhabitants declared themselves as German-speaking and 1.54% as Italian speaking [21].

After the collapse of the mining industry in the eighteenth century, animal husbandry and irrigated agriculture became the primary sources of income. At this time, villagers would spend time in fields and pastures but avoided the glaciers, as dangerous, hostile, frightening places. At the end of the nineteenth century, farming became increasingly difficult due to land tenure changes and the system of partible inheritance that fragmented plots [22]. At the same time, mountain tourism started growing, with a strong focus on the glaciers, drawing wealthy and titled people from all over Europe [23] until World War I forced a hiatus. During this war, local men were pressed into service as mountain guides by the opposing forces of Italy and Austria-Hungary, drawing on the experience they had gained in the previous decades. They were sometimes caught in cross-fire and would hide in crevasses and tunnels they dug into the glaciers. In 1919 South Tirol became part of Italy, and the 200 000 inhabitants of the province became a cultural minority within the Italian nation [24].

Tourism became more lucrative in the interwar period but declined dramatically again during World War II. During the war, residents were forced to make the wrenching decision known as "the Option," choosing between remaining in Italy with Mussolini's policy of forced Italianization or moving to Germany (to which Austria had been annexed in 1938) with the possibility of receiving lands in Ukraine after a German victory in the war. Many of those who left the area died during the war; others survived, and some of them returned to their villages [25].

Tourism rebounded in the 1950s, with the glaciers of Stilfser Joch becoming popular ski destinations in summer as well as winter. Today, Sulden and Trafoi are small but well-established tourist destinations for skiing and hiking, with many small guesthouses and restaurants run as family enterprises. In recent years migrants from Eastern Europe have come to work during the busy seasons of the year; the local owners welcome the additional hands, but fear an erosion of local identity and of the authenticity of the visitors' experiences. Tourism is the economic activity that generates the most taxes in the municipality. Stilfs, the third village of the site, draws fewer tourists and considers itself a village of academics and intellectuals.

6.4.1 *The role of glaciers for community, identity, and self-reliance in Stilfs*

Residents can see glaciers from the villages, and glaciers features prominently in local narratives. They are said to represent a feeling of "home." As one villager put it,

If you are outside in spring ... then you can always hear some sound. The glacier is steadily working. In wintertime less, because everything is frozen. [But] now there is life in it, the whole mountain is alive together with the glacier ... because you cannot only see it, you can also hear it, day and night. You can hear each change of the weather, of the temperature. ... And this is home, yes, when you hear this sound.

The glaciers' retreat and the "black" spots in summertime – areas formerly covered with ice – evoke sadness. This sadness, mentioned by almost all interviewees, goes beyond an aesthetic loss. The residents began visiting glaciers when tourism expanded in the late nineteenth century. The tourism allowed them to survive as a group under the Italian fascist regime with a strong sense of *community*, with an identity based on their links to a broader German language and culture, heavily repressed in Italy during that time. The links of community and identity came under intense scrutiny when some families "opted" for Germany and left the municipality, while other refused the Option and remained – a tense issue still raised in present-day conversations.

Recently, glacier melt has led to the discovery of World War I artifacts in the ice. These are avidly gathered by many local young men who participate in the association of World War I memorabilia collectors, which organizes presentations about the history of the glaciers during that time. These shared memories evoke a period of unity, when the communities all suffered under the pressures of living at battlefronts; these memories overcome some of the troubled recollections of divisions at the time of the Option. The glaciers protected local men when they hid in the tunnels, and also protected them by marking the distance from political centers, which also contributed to their survival as a community.

In the present, the glaciers continue to be linked to the *identity* of the residents as Bergler (mountain people), distinguishing them from the lowlanders, often

Italian-speakers, who are perceived to be less able to cope with the harsh (though beautiful) Alpine environment, or even walk on the mountain paths. Many young men participate in the local rescue service, working together to save tourists who encounter hazards. Membership in the rescue service allows men to strengthen local social networks and be part of the community. However, not everyone has regular physical contact with the glaciers. Most of the women interviewed came from other villages in the province and have had little direct contact with the glaciers; instead, they are oriented to domestic life and their husbands' tourist enterprises. Women tended to mention the dangers of glaciers and the fear they felt when their husbands, sons, and friends were on a rescue operation. Rescue service operations are often difficult to carry out. The members need to have excellent climbing ability, and also need to be physically strong to carry out the often dangerous rescue service operations. Women know about these dangers and efforts of the local men, and they worry when men go on rescue operations.

The importance of tourism is undisputed among the residents. Even more than the autonomous status of Stilfs [26], of which the residents are quite proud, the household-based tourist guesthouses and restaurants are fundamental to the communities' *self-reliance* and their stance of separation from the Italian state. Residents are conscious of the importance of tourism – based in large part on the glaciers – for the municipality's survival, and they recognize the aesthetic and recreational value of the glaciers. Glacier retreat thus threatens local livelihoods and the survival of local communities.

Residents have developed some strategies to reduce their dependence on the glaciers. For example, the district heating station relies not on hydropower – which has become less reliable as glacier retreat has reduced water supplies– but on locally produced wood chips. Nevertheless, glacier retreat is a serious threat for them as the following quote from an interview shows:

I am too attached to this village to offer a negative outlook of its future. . . . It's difficult, though. You don't know . . . you feel sometimes that you are ill, but you don't really believe it. And this could be the case here, you will not act until an illness is really severe. If it is not severe, then it might only be a mild case of the flu. . . . We are having a flu epidemic, that's for sure.

This uncertainty about the future conveys the sense of rupture with a past time of well-being for these communities, with strong senses of identity and self-reliance.

6.5 Case study 2: Siete Imperios, Cordillera Blanca, Peru

The Cordillera Blanca of Peru is home to the largest contiguous area of tropical glaciation in the world, with glaciers spanning 507 km^2 on 722 peaks. Glacier meltwater drains into the Rio Santa watershed on Peru's desert coast, providing water for domestic needs, agricultural irrigation, hydroelectric power, large- and small-scale mining operations, commercial agricultural operations, and growing cities. From a 1970 baseline, the national inventory of Peruvian glaciers shows that glaciers in the Cordillera Blanca have shrunk by 27% in the last 30 years [27–29]. Projected temperature increases will lead to further mass loss in coming decades [29].

Siete Imperios, a *comunidad campesina* (peasant community) lies at an altitude of 3300 meters, just under the Copa glacier. It is located in the rural district of Marcará about 34 km from the regional administrative city of Huaráz. Siete Imperios spans a wide range of elevations, from the foot of the glaciers down through pastures, woodlots, and agricultural fields to dry slopes low in the Rio Santa valley. The community was established by the agrarian reform in the 1970s that dissolved large private estates and deeded the land to the families who had lived and worked there. The administrative center of the community lies at the highest point in the district before reaching the boundaries of Huascarán National Park. Tributaries to the Rio Santa form above Siete Imperios from glacial melt-water. These tributaries form the northern and southern borders of the community and are diverted into irrigation canals or potable water reservoirs before joining the Rio Santa below.

Siete Imperios is composed of seven sectors or population centers, totaling about 800 people. The five sectors that make up the administrative center of the community are relatively close to each other, in a small valley between the glacier and several foothills. The other two sectors are found on the lower, more gently sloped foothills just above Marcará. Primarily subsistence agropastoralists, each family has one or several fields where they grow maize, potatoes, and other crops nearly year-round, with glacial meltwater permitting agriculture during the dry season. Glacier retreat threatens meltwater availability, exacerbating the loss of water due to recent diversions from community streams for the potable water system of Marcará.

6.5.1 The role of glaciers and mountains for community, identity, and self-reliance in Siete Imperios

Glaciers are an important element of the landscape in Siete Imperios. From almost every house and farming or grazing parcel, Copa and other dramatic glaciated peaks are visible. The management of the canals that bring meltwater to fields is a key task of the *community* as an institution.

Glaciers are places to visit and harvest with other members of the community. The inhabitants of Siete Imperios point out places much lower on the mountain where they once went to pasture animals and gather wildflowers with medicinal, culinary, or aesthetic qualities that grew just under the ice fields. They also note places, now ice-free, from which they harvested ice to sell in the valley towns or to use at home to prepare treats, known locally as *raspadilla* or *shika-shika*, by mixing shaved glacier ice with sugar and flavorings. The walk from nearby pastures, once easy and enjoyable, is now a several-hour trek, made only infrequently. For those who still visit, "singing" before entering the glacial field continues to be common to encourage any close-to-crumbling edges of the glacier to fall ahead of their arrival.

Though older and younger generations usually do not travel to the glacier together, intergenerational ties are strengthened through oral histories of past experiences with the ice, stories of place-names, tales of encounters with supernatural beings at periglacial lakes, and cautionary practices of how to approach the glacier safely, all contributing to the place-based *identity* as *serranos*, indigenous mountain agriculturalists. Trips to the glaciers are also part of important rituals, cosmologically connected to a larger Andean history. The most significant is the Fiesta de San Juan on June 24, which celebrates the birth of Saint John the Baptist and the summer solstice. Traditionally, communities observe San Juan by trekking to the glacier and burning offerings for the continued fertility of their livestock. These rituals persist but are changing with glacier retreat, not only because of the difficult terrain left by recent recession, but also due to restrictions on burning within the borders of the national park.

Beyond being a social space, the glaciers have long been important to the community's sense of *self-reliance*. The residents of Siete Imperios who witnessed the agrarian reform recount that their first act as an established community after long-established domination by private estates was to improve the canal system delivering glacial meltwater. Families initially farmed communally but fairly quickly changed to communal management of individual parcels so individual households could determine their own cropping preferences. As mentioned earlier, the glacier meltwater permitted irrigation, allowing the production of crops for home consumption and sale throughout the year, which is essential to household livelihoods.

For community members in Siete Imperios, the current water scarcity is due at least in part to institutional factors. Water conflicts, likely driven by awareness of projected physical scarcity, have arisen as a result of the recent Marcará diversion. In addition, the expansion of potable water systems within the community has led to tensions between sectors over access to these systems. The introduction of such systems in neighboring communities has resulted in inter-community border disputes. The community administration anticipates that such disputes will continue to grow. A community official mentioned that the current Siete Imperios administration was considering loosening the current rules which establish a minimum landholding area for membership in water management associations. This shift would allow the community to show higher population numbers for the district water allocation, favoring them in competition with neighboring communities, but placing greater demands on local land and water, and creating the risk of greater conflicts within the community.

Although the current stress stems more from increasing demands for water and overextension of service networks than from environmental change or related glacial loss, when asked about future concerns related to observed glacial loss, water availability alone was the focus. "We will die without water," "we will die of thirst," or "what will we eat without water?" were immediate responses to queries about life when glacier retreat would lead to complete loss of glaciers. Most scoffed at the idea of moving to another place, stating that these glaciers are the source of all water in Peru. Many feel as though there are no solutions to losing the glaciers.

6.6 Case study 3: Glacier and Concrete, North Cascades, USA

The small towns of Glacier and Concrete are located in the North Cascades in Washington State, on the Nooksack and at the confluence of the Baker and Skagit Rivers, respectively. These are the most densely glaciated mountains in the contiguous United States, although few peaks are more than 3000 m in elevation. Although abundant winter precipitation on the west slopes produces a heavy snowpack above 1000 m, rising temperatures have contributed to a 40–60% decrease in the volume of North Cascades glaciers since 1984 [31]. Snowmelt keeps mountain streams running through the summer and feeds numerous large rivers and lakes. Rivers are subject to flooding in the winter when extreme precipitation events combine with unseasonably warm temperatures that bring rain rather than snow to the slopes.

At the time of contact with Euro-Americans, indigenous land use in the region was based on fishing, hunting, and gathering. Euro-American settlers were initially attracted to the region in the late eighteenth century by the discovery of precious metals and the rich supply of timber. The discovery of lime deposits led to the flourishing of the cement industry in Concrete, incorporated in 1909. Glacier was founded in the same year. Around that time, logging also became industrial in scale, with water resources used to power machinery and transport timber. Farmers and ranchers developed the logged-off lands into productive agricultural enterprises. Federal administration of the forest lands began in the 1890s. The construction of dams along the Skagit and Baker Rivers in the first half of the twentieth century provided employment in the hydropower industry. These industries attracted recent immigrants and workers from across the United States.

A highway was built in the 1920s, providing access to ski slopes beyond, which eventually developed into the Mt. Baker Ski Area, providing seasonal income.

The closing of the cement plants in the 1960s and the reduction in logging since the 1970s led to economic decline and depopulation in the region. Today the population of Concrete hovers around 800, and that of unincorporated Glacier around 500. Population is increasing again as new residents are attracted to the area for its scenic beauty, recreational opportunities, rural lifestyle, and lower cost of living. Most of the Baker, Upper Skagit, and North Fork of the Nooksack basins lie within the Mt. Baker–Snoqualmie National Forest, where they are currently managed as wilderness and roadless areas, or in North Cascades National Park, created in 1968. With these designations and the resulting restrictions on resource development, today residents rely mainly on recreation and tourism for income.

6.6.1 The role of glaciers and mountains for community, identity, and self-reliance in Glacier and Concrete

Because the Cascades are a much younger mountain range than the Cordillera Blanca and the Alps, the terrain is too steep and rugged for settlements to be established in proximity to glaciers. Dense forestation further obscures views of glaciers from the study communities and makes them difficult to reach. Residents of Concrete and Glacier are aware of melting glaciers, but only from a distant view of them, others' stories, repeat photography, or awareness of the glacier monitoring research in North Cascades National Park. In addition, many see glacier retreat as part of natural cycles of advance and retreat. Few express concern that glacier retreat would affect them personally, although some expressed concern about the impacts in other places and for the world as a whole. Others expressed concern about how changes in snowfall might affect water runoff and winter recreation. We note as well that a major public agency recently ignored glacier retreat for a locally important matter. When the Federal Energy Regulatory Commission in 2008 issued Puget Sound Electric a 50-year operating license on the hydroelectric facility at Lower Baker Dam on the Baker River, it made no mention of the potential impact of glacier retreat on Mount Baker, the major source of water for the river. Had this matter been discussed, it might have had some influence on local perceptions in Concrete. The facility is well-known in the town, even though it does not employ many locals.

Concrete and Glacier are younger than the other communities discussed in this chapter, but the cultural value of *community* is still important for them. Until recently, community identity centered on residents' shared participation in resource-producing industries. Many interviewees were descendants of the original settlers or arrived to participate in the booming industries up until the 1970s. They emphasized the values of hard work, self-reliance, and community solidarity, which were necessary to subsist and to sustain families in the remote mountain environment, and expressed concern about the loss of environmental knowledge and skills. They also spoke nostalgically about the activities in which people participated as a community, such as ice skating on the Skagit River when it froze (pointing out that it no longer does), and the ways in which everyone worked together in times of need.

Today this shared community identity is changing as employment in resource-producing industries has declined and new residents are moving in. Some long-time residents expressed concern about the new arrivals. On the one hand, some come from more affluent, urban backgrounds and bring a different set of values; on the other hand, some come from more disadvantaged backgrounds and bring increasing problems with crime and drugs. In both cases they threaten the established community identity.

However, in both Concrete and Glacier residents are actively involved in activities that contribute to recreating community identity; at the same time they encourage recreation and tourism and attract new residents and businesses. For example, one new Concrete resident has resuscitated the defunct local newspaper, and one who was born in Concrete, moved away, and returned purchased and refurbished the long-closed Concrete Theater to show movies and host community events. The Skagit Eagle Festival attracts many visitors each winter to see the largest convergence of bald eagles in the contiguous United States. In Glacier, summer tourism and recreation, including a late summer bicycle race, have outpaced that of winter. As community identity is reimagined, many of the values associated with the older community identity are incorporated.

While the residents of Glacier and Concrete do not perceive the glaciers as essential to their lives, mountains more generally are important for their *identity*. The site of earlier economic activities, mountains are where residents hunt and fish, and still serve to keep the communities relatively isolated from urban centers to the west. Mountain vistas are painted on buildings in Concrete, and Sauk Mountain is depicted on T-shirts sold at local festivals. A hike to its summit is considered de rigueur by many Concrete residents. In Glacier, mountains have played a more explicit role in community identity because of the proximity of the Mt. Baker Ski Area and seasonal employment associated with it.

Self-reliance and the related value of independence are important to the Concrete and Glacier community identities, and a source of pride for residents. In the past, these were necessary for survival in the remote mountain environment. Today, these communities are still somewhat isolated (floods and debris flows can interrupt road traffic) and lacking in amenities. For example, a Glacier resident described it as a wonderful place to live. . . . It's a very small community. We have no cell phone service. We have no high-speed Internet. . . . We're thirty-five miles away from medical attention. I'm not saying it's a hard life up here, but you really have to be organized, you really have to think ahead. . . . I love it here. The people that live here are probably some of the most friendly people that you'll ever run across. It's a small town atmosphere of people that like to do what they do and not be told what to do.

In addition, local residents are aware of their political marginality in relation to larger cities, and assert their desire not to be dominated or dictated to by outsiders.

Overall, the interviewees do not see a direct impact of glacier retreat on their lives. For them the greatest challenge is how to maintain community viability in the face of economic decline.

6.7 Discussion and conclusions

We now turn to a comparison and discussion of the cultural values of glaciers in the three cases, drawing on both senses of the term "value," as a standard of assessment of worth and as an application of such a standard to a specific entity. In this chapter we have considered the standards of worth in three distinct communities and have found that they share three of them: the belief that the local *community* should receive recognition and support; the notion that the specific local *identity* warrants protection and promotion; the idea that *self-reliance* is a meritorious form of action. However, these standards are expressed in somewhat different forms in each setting. The specific historical experiences and memories are distinct, as are the hopes and concerns for the future. The physical characteristics of the communities are important as well, since the glaciers are visible in Stilfs, Italy and Siete Imperios, Peru, dominating the landscape, whereas they can be seen from only a few vantage points in Glacier and Concrete, United States. We have found also that these standards of worth apply differently to mountains and to glaciers in each case.

In all three cases, the communities have a high degree of historical self-awareness. The twentieth century looms large for all of them, particularly the intense conflicts of the two World Wars in Stilfs, the agrarian reform in Siete Imperios, and the period of resource extraction in Glacier and Concrete, though all of them are aware of an earlier history, even in the United States, where settlement in the Pacific Northwest expanded in the nineteenth century. They have all experienced recent transformations – the decline of traditional livelihoods and the rise of tourism in Stilfs and in Glacier and Concrete, and the shifts in water governance in Siete Imperios. And they all face environmental, economic, and social uncertainties. The glaciers are key sites of memory in Stilfs (where glaciers serve as important sites of warfare, tourism, and male bravery and prowess) and in Siete

Imperios (where glaciers remain a site of recreation and enjoyment). In these two cases they are sites of anticipation as well, since their retreat threatens the tourism that supports Stilfs and the irrigation supplies that are crucial in Siete Imperios. By contrast, the glaciers play a minor role in Glacier and Concrete; the tourism is associated more broadly with mountains and with snow.

In all three cases, identities are linked to mountains and the countryside. The term *Bergler* used in Stilfs is very close to the term *serrano* used in Siete Imperios, while people in Glacier and Concrete speak more generally of life in the country and near mountains. In all these cases there are oppositions – to the lowland areas of Italy where Italian, rather than German, is spoken; to the coastal desert lowlands of Peru; to the more densely populated lowlands in Washington State, United States. The glaciers are valued more specifically in local identity in Stilfs, where the collection of artifacts from World War I, exposed by glacier retreat, serve a unifying function, helping to overcome the darker past of division during the Option. In Siete Imperios, the glaciers are linked to memories of an earlier and possibly simpler time. Residents offer stories which are filled with many carefree images of time spent in the lands above the villages, close to the glaciers. These identities have a certain element of performance or enactment, both by the young men in Stilfs who show their strength and courage in the rescue service, and by the residents of Siete Imperios who gather wildflowers and take ice for *raspadilla*. These identities have an emotional character as well – the pleasure that people in Stilfs take in the beauty and peace of the glaciers, the enjoyment in Siete Imperios of the times at the foot of the glacier. In this case, Concrete and Glacier are once again distinct, with identities linked more generally to mountains, and, in the case of Concrete, to a closer, unglaciated peak, Sauk Mountain, rather than the taller Mt. Baker.

Glaciers play different roles in the expression of the shared value of self-reliance as well. In Stilfs, glaciers serve as the site of "tunnels" in which men, caught between enemy forces during World War I, could take refuge to survive. In more recent decades they have been a crucial element for tourism. In Siete Imperios, glacial meltwater allows the communities to maintain year-round agriculture, supporting both food security for domestic needs and income generation. In Concrete and Glacier, self-sufficiency comes more generally from the mountains: the abundant forests and streams that provide timber and fish, the deep canyons which isolate them from cell phone networks, the sudden floods which cut off roads.

Two more general points emerge from this discussion of cultural values of glaciers, both as standards of assessment within glacier communities and as applications of these standards to glaciers. The first is that economic and cultural values overlap but are distinct. In all three cases, residents are keenly aware of the importance of monetary income (even in Siete Imperios, where households produce a good portion of their foodstuffs, cash is necessary for basic purchases). And they recognize the links between glaciers and income, particularly in Stilfs and in Siete Imperios. However, the livelihoods are valued not merely by an economic standard, in which value is proportional to income, but rather by cultural standards, in which community survival, the upholding of identities, and the promotion of self-reliance are crucial. The mountain rescue service in Stilfs is not merely a means to meet the economically valued end of income generation; it expresses the identity of the community by showing the character of the young men and by affirming the continuity with the past. Similarly, the production of *raspadilla* in Siete Imperios is not just a way of producing a food item, of economic value, but, in a parallel fashion, shows the character of residents who can climb upslope to the glacier front and affirms continuity. In this way, hunting and fishing in Concrete and Glacier play a similar role in relation to mountains, simultaneously producing economically valued foodstuffs and affirming identity and self-reliance.

The second is that the cultural values apply not only to glacier retreat but also to other social, economic, and political factors. In Stilfs, the future of tourism is affected not only by glacier processes but also by the special status of South Tirol as an Autonomous Province, which grants it certain concessions, and by European Union policies on migration which affect the availability of workers for tourist enterprises. In a similar fashion, water issues in Siete Imperios revolve around investment in infrastructure and shifts in governance, as much as changes in melt-water availability. The character of Glacier and Concrete as mountain communities is influenced as much by new patterns of migration as by environmental dynamics.

This chapter has shown that responses to climate change risks need to be seen and developed within a context of social, cultural, economic, and political change, taking into account the diversity and limits of adaptation. It is important to concentrate not only on economic valuation that can attach a price to water and a cost to hazards, but also on cultural values. The cases in this chapter demonstrate the importance of such values to local residents. These cultural values have a great capacity to mobilize people into action, whether to operate a rescue service that supports tourism, to maintain irrigation canals, or to sponsor new festivals. A fuller understanding of such values can promote a fuller incorporation of glacier communities in the formulation and implementation of plans to address glacier retreat.

Acknowledgments

We would like to thank the residents of Stilfs, Italy, Siete Imperios, Peru, and Concrete and Glacier in Washington, United States, who gave their time to make this research possible.

References

1. KL O'Brien, J Wolf, A values-based approach to vulnerability and adaptation to climate change. *Wiley Interdisciplinary Reviews: Climate Change*, **1** :2 (2010), 232–242.
2. UNFCCC. Summary note: outcomes of the work programme to consider approaches to address loss and damage associated with climate change impacts in developing countries that are particularly vulnerable to the adverse effects of climate change – (AC/2013/8), (2013). http://unfccc.int/files/adaptation/cancun_adaptation_framework/adaptation_committee/application/pdf/l_and_d_summary_25_feb.pdf.
3. J Morrissey, A Oliver-Smith, Perspectives on non-economic loss and damage. (2013). www.loss-and-damage.net/download/7213.pdf.
4. D Graeber, *Toward an Anthropological Theory of Value: The False Coin of Our Own Dreams* (Basingstoke: Palgrave Macmillan, 2001).
5. A Appadurai, *The Social Life of Things: Commodities in Cultural Perspective* (Cambridge: Cambridge University Press, 1986).
6. S Graham, J Barnett, R Fincher, A Hurlimann, C Mortreux, E Waters, The social values at risk from sea-level rise. *Environmental Impact Assessment Review*, **41** (2013), 45–52.
7. UNESCO, *Operational Guidelines for the Implementation of the World Heritage Convention* (Paris: World Heritage Centre, 2013).
8. B Orlove, E Wiegandt, BH Luckman (eds.), *Darkening Peaks: Glacier Retreat, Science, and Society* (Berkeley, CA: University of California Press, 2008).
9. M Carey, The history of ice: How glaciers became an endangered species. *Environmental History*, **12**: 3 (2007), 497–527.
10. J Cruikshank, *Do Glaciers Listen? Local Knowledge, Colonial Encounters, and Social Transformation* (Vancouver: University of British Columbia Press, 2005).
11. F Toennies, *Gemeinschaft und Gesellschaft* (Leipzig: Fues Verlag, 1887).
12. F Barth, *Ethnic Groups and Boundaries* (Boston, MA: Little, Brown, 1969).
13. A Cohen, Community. In *Social and Cultural Anthropology: The Key Concepts*, eds. N. Rapport, J. Overing (London and New York: Routledge, 2002), pp. 60–64.
14. A Escobar, Culture sits in places: reflections on globalism and subaltern strategies of localization. *Political Geography* **20**: 2 (2011), 139–174.
15. M Fried, Continuities and discontinuities of place. *Journal of Environmental Psychology*, **20** (2000), 193–205.
16. M Foucault, *Les Mots et les choses* (Paris: Editions Gallimard, 1966).
17. M Sökefeld, Debating self, identity, and culture in anthropology. *Current Anthropology*, **40**: 4 (1999), 417–448.
18. P Devine-Wright, Think global, act local? The relevance of place attachments and place identities in a climate changed world. *Global Environmental Change*, **23**: 1 (2013), 61–69.
19. R Messner, *König Ortler* (Lana, Italy: Tappeiner Verlag, 2004).
20. L Carturan, R Filippi, R Seppi, *et al.*, Area and volume loss of the glaciers in the Ortles-Cevedale group (Eastern Italian Alps): controls and imbalance of the remaining Glaciers. *The Cryosphere*, **7** (2013), 1339–1359.
21. ASTAT, *Südtirol in Zahlen-Alto Adige in cifre* (Bozen, Italy: Südtirol Landesinstitut für Statistik, 2011).
22. J Rampold, *Vinschgau*, 2nd edn. (Bozen, Italy: Verlagsanstalt Athesia, 1997).
23. J Hurton, *Sulden: Geschichte, Land, Leute und Berge*, 7th edn. (Bozen, Italy: Eigenverlag, 2004).
24. G Grote, *The South Tyrol Question, 1866–2010: From National Rage to Regional State* (Bern: Peter Lang, 2012).

25. M Zappe, *Das ethnische Zusammenleben in Südtirol : Sprachsoziologische, sprach-politische und soziokulturelle Einstellungen der deutschen, italienischen und ladinischen Sprachgruppen vor und nach den gegenwärtigen Umbrüchen* (Frankfurt am Main: Peter Land, 1996).

26. ASTAT, *Das neue Autonomiestatut*, 14th edn. (Bozen, Italy: Südtirol Landesinstitut für Statistik, 2009).

27. M Beniston, DG Fox, Impacts of climate change on mountain regions. In *IPCC 1995: Climate Change: Impacts, Adaptations and Mitigation of Climate Change: Scientific-Technical Analysis (Working Group II)* (Cambridge: Cambridge University Press, 1996), pp. 191–213.

28. E Ramírez, B Francou, P Ribstein, *et al.*, Small glaciers disappearing in the Tropical Andes: a case study in Bolivia; the Chacaltaya Glacier (16°). *Journal of Glaciology.* **47** :157 (2001), 187–194.

29. G Kaser, H Osmaston, *Tropical Glaciers* (Cambridge: Cambridge University Press, 2002).

30. R Bradley, SM Vuille, HF Daz, W Vergara, Threats to water supplies in the tropical Andes. *Science*, **312** (2011), 1755–1756.

31. MS Pelto. Impact of climate change on North Cascade alpine glaciers, and alpine runoff. *Northwest Science*, **82** :1 (2008), 65–75.

Part II
Processes

7

Implications for hazard and risk of seismic and volcanic responses to climate change in the high-mountain cryosphere

BILL MCGUIRE

7.1 Introduction

The climate and the solid Earth are not isolated geophysical systems, but are bound in such a way that a significant change occurring within one can result in a clear and measurable response in the other. Notably, the exceptional swings in global climate that characterised much of the Quaternary are associated with dynamic responses from the geosphere, involving the adjustment, modulation or triggering of a broad range of surface and crustal phenomena that include seismic and volcanic activity [1]. This response is especially strong in areas of diminishing cryosphere, where major changes in ice mass are linked to large variations in the load exerted upon the crust and underlying mantle in general, and on active faults and extant volcanic systems in particular. Following the Last Glacial Maximum (LGM) $c.20\,000$ years ago, the biggest responses occurred in areas where major ice sheets were losing mass rapidly, resulting in a rapid diminution of load on the underlying crust; for example triggering $M = \sim 8$ earthquakes on faults in Scandinavia [2–4] and promoting a wholesale increase in the level of volcanic activity in Iceland [5,6]. Rapid warming during the Late Pleistocene and the Holocene also resulted in significant and widespread modifications to the high-mountain cryosphere, which was dramatically reduced in both extent and thickness. The post-glacial reduction in area of the European Alpine Ice Cap has, for example, been implicated in coincident raised levels of seismic activity [7–9], as has the loss of ice cover across the Teton Range (Wyoming and Idaho, United States) during the Late Pleistocene [10,11]. Similarly, ice recession over the same period, at mountain volcanoes as widely distributed as the western United States, southern Kamchatka (Russia), the Andes, Mexico and New Zealand, has been charged with accelerating eruptive activity or promoting edifice collapse [12,13].

The High-Mountain Cryosphere, ed. Christian Huggel, Mark Carey, John J. Clague and Andreas Kääb. Published by Cambridge University Press. © Cambridge University Press 2015.

As anthropogenic climate change accelerates, forecasts for the survival of the high-mountain cryosphere are bleak. In its 5th Assessment Report [14], the IPCC observes that is it very likely that between 1993 and 2012 the average rate of ice loss from the world's glaciers (excluding those marginal to the ice sheets) was 275 (140–410) Gt yr^{-1}. By the end of the century, global glacier volume (excluding glaciers around the margins of Antarctica) is projected to lessen by up to 85%. The resulting combination of diminishing load on the crust, groundwater pore pressure increases around meltwater lakes, the removal of buttressing ice from the flanks of mountain volcanoes, and increased opportunities here for magma and meltwater to come into contact, supports a potential increase in seismic and volcanic activity in glaciated mountainous terrain (Table 7.1). In this regard, it is significant, and concerning, that with just ~0.85 (0.65–1.06) °C of warming over the period 1880–2012, a measurable seismic response to ice-mass loss has already been recognised in southern Alaska [15]. Looking ahead, continued rapid thinning of glacier ice across tectonically active regions including Alaska and the western United States, the Andes, the European and Southern Alps and the Himalayas, and on glaciated volcanoes such as Mount Rainier (Washington State, United States), Katla (Iceland) and Sollipulli (Chile), raises the prospect of tectonic and volcanic responses capable of impinging locally or regionally upon human society and economy.

7.2 Ice retreat and earthquakes

A correlation between times of significant ice retreat and a seismic response from the crust is well established, both in relation to Late Pleistocene and Holocene ice sheet deglaciation in Scandinavia [16,17] and to a lesser extent eastern Canada [18], and to more recent ice-mass loss in southern Alaska [15]. The mechanism whereby ice retreat promotes earthquakes is also well constrained and can be summarised in terms of loading and flexure of the lithosphere by ice, reducing the slip rates on proximal faults, whereas the unloading and rebounding of the lithosphere promotes the acceleration of accumulated slip on the same faults [10,11], resulting in earthquakes with larger than expected (in the absence of glaciation) magnitudes.

The coincidence in space and time of major, receding, ice sheets and palaeo-earthquakes was recognised in the Lapland Fault Province of northern Fennoscandia as long ago as the 1970s [e.g. 2,3]. Here, in the heart of a supposedly tectonically stable craton, exposed neotectonic structures, taking the form of fault scarps, some in excess of 150 km long and 15 m high, testify to the occurrence of earthquakes as large as M_w (moment magnitude) = 8 [19,20]. In fact, geomorphological and stratigraphical evidence supports the idea that all the main post-glacial

Table 7.1 *Principal cryosphere retreat phenomena, potential seismic and volcanic responses, and associated hazards*

Cryosphere retreat phenomenon	Potential seismic and volcanic response	Resulting hazards
Ice thickness reduction	Increased seismicity due to pressure release; elevated levels of volcanic activity; increased magma production in asthenosphere	Seismic shaking; remote ground liquefaction and/or seismic wave amplification; rock and ice falls; debris avalanches; GLOFs; debris flows; increase in explosive volcanic activity; volcanic debris flows and floods
Loss of buttressing ice	Volcanic edifice destabilisation and structural failure due to reduced mechanical support	Rockfalls; debris avalanches; wholesale lateral collapse; possible triggering of eruptions
Elevated water tables and meltwater saturation of volcanic edifices	Volcanic edifice destabilisation and structural failure due to increased pore water pressures; incorporation of saturated wall rock into stored magma	Rockfalls; debris avalanches; wholesale lateral collapse; possible triggering of eruptions; raised potential for explosive eruptions
Exposure of volcanic debris fields due to ice retreat	Precipitation/meltwater mobilisation of surface material	Volcanic debris flows
Expanding glacial meltwater lakes	Increased seismicity due to increased pore water pressures	Seismic shaking; remote ground liquefaction and/or seismic wave amplification; rock and ice falls; debris avalanches; GLOFs; debris flows
Catastrophic draining of meltwater lakes	Increased seismicity due to pressure release	Seismic shaking; remote ground liquefaction and/or seismic wave amplification; rock and ice falls; debris avalanches; GLOFs; debris flows

faults in the region formed in concert with deglaciation [17]. Examples include northern Sweden's 50 km long Lansjärv Fault, which appears to have ruptured just years to decades after being uncovered by retreating ice [21], and the 165 km long Pärvie Fault (Figure 7.1), also in northern Sweden, which developed while part of its surface trace was still covered by ice [22]. End-glacial seismicity is also recorded in association with the melting of smaller ice bodies, such as the Yellowstone Ice Cap and late Pleistocene glaciers in the western United States,

Figure 7.1 The surface trace of the end-glacial Pärvie Fault between Lake Kamasjaure and Mount Tsåktso, 70 km north of Kiruna in northern Sweden. Source: R. Lagerbåck, with permission.

which buried the northern portion of the Teton normal fault in the Basin and Range Province. Here, a burst of earthquake activity is recorded between 8000 and 14 000–16 000 years ago, following deglaciation, involving 70% of the post-glacial slip on the Teton Fault. Modelling strongly supports a mechanism for significantly increased slip over this interval involving ice unloading [10,11].

Beyond ice sheet margins, Muir-Wood [17] proposes that elevated levels of post-glacial seismicity reflect the interplay of tectonic forces and an outward-migrating wave of strain release arising from the progressive collapse of a so-called 'forebulge' – the compensatory upward-flexed zone peripheral to areas of glacially depressed lithosphere. This mechanism is advocated [17] as underpinning patterns of post-glacial seismicity in the UK and northern Europe, and explaining large intraplate seismic events in eastern North America that occurred several thousand years after decay of the Laurentide Ice Sheet; notably, the $M = 7+$ early nineteenth-century earthquakes in the New Madrid Seismic Zone (Missouri), and an $M = \sim 7$ shock that struck Charleston (South Carolina) in 1886.

Evidence for a post-glacial seismic response to a diminishing cryosphere is also recognised in high-mountain terrain. The European Alps, for example, hosted a 2 km thick ice cap at the LGM and earthquake catalogues for both the French and Swiss sectors support a period of elevated seismicity following the LGM [7,8], which is likely the result of deglaciation-related rebound. More specifically, the melting of thick valley glaciers has been charged with the formation of unusual

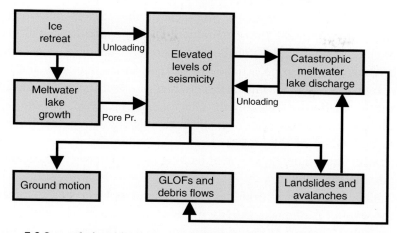

Figure 7.2 Interrelationships between high-mountain cryosphere retreat, elevated levels of seismicity and associated hazards. Pore Pr. = pore water pressurisation.

~10 m high uphill-facing fault scarps in the central Swiss Alps, which parallel the valley axes, with the faults arising from the more rapid rebounding of the valley floors in comparison to the margins [23]. Accelerated post-glacial slip is also recorded on a normal fault in the southern part of the Upper Rhine Graben, which has been implicated in the $M = \sim7$ earthquake that destroyed Basel in 1356, and linked to the retreat and melting of glaciers in the Alps, the southern Vosges (France) and southern Black Forest region (Germany) [24].

Earthquakes occurring in high mountains are capable of triggering a range of hazardous phenomena with the potential not only to have local impacts, but also to cause damage, loss of life and disruption at lower altitudes (Figure 7.2). In addition to the effects of strong ground motion on local buildings and other structures, these impacts include the triggering of major rock and/or ice falls, which can transform into debris avalanches or mix with meltwater or ice and snow to form far-reaching debris flows. The seismically triggered collapse of rock faces into meltwater lakes may also cause over-topping, resulting in potentially devastating glacial lake outburst floods (GLOFs). Strong seismic shaking may itself directly cause the failure of the natural dams impounding meltwater lakes.

The link between historical ice retreat and increased levels of seismicity in high-mountain regions is best established for southern Alaska, where more than 3000 km^3 of ice has been lost from the Glacier Bay area since 1770, and total thinning, locally, has been as great as 1.5 km [25]. Sauber and Molnia [15] evaluated the impact of ice-mass changes in the region between the 1899 Yakataga and Yakutat earthquakes (for both $M_w = 8.1$) and the 1979 St. Elias earthquake ($M_s = 7.2$), during which time tectonic strain continued to accumulate. They concluded that the resulting cumulative reduction in fault stability at seismogenic depths was up to ~2 MPa. This, the authors determined, was sufficiently high to promote thrust faulting of the type associated with the 1979 St. Elias earthquake and its aftershocks. Short-term changes in seismicity in southern Alaska have also been linked to changes in ice thickness and hence in the load exerted by ice on the crust. Observations of seismic shocks ($M \geq 2.5$) and ice-thickness variations of the Bering Glacier [26,27] reveal that during surge events, the glacier reservoir area was more seismically active than its receiving area. The orientation of the crustal stress field within the ice reservoir area is consistent with thrust faulting, a style of faulting that would be enhanced due to ice thinning during surges [27].

Accumulating meltwater from retreating high-mountain glaciers may also elevate the level of earthquake activity by increasing pore-water pressures acting on active faults marginal to large meltwater lakes. The role of elevated pore pressures in triggering earthquakes is well established in the context of filling man-made reservoirs, such as at Koyna (India) in 1967 [28] and, more contentiously, at Zipingpu (China) in 2008 [29]. Based on observations of seismicity in southeast Germany, Hainzl *et al.* [30] demonstrate that the crust can sometimes be so close to failure that even tiny (less than 1 kPa) pore-pressure variations can trigger earthquakes in its top few kilometres. Additionally, Christiansen *et al.* [31] speculate that pore-pressure or load-related changes, on the order of 2 kPa, might modulate seismicity on a creeping section of the San Andreas Fault in the vicinity of Parkfield. Looking ahead, it is worth noting that meltwater lake depth changes of just 10–20 cm would be sufficient to promote variations in load and pore pressure comparable to those implicated above.

In a positive feedback, the seismic threat to natural dams of glacial debris that typically impound growing pro-glacial meltwater lakes may itself be increased as a consequence of rising pore pressures around the expanding lake margins. Resulting earthquakes may promote dam failure as a consequence of the strong ground motions or due to shake-induced avalanches or landslides displacing lake waters that in turn cause dam over-topping or erosion and the formation of potentially devastating GLOFs and debris flows. Within this scenario, growing meltwater lakes may contain within themselves the seeds of their own catastrophic failure. In the Swiss Alps, strong ground motion arising from earthquake activity has been proposed as the trigger of the 2.5 km^3 volume Late Pleistocene outburst of Lake Zurich [32].

In a further complication, the sudden draining of glacial meltwater lakes in mountain landscapes may itself promote earthquakes, this time on active faults directly beneath the water bodies. Sudden unloading of the crust due to reductions in water loads has been implicated in promoting earthquake activity at a number of locations and at a range of scales, including at Glen Roy in the Scottish Highlands at the end of the Younger Dryas [33] and in association with the catastrophic draining of the great Late Pleistocene lakes, Bonneville and Lahontan, in the Basin and Range Province of the western United States [10,11].

7.3 The hazard consequences of ice retreat in volcanic landscapes

There is compelling evidence for a causative link between episodes of Late Pleistocene and Holocene deglaciation and increased levels of volcanic activity, related either to surface load reduction due to ice cover thinning [34,35] or to the effects of rising global sea level on the stress regimes of ocean margins and the volcanoes located thereon [36]. The most convincing example is provided by the Icelandic volcanoes, which were buried beneath ~1 km of ice at the LGM. In response to rapid melting, the level of volcanic activity rose by between 30 and 50 times, starting around 12 000 years ago and continuing for 1500 years [34]. The response was in large part a function of decompression melting within the underlying asthenosphere, as the formerly depressed lithosphere rebounded by around 0.5 km. As a consequence, magma production is estimated to have increased 30-fold [34,35].

Elsewhere, accelerated deglaciation has been proposed as a driver of increased levels of volcanic activity in the Eifel Mountains (Germany) and Massif Central (France) [37]; eastern California [38]; southern Kamchatka (Russia) [39]; Mount Mazama (Oregon State, United States) [40]; and at the Nevados de Chillan [41], Lascar [42] and Puyehue [43] volcanoes in the Chilean Andes. Globally, Huybers and Langmuir [44] find a statistically significant correlation, over the past 40 000 years, between deglaciation and the number of recorded volcanic eruptions greater than 2 on the Volcanic Explosivity Index.

Deglaciation has also been linked to edifice failure and flank collapse at mountain volcanoes, leading to the formation of massive landslides. Capra [45] attributes collapses at 24 volcanoes, located mainly in the Andes, Mexico, the Cascade Range and New Zealand, to episodes of rapid ice cover thinning over the past 30 000 years. A range of causes are proposed for destabilisation and ultimate structural failure, notably mechanical debuttressing due to ice loss and increases in fluid circulation and precipitation. A much larger study by Lowe [46] suggests that over the last 40 000 years, the incidence of volcano flank failure and giant landslide formation increases when the climate becomes warmer and wetter (Figure 7.3) following drier, colder conditions, including after the Younger Dryas and other short-lived colder intervals during the Holocene.

As a consequence of anthropogenic warming, the mountain cryosphere supported by active volcanoes is rapidly diminishing, both in terms of thickness and area [12,13]. At the largest scale, Iceland's Vatnajökull Ice Cap, which buries or

Figure 7.3 The Valle del Bove lateral collapse amphitheatre, excavated from the east flank of the volcano, was formed around 7000 years ago. Magma intrusion into a water-saturated edifice has been proposed as a likely cause [47].

partly buries several active volcanoes, including Grimsvötn and Katla, lost 435 km^3 of ice between 1890 and 2004 [48] and continued to thin by 0.8 m per year between 1995 and 2008 [49]. At individual volcanoes too, ice cover is quickly dissipating. The glaciated area on Cotopaxi (Ecuador) reduced from a little over 19 km^2 in 1976 to 13.4 km^2 in 1997 [50], while ice cover on Nevado del Ruiz (Colombia) shrank from between 19 and 25 km^2 in 1985 to just over 10 km^2 in 2002–2003 [51,52].

Based upon current rates, Tuffen [12,13] predicts that ice cover on many glaciated volcanoes is likely to thin by around 50–150 m by 2100. This may raise the potential for increased hazards due to: (1) a reduction in surface load; (2) the formation of increased volumes of meltwater; (3) the elevation of water tables; (4) exposure of new areas of unconsolidated volcanic debris; and (5) the development of surface water bodies in summit craters, calderas and other depressions.

The reduction in surface load is likely to have the greatest impact on the mountain volcanoes beneath Iceland's Vatnajökull Ice Cap, where glacial rebound is already occurring at rates exceeding 20 mm per year [53,54], and for which a 25% ice volume reduction has been forecast by 2060 [55]. Here, modelling predicts a significant pressure release due to ice retreat, resulting in an annual increase in magma formation in the underlying mantle of 0.014 km^3 [49], and leading to more voluminous or more frequent eruptions [53]. It is estimated that future ice retreat will result in 0.07–0.17 km^3 per year of additional magma across all the island's rift zones [54].

Ice thinning at volcanoes buried beneath thick cryosphere may also promote more explosive eruptions as the ice load pressures are reduced. According to Tuffen [12,13], such a response is most likely due to occur if ice cover of more than 300 m thins, as is the case, for example, in the deep, ice-filled calderas of Sollipulli (Chile) and Katla (Iceland). Tuffen [12,13] also notes, however, that 100 m or more of thinning of ice in excess of 150 m thick will ensure that the probability of more hazardous explosive eruptions increases. Furthermore, Geyer and Bindeman [56] have proposed that the incorporation of water-saturated wall rock into stored magma may raise the volatile content, thereby increasing the potential for explosive eruptions of volcanoes that host melting ice cover.

As ice retreat accelerates, structural failure and lateral collapse of mountain volcanoes may be promoted, either due to the loss of mechanical support [45,56] or to the pressurisation of shallow hydrothermal systems as a consequence of perco-lating meltwater and elevated water tables; this, in turn, promotes lubrication and failure along existing weak zones [45].

Ice retreat on the summit or flanks of mountain volcanoes may expose tracts of unconsolidated debris that can serve as sources for debris flows arising either from the eruption of hot material across remaining ice or as a consequence of intense rainfall. Loss of life due to rainfall mobilisation of volcanic debris occurred at Sarno (Campania, Italy) in 1998 [57] and at Nicaragua's Casita volcano later the same year [58]. Additionally, meltwater may pond in surface craters, calderas or other depressions and break out to generate meltwater floods once water levels reach low points in the bounding walls or when the water mass, or pore pressurisa-tion of the retaining walls, causes mechanical failure. Such an event occurred at Ruapehu (New Zealand) in 2007 [59].

7.4 Future risk to society and economy and potential impacts

Close to half a billion people now live in seismically vulnerable settings [60]. Coincidentally, about the same number live within the danger zones around active volcanoes [61], and many hundreds of millions more are susceptible to the broader impacts of large, explosive eruptions. More than 85% of all the world's recorded earthquake fatalities have been in the Alpine–Himalayan Collision Zone. The eastern margin of the Pacific, too, is especially seismically active, hosting 12% of catalogued fatal earthquakes [60]. Both regions are characterised by high mountains and/or active volcanoes.

7.4.1 The seismic threat

So far, a clear link between historical ice retreat and increased levels of seismicity has only been established for southern Alaska [15,26,27]. As rising global temperatures accelerate the shrinking of mountain cryosphere, however, it is likely that similar relationships will be established for other parts of the world where ice thinning is occurring on a significant scale. These may be areas such as the Southern Alps, where levels of seismic strain are high and driven by plate tectonics, or regions that have previously been regarded as aseismic, or largely so. The impact of rapid ice thinning is particularly pertinent to the Greenland and Antarctic ice sheets, which are areas of very low seismicity. It may be that the crust in these regions has simply adjusted to the enormous glacial loads exerted upon them, and the rates of low seismic strain release reflect their intraplate setting. The alternative is that – like northern Scandinavia at the LGM – these regions host seismogenic faults that have been accumulating strain for perhaps hundreds of thousands of years and that would be available for release given sufficient future ice unloading. Indeed, modelling by Hampel *et al.* [10,11] supports the idea that the seismic calm that characterises Greenland and Antarctica is a reflection of the presence of the large ice sheets and suggests that an increase in seismic activity in these regions can be expected as melting accelerates, perhaps on timescales as short as decades to centuries.

Away from the great polar ice sheets, the load reductions on active faults hidden beneath ice cover in the high-mountain crysosphere will – in most cases – be significantly smaller. Nonetheless, glacier thicknesses of 1 km are encountered in many of the world's high-mountain ranges, including the European Alps, Southern Alps, Saint Elias Mountains (Alaska), Andes and Himalayas. As demonstrated for southern Alaska [26], an ice thickness reduction of just 50 m results in a 0.5 MPa fall in vertical compressive stress; an amount likely to enhance slip on thrust faults. It is noteworthy that such faults are characteristic of the compressive tectonics typically associated with the formation of high-mountain ranges.

Looking ahead, therefore, an increased role for a warming climate in triggering or accelerating seismic activity in deglaciating high-mountain landscapes can be expected, through a combination of ice unloading, pore-pressure increases in the vicinity of developing meltwater lakes and the catastrophic draining of these lakes. Where resulting pressure changes are large, and faults of appropriate geometry and length sufficiently strained, earthquakes of $M \geq 7$ are possible. Given relatively low population densities in high-mountain areas, any resulting damage and loss of life will, inevitably, be lower than for a comparably sized earthquake in an urbanised region. Nonetheless, for larger magnitude events, shaking intensities may be damaging and lethal to distances of hundreds of kilometres, particularly to communities built on soft sediment or landfill, which can liquefy and/or amplify shaking. Strong shaking may also cause the breaching of high-mountain hydropower dams, resulting in far-reaching catastrophic floods. Consequently, an earthquake triggered by cryosphere loss at high elevations may impinge detrimentally on higher population densities at lower altitudes.

The geographical reach of earthquakes forced by climate change, in relation to impacts on society and economy, may also be significantly increased as a consequence of secondary hazards, including rock avalanches, debris flows and GLOFs, triggered by strong ground motion and/or the breaching or overtopping of meltwater lakes, or adjacent slope instabilities imposed by rising lake levels. Rock avalanches and, in particular, debris flows, can remain destructive over considerable distances. The 2002 Kolka-Karmadon rock and ice avalanche (Caucasus), which took 125 lives, travelled 32 km, while most of the 23 000 people killed in 1985 by debris flows triggered on Colombia's Nevado del Ruiz volcano resided 70 km from the volcano. Valley-constrained rock avalanches, debris flows and floods are able to maintain high velocities and volumes over distances of 100 km or more [52].

Globally, hazards triggered as a direct or indirect consequence of earthquakes promoted by ice retreat in high mountains have the potential for widespread impacts. Most at risk are high-elevation communities and resorts, alongside population centres sited within the valleys that drain mountain ranges and at their mouths. For earthquakes of sufficient magnitude, liquefaction and amplification effects may extend resulting damage to distances in excess of 100 km from the source. The greatest risk lies in those high-mountain environments where seismogenic faults are known to exist or may be present; that host significant numbers of communities or have valleys that drain into areas of higher population density; that are characterised by the rapid melting of thick (more than hundreds of metres) glaciers and ice fields; and where growing meltwater lakes are commonplace (Figure 7.4). Candidates include Nepal, Bhutan, Kyrgyzstan, the Himalayan regions of Pakistan, India and China, the Peruvian Andes, Chilean and Argentinian Patagonia and the Southern Alps (New Zealand).

7.4.2 The volcanic threat

The eruption of additional magma at deglaciating mountain volcanoes must translate either to more frequent or larger eruptions, or a combination of the two. This, in turn, would lead to heightened risk to nearby populations and infrastructure, and even, in the case of explosive eruptions, to a raised threat of regionally disruptive events. Constraining the level of elevated risk may, however, prove to be difficult, if not impossible, and there is, as yet, no concrete evidence for an increase in either the frequency or size of eruptions during the twenty-first century at ice-supporting volcanoes. Any response from deglaciating volcanoes may, in fact, take significantly longer to make itself apparent. Jellinek *et al.* [38] suggest that the delay between unloading and eruption, following past ice retreat episodes in eastern California, may have been as long as thousands of years. Looking ahead,

Figure 7.4 Rapid melting of Himalayan glaciers is resulting in the formation and expansion of pro-glacial meltwater lakes in Tibet (top of image) and Bhutan (bottom). Earthquakes linked to ice retreat may promote catastrophic lake discharges, either due to strong ground motions or to the triggering of rockfalls or landslides into the lakes.
Source: NASA.

Sigmundsson *et al.* [54] note that for the Vatnajökull volcanoes, new magma generated in the underlying mantle due to ice-mass loss may take longer than decades or centuries to reach the surface. While ice retreat may result in a ~1% increase in melt production beneath the Vatnajökull volcanoes over the twenty-first century, therefore, it may be that this magma will only be erupted in following centuries [12,13]. It is also the case that the episodic nature of volcanic activity may make it impossible for any increase in activity to be statistically distinguishable from background.

Nonetheless, evidence for the extraordinary sensitivity of some active volcanic systems to surface load changes supports the idea that some volcanoes at least may respond rapidly to ice retreat. Modelling [53,54] suggests, for example, that eruptions at Katla, which seem to occur preferentially in the summer months [62], are triggered by the loss of seasonal snow cover that varies by 6 m from winter to summer, despite the resulting load reduction being just a few kPa. Similarly, eruptions of Pavlof volcano (Alaska) seem to have been modulated by surface loading arising from seasonal changes in local sea level amounting to less than 20 cm, equivalent to 2 kPa [63,64]. Furthermore, the recognition of a global annual volcano 'season' [65], attributed to minuscule Earth surface deformation driven by yearly variations in the planet's hydrological cycle, argues for many active volcanoes displaying extreme sensitivity to very small changes in their external environment.

Distinguishing eruptions triggered at deglaciating volcanoes from those that would have happened anyway may prove impossible in most circumstances, but linking other hazardous events at active volcanoes to ice retreat is likely to be more straightforward. These events might include catastrophic flood generation due to the breaching of meltwater-filled calderas or craters, debris flow formation as a consequence of heavy rains falling on unconsolidated debris fields exposed by ice retreat, and debris avalanche formation occurring as a result of the collapse of a formerly ice-buttressed flank.

Perhaps the largest increase in risk involves eruptions promoted by ice retreat at mountain volcanoes that still maintain significant levels of ice cover and/or glacial meltwater. Evidence for the potentially lethal and destructive nature and disruptive capacity of such events is provided, in particular, by two eruptions in the last half-century. In 1985, the second largest volcanic disaster death toll of the twentieth century (23 000) resulted from debris flows fed by a meltwater flood caused by the passage of pyroclastic flows across ice fields capping Colombia's Nevada del Ruiz volcano [66]. In 2010 the explosive eruption of Iceland's Eyjafjallajökull volcano generated an ash column that resulting in the cancellation of more than 100 000 flights and losses to the aviation industry totalling €1.3 billion [67]. Further damage and loss of life arising from the interplay of ice and magma occurred at Cotopaxi (Ecuador) in 1877, Tokachi-dake (Japan) in 1926 and Villarrica (Chile) in 1971, while the 2011 eruption of Grimsvötn resulted in further – if less extensive – aviation disruption.

In some circumstances, however, ice retreat may reduce the risk of destructive or disruptive hazards. For Katla (Iceland) volcano, for example, Sigmundsson *et al.* [54] demonstrate that certain peculiarities of ice unloading and magma reservoir form could mean that ice retreat might actually inhibit eruptive activity. More generally, the complete or near-complete loss of ice cover at mountain volcanoes would result in a significant fall, ultimately to zero, in the risk of debris flows triggered by meltwater discharges or the eruption of hot material across ice fields. On some volcanoes, at least, this might be partly compensated for by the raised risk of debris flows sourced by heavy rain falling on newly exposed slopes of ash and debris.

Looking ahead, some specific glaciated volcanoes can be identified as presenting a potentially significant level of increased risk to adjacent population centres and/or infrastructure as a consequence of the ramifications of ice retreat. Notably, Redoubt volcano (Alaska); the ice-supporting peaks of the Cascade Range in the western United States, in particular Mount Rainier (Figure 7.5) and Glacier Peak (both Washington State); the high Andean volcanoes of South America, including Cotopaxi (Ecuador), Sollipulli, Llaima and Villarrica (all Chile) and Nevado del Ruiz (Colombia); and Ruapehu (New Zealand). In addition, the volcanic peaks

Figure 7.5 Loss of glacier mass on the Mount Rainier volcano (Cascade Range, western United States) will increase the potential for landslide and debris flow formation, threatening adjacent communities including Tacoma.
Source: USGS/Cascades Volcano Observatory.

covered by Iceland's Vatnajökull Ice Cap – including Katla, Bárdarbunga, Kverkfjöll and Öraefajökull – present a two-fold threat should the frequency or size of future eruptions be forced by thinning ice cover; first, as a consequence of the formation of jökulhlaups (glacial outburst floods); and second, and potentially more disruptively, through the formation of ash columns that extend across European and North American airspace.

7.5 Conclusions

The evidence that ice retreat due to past climate change elicited elevated levels of seismic and volcanic activity is incontrovertible, and the underlying mechanisms are well understood. The rapid melting of the high-mountain cryosphere is already driving increased seismicity in southern Alaska, and it is not unreasonable to suppose that a comparable response may become evident across mountain landscapes, wherever the substantial reduction of thick ice cover acts to unload subglacial seismogenic faults. This seismic threat can be expected to be compounded by raised pore-pressures associated with expanding meltwater lakes, and further crustal unloading due to lake outbursts. Taken together, the resulting hazards, potentially including strong ground motion, GLOFs, debris flows and ice and debris avalanches, present an increased level of risk to mountain communities and those living at lower elevations within the local watersheds. Ice thinning and retreat on glaciated volcanoes raises the prospect of an increased threat of explosive eruptions and flank collapse, in addition to smaller-scale mass movements linked to eruptions or slope instabilities. Where the future loss of ice cover is substantial (e.g. Iceland's Vatnajökull Ice Cap), an eventual rise in magma production in the underlying mantle and an increase in the volume of magma erupted at the surface may be the result, although this may not be statistically distinguishable from a background level of activity that is, in any case, highly variable.

References

1. WJ McGuire, Potential for a hazardous geospheric response to projected future climate changes. *Philosophical Transactions of the Royal Society A*, **368** (2010), 2317–2346.
2. J Lundquist, R Lagerbäck, The Pärve fault: a lateglacial fault in the Precambrian of Swedish Lapland. *Geologiska Föreningens i Stockholm Förhandlingar*, **98** (1976), 45–51.
3. N-A Mörner, Faulting, fracturing and seismic activity as a function of glacial-isostasy in Fennoscandia. *Geology*, **6** (1978), 41–45.
4. R Lagerbäck, Neotectonic structures in northern Sweden. *Geologiska Föreningens i Stockholm Förhandlingar*, **100** (1979), 271–278.
5. M Jull, D McKenzie, The effect of deglaciation on mantle melting beneath Iceland. *Journal of Geophysical Research*, **101** (1996), 21815–21828.
6. J Maclennan, M Jull, D McKenzie, L Slater, K Gronvold, The link between volcanism and deglaciation in Iceland. *Geochemistry, Geophysics, Geosystems*, **3** (2002), doi:10.1029/2001GC00082.
7. C Beck, F Manalt, E Chapron, PV Rensbergen, MD Batist, Enhanced seismicity in the early post-glacial period: evidence from the post-Würm sediments of Lake Annecy, northwestern Alps. *Journal of Geodynamics*, **22** (1996), 155–171.
8. A Becker, M Ferry, K Monecke, M Schnellmann, D Giardini, Multiarchive palaeoseismic record of late Pleistocene and Holocene strong earthquakes in Switzerland. *Tectonophysics*, **400** (2005), 153–177.
9. M Ferry, M Meghraoni, B Delouis, D Giardini, Evidence for Holocene palaeoseismicity along the Basel–Reinach active normal fault (Switzerland): a seismic source for the 1356 Basel earthquake in the Upper Rhine Graben. *Geophysical Journal International*, **160** (2005), 554–572.
10. A Hampel, R Hetzel, G Maniatis, Response of faults to climate-driven changes in ice and water volumes on the Earth's surface. *Philosophical Transactions of the Royal Society A*, **368** (2010), 2501–2518.
11. A Hampel, R Hetzel, G Maniatis, Response of faults to climate-driven changes in ice and water volumes on the Earth's surface. In *Climate Forcing of Geological Hazards*, ed. B. McGuire, M. Maslin (Chichester: John Wiley & Sons, 2013), pp. 124–142.
12. H Tuffen, How will melting of ice affect volcanic hazards in the twenty-first century? *Philosophical Transactions of the Royal Society A*, **368** (2010), 2535–2578.
13. H Tuffen, Melting ice and volcanic hazards in the 21st century. In *Climate Forcing of Geological Hazards*, ed. B. McGuire, M. Maslin (Chichester: John Wiley & Sons, 2013), pp. 78–107.
14. Intergovernmental Panel on Climate Change (IPCC), *Climate Change 2013: The Physical Basis* (Cambridge: Cambridge University Press, 2014).

15. JM Sauber, BF Molnia, Glacier ice mass fluctuations and fault instability in tectonically active southern Alaska. *Global Planetary Change*, **42** (2004), 279–293.

16. IS Stewart, J Sauber, J Rose, Glacio-seismotectonics: ice sheets, crustal deformation and seismicity. *Quaternary Science Reviews*, **19** (2000), 1367–1389.

17. R Muir-Wood, Deglaciation seismotectonics: a principal influence on intraplate seismogenesis at high latitudes. *Quaternary Science Reviews*, **19** (2000), 1399–1411.

18. C Fenton, Postglacial faulting in eastern Canada: an annotated bibliography. Geological Survey of Canada Open-File Report 2774 (1994).

19. R Arvidsson, Fennoscandian earthquakes: whole crustal rupturing related to postglacial rebound. *Science*, **274** (1996), 744–746.

20. JF Dehls, O Olesen, L Olsen, L Harald-Blikra, Neotectonic faulting in northern Norway; the Stuoragurra and Nordmannvikdalen postglacial faults. *Quaternary Science Reviews*, **19** (2000), 1447–1460.

21. R Lagerbäck, *Postglacial Faulting and Palaeoseismicity in the Lansjärv Area, Northern Sweden* (Stockholm: Swedish Nuclear Fuel and Waste Management Co., 1988).

22. R Lagerbäck, F Witscard, *Neotectonics in Northern Sweden: Geological Investigations.* (Stockholm: Swedish Nuclear Fuel and Waste Management Co., 1983).

23. M Ustaszewski, A Hampel, OA Pfiffner, Composite faults in the Swiss Alps formed by the interplay of tectonics, gravitation and postglacial rebound: an integrated field and modelling study. *Swiss Journal of Geoscience*, **101** (2008), 223–235.

24. J Ehlers, P Gibbard, *Quaternary Glaciations: Extent and Chronology* (Amsterdam: Elsevier, 2004).

25. CF Larsen, RJ Motyka, JT Freymueller, KA Echelmeyer, EI Ivins, Rapid viscoelastic uplift in southeast Alaska caused by post-Little Ice Age glacial retreat. *Earth and Planetary Science Letters*, **237** (2005), 548–560.

26. J Sauber, BF Molnia, Glacier ice mass fluctuations and fault instability in tectonically active Southern Alaska. *Global Planetary Change*, **42** (2004), 279–293.

27. D Doser, KR West, J Sauber, Seismicity of the Bering Glacier region and its relation to tectonic and glacial processes. *Tectonophysics*, **439** (2007), 119–127.

28. P Talwani, On the nature of reservoir-induced siesmicity. *Pure and Applied Geophysics*, **150** (1997), 473–492.

29. S Ge, M Liu, N Lu, JW Godt, G Luo, Did the Zipingpu Reservoir trigger the 2008 Wenchuan earthquake? *Geophysical Research Letters*, **36** (2009), doi:10.1029/2009GL040349.

30. S Hainzl, T Kraft, J Wassermann, H Igel, Evidence for rain-triggered earthquake activity. *Geophysical Research Letters*, **33** (2006), L19303.

31. LB Christiansen, S Hurwitz, S Ingebritsen, Annual modulation of seismicity along the San Andreas Fault near Parkfield, CA. *Geophysical Research Letters*, **34** (2007), L04306.

32. M Strasser, C Schindler, FS Anselmetti, Late Pleistocene earthquake-triggered moraine dam failure and outburst of Lake Zurich, Switzerland. *Journal of Geophysical Research* **113** (2008), doi:10.1029/2007JF000802.

33. AP Palmer, J Rose, J Lowe, A Macleod, Annually resolved events of Younger Dryas glaciation in Lochaber (Glen Roy and Glen Spean), western Scottish Highlands. *Journal of Quaternary Science*, **25** (2010), 581–596.

34. M Jull, D McKenzie, The effect of deglaciation on mantle melting beneath Iceland. *Journal of Geophysical Research*, **101** (1996), 21815–21828.

35. J Maclennan, M Jull, DP McKenzie, L Slater, K Gronvold, The link between volcanism and deglaciation in Iceland. *Geochemistry, Geophysics, Geosystems*, **3** (2002), 1–25.

36. WJ McGuire, RJ Howarth, CR Firth, *et al.*, Correlation between rate of sea level change and frequency of explosive volcanism in the Mediterranean. *Nature*, **389** (1997), 473–476.

37. D Nowell, C Jones, D Pyle, Episodic Quaternary volcanism in France and Germany. *Journal of Quaternary Science*, **21** (2006), 645–675.

38. AM Jellinek, M Manga, MO Saar, Did melting glaciers cause volcanic eruptions in eastern California? Probing the mechanics of dike formation. *Journal of Geophysical Research*, **109** (2004), B09206.

39. G Bigg, C Clark, A Hughes, A last glacial ice sheet on the Pacific Russian coast and catastrophic change arising from coupled ice–volcanic interaction. *Earth and Planetary Science Letters*, **265** (2008), 559–570.

40. C Bacon, M Lanphere, Eruptive history and geochronology of Mount Mazama and the Crater Lake region, Oregon. *Geological Society of American Bulletin*, **118** (2006), 1331–1359.

41. K Mee, H Tuffen, JS Gilbert, Snow-contact volcanic facies at Nevados de Chillan volcano, Chile, and implications for reconstructing past eruptive environments. *Bulletin of Volcanology*, **68** (2006), 363–376.

42. MC Gardeweg, RSJ Sparks, SJ Matthews, Evolution of Lascar volcano, northern Chile. *Journal of the Geological Society*, **155** (1988), 89–104.

43. BS Singer, BR Jicha, MA Harper, JA Naranjo, LE Lara, H Moreno, Eruptive history, geochronology, and magmatic evolution of the Puyehue-Cordón Caulle volcanic complex, Chile. *Geological Society of America Bulletin*, **120** (2008), 599–618.

44. P Huybers, C Langmuir, Feedback between deglaciation, volcanism, and atmospheric CO_2. *Earth and Planetary Science Letters*, **286** (2009), 479–491.

45. L Capra, Abrupt climatic changes as triggering mechanisms of massive volcanic collapses. *Journal of Volcanology and Geothermal Research*, **155** (2008), 329–333.

46. R Lowe, Volcano lateral collapse events in the Quaternary. Masters thesis (University College London, 2007).

47. KR Deeming, B McGuire, P Harrop, Climate forcing of volcano lateral collapse: evidence from Mount Etna, Sicily. *Philosophical Transactions of the Royal Society A*, **368** (2010), 2559–2578.

48. C Pagli, F Sigmundsson, B Lund, *et al.*, Glacio-isostatic deformation around the Vatnajökull ice cap, Iceland, induced by recent climate warming: GPS observations and finite element modeling. *Journal of Geophysical Research*, **112** (2007), B08405.

49. C Pagli, F Sigmundsson, Will present day glacier retreat increase volcanic activity? Stress induced by recent glacier retreat and its effect on magmatism at the Vatnajökull ice cap, Iceland. *Geophysical Research Letters*, **35** (2008), L09304.

50. M Vuille, B Francou, P Wagnon, *et al.*, Climate change and tropical Andean glaciers: past, present and future. *Earth Science Reviews*, **89** (2008), 79–96.

51. JL Ceballos, C Euscátegui, J Ramírez, *et al.*, Fast shrinkage of tropical glaciers in Colombia. *Annals of Glaciology*, **43** (2006), 194–201.

52. C Huggel, JL Ceballos, B Pulgarín, J Ramírez, J-C Thouret, Review and reassessment of hazards owing to volcano–glacier interactions in Colombia. *Annals of Glaciology*, **45** (2007), 128–136.

53. F Sigmundsson, V Pinel, B Lund, *et al.*, Climate effects on volcanism: influence on magmatic systems of loading and unloading from ice mass variations with examples from Iceland. *Philosophical Transactions of the Royal Society A*, **368** (2010), 2519–2534.

54. F Sigmundsson, F Albino, P Schmidt, *et al.*, Multiple effects of ice load changes and associated stress change on magmatic systems. In *Climate Forcing of Geological*

Hazards, ed. B. McGuire, M. Maslin (Chichester: John Wiley & Sons, 2013), pp. 108–1123.

55. H Björnsson, F Pálsson, Icelandic glaciers. *Jökull*, **58** (2008), 365–386.
56. A Geyer, I Bindeman, Glacial influence on caldera-forming eruptions. *Journal of Volcanology and Geothermal Research*, **202** (2011), 127–142.
57. F Brondi, L Salvatori, The 5–6 May (1998) mudflows in Campania, Italy. In *Lessons Learned from Landslide Disasters in Europe*, ed. J. Hervás (Brussels: European Commission Joint Research Centre, 2003), pp. 5–16.
58. KM Scott, JW Vallance, N Kerle, JL Macías, W Strauch, G Devoli, Catastrophic precipitation-triggered lahar at Casita volcano, Nicaragua: occurrence, bulking and transformation. *Earth Surface Processes and Landforms*, **30** (2005), 59–79.
59. JL Carrivick, V Manville, SJ Cronin, A fluid dynamics approach to modelling the 18th March (2007) lahar at Mt. Ruapehu, New Zealand. *Bulletin of Volcanology*, **71** (2009), 153–169.
60. R Bilham, The seismic future of cities. *Bulletin of Earthquake Engineering*, **7** (2009), 839–887.
61. L Siebert, T Simkin, P Kimberly, *Volcanoes of the World*. (Berkeley, CA: Smithsonian Institution and University of California Press, 2010).
62. J Eliasson, G Larsen, MT Gudmundsson, F Sigmundsson, Probabilistic model for eruptions and associated flood events in the Katla caldera, Iceland. *Computational Geosciences*, **10** (2006), 179–200.
63. S McNutt, R Beavan, Eruptions of Pavlof volcano and their possible modulation by ocean load and tectonic stresses. *Journal of Geophysical Research*, **92** (1987), 11509–11523.
64. S McNutt, Eruptions of Pavlof Volcano, Alaska, and their possible modulation by ocean load and tectonic stresses: re-evaluation of the hypothesis based on new data from 1984–1998, *Pure and Applied Geophysics*, **155** (1999), 701–712.
65. B Mason, D Pyle, W Dade, T Jupp, Seasonality of volcanic eruptions. *Journal of Geophysical Research*, **109** (2004), doi:10.1029/2002JB002293.
66. B Voight, The 1985 Nevado del Ruiz volcano catastrophe: anatomy and retrospection. *Journal of Volcanology and Geothermal Research*, **42** (1990), 151–188.
67. BBC News, Flight disruptions cost airlines $1.7 billion, says IATA. April 21, 2010. Archived from the original on May 12, 2011. Retrieved March 12, 2014.

8

Catastrophic mass wasting in high mountains

OLIVER KORUP AND STUART DUNNING

8.1 Introduction

Much of the decay of Earth's mountains is initiated by mass wasting, a natural hazard caused and triggered by, among other things, high topographic relief, steep hillslopes, and strong precipitation gradients [1]. The myriad mass-wasting processes in mountain belts is impressive, ranging from rockfalls to soil slides, snow and ice avalanches, debris flows, rockslides and rock avalanches to large deep-seated gravitational slope deformation that slowly denudes entire valley flanks. Even Earth's highest peaks may not have escaped catastrophic rock-slope failure [2]. In this chapter we briefly summarise key characteristics of catastrophic slope failures in high-mountain environments, marked by the intersection of high topographic relief, steep bedrock hillslopes, permanent snow cover, widespread glacierisation, and alpine permafrost. We focus on larger ($>10^6$ m^3) rockfalls, rock slides, and rock avalanches for several reasons. Larger rock-slope failures are commensurately rarer than smaller ones, but their geomorphic and sedimentary evidence is likely to persist much longer in landscapes near the permanent snow line. Moreover, larger mass-wasting events have a much higher destruction potential. Yet the rare occurrence of extreme rock-slope failures means that their detection and prediction remain a major challenge in quantitative hazard assessments.

In the high-mountain environment both glacial and mass-wasting processes may show a systematic response to changes in topographic relief [3]. A common view is that many large and catastrophic rock-slope failures have been conditioned indirectly by past glacial and periglacial processes. This concept of a paraglacial cycle [4] has been influential and persistent in deducing the causes and triggers of past, mostly post-glacial, catastrophic rock-slope failures in many mountain belts [5]. Testing this concept has become highly relevant, given the number of studies that

The High-Mountain Cryosphere, ed. Christian Huggel, Mark Carey, John J. Clague and Andreas Kääb. Published by Cambridge University Press. © Cambridge University Press 2015.

predicted significant changes in the frequency and magnitude of rock-slope failures in the wake of contemporary global warming.

Accordingly, we direct our focus to recent studies that have tried to quantify effects of climate change on the occurrence of large catastrophic landslides in high mountains. We review current developments on how such landslides interact with glaciers and cast a glance at potentially highly sensitive mountain regions in polar regions that have so far seen little attention in terms of this theme. We conclude with a number of recommendations for future research.

8.2 More mass wasting because of climate change?

Slope instability in high mountains has seen increased research attention in recent years, owing to projections of climate change and global warming [6–8]. Recent research has pointed out the potential control of climate in general, and trends of global warming in particular, on the apparent increase of rock-slope failures in mountain belts worldwide [9]. Studies have documented increases in the observed magnitude and frequency of mass-wasting events in glacierised or permafrost terrain; for example, enhanced rockfall and debris-flow activity during the exceptionally warm summers of 2003 and 2010 [7]. Several research groups are now investigating the mechanistic framework that links degradation of alpine permafrost to rock-slope stability [10,11]. Time series of degrading alpine permafrost have led to much argument about whether warmer conditions may compromise the future stability of mountain slopes. Using repeat photogrammetry, Fischer and her colleagues [12] estimated a net bedrock and ice loss of $25 \times 10^6 \, \text{m}^3$ from the east face of Monte Rosa between 1988 and 2007, and inferred an alarming order-of-magnitude increase in mass loss rates in recent decades. The few other available systematic inventories of reliably dated rock-slope failures indicate a clear increase in frequency, with a concentration of detachment in high-elevation parts of the Central Alps [13]. In the Southern Alps of New Zealand, Allen and co-workers [8] reported that 19 out of 20 documented rock avalanches since 1950 occurred within a thin elevation band on ice-covered bedrock slopes; 13 of the 20 events sourced less than 300 m away from the estimated lower limit of permafrost. In places such as the Himalayas, however, obtaining detailed data is exacerbated by the sheer size of the mountain belt, and most remote sensing-based monitoring studies have focused on the hazards from glacial lake outburst floods (GLOFs) rather than assessing the potential for catastrophic rock-slope failures.

8.3 The search for causes and triggers

Despite the intuitive notion that global warming may give rise to more frequent mass wasting in high mountains, most rigorous statistical tests of the idea have so

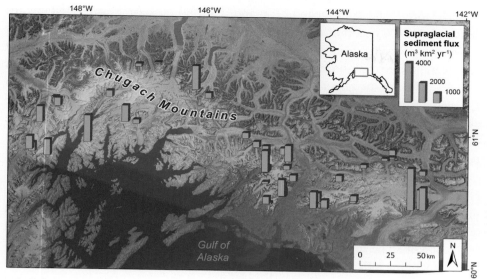

Figure 8.1 Estimates of average multi-year supraglacial sediment flux [m^3 km^{-2} yr^{-1}] resulting from rock avalanches onto valley glaciers in the Chugach Mountains, Alaska, between 1972 and 2008. This sample is limited to deposits that originally covered >0.1 km^2 of ice.
Source: after [15].

far been limited to data sets that are too short, too local, or biased. The data density and quality drop sharply as we look further back in time. Thus, comprehensive chronologies are rarely available for even single catchments. For example, in an exhaustive study Guthrie and his colleagues [14] were able to recognise 25 post-glacial landslides and debris flows greater than 0.5×10^6 m^3 that mobilised a total of 2.47×10^9 m^3 from the volcanic Mount Meager Complex, British Columbia, Canada during the last 8000 years. On the other hand, Uhlmann and co-workers [15] provided a comprehensive inventory of more than 120 historic rock avalanches that fell onto glaciers in the Chugach Mountains, Alaska (Figure 8.1), and noted the high resulting supraglacial sediment fluxes that generally match those recorded in mountain rivers throughout the world. The mountain ranges of Alaska are a prime area to study the frequent occurrences of rockslides and rock avalanches, many of which fall onto glaciers. The advantage of this natural laboratory is offset by the problem of disentangling seismic triggers from other controls, notably glacial debuttressing and rapid glacio-isostatic rebound of up to 30 mm yr^{-1} and more in places [16]. There is the additional problem of distinguishing potential volcanic triggers [17], a common phenomenon in the high mountains bordering subduction zones [18]. The 2002 $M_w = 7.9$ Denali Fault earthquake, Alaska, provided an unparalleled opportunity to investigate using remote sensing data the regional rock-slope response to strong seismic shaking in glacierised terrain. Gorum and colleagues [19] mapped ~1580 co-seismic slope failures, of which some 20% occurred above large valley glaciers, including several large rock avalanches. Results from this study indicate that fault geometry and rupture propagation exert first-order controls on the catastrophic delivery of sediment to glaciers, whereas the effect of glacial cover itself was considered to be minor.

Establishing causes and triggers of prehistoric events remains difficult. For example, Ballantyne and Stone [20] concluded from their absolute age dating study of 17 large post-glacial rock-slope failures in the Scottish Highlands that: 'Neither permafrost degradation (thaw of ice in joints) nor seismic activity offers a general explanation for triggering ..., but could be contributory in some cases.' They concluded that catastrophic failures are likely to occur again in the future. Despite these limitations, a prevalent view holds that the Quaternary geological record corroborates the hypothesis that climatic change modulates the magnitude and frequency of slope instability in mountainous terrain. The underlying argument is that many of the hitherto-documented catastrophic rock-slope failures had occurred during or following the termination of Pleistocene glaciation. However, a number of studies have now begun to document the prevalence of large rock-slope failures throughout the Holocene [21]. Explanations to reconcile this trend with the concept of a paraglacial cycle centre on a delayed response owing to gradual redistribution of hillslope stresses to retreating ice sheets and valley glaciers [5].

8.4 The rockslide–glacier couple

Recent research on mass wasting in high mountains has focused on measuring geomorphic feedbacks of supraglacial rockslides and rock avalanches on glacier dynamics. Reznichneko and co-workers [22] proposed that rock avalanches emplaced on glaciers might result in a glacier tongue being out-of-phase with climate, with the insulated ice stagnating and downwasting. A hotly debated topic is the possibility that distinct moraine ridges such as the Franz Josef Glacier's Waiho Loop in the Southern Alps of New Zealand – a prime location for gauging Southern Hemisphere climate oscillations during the Pleistocene–Holocene transition – owe their formation to glacier advances assisted by supraglacial rock avalanches [23]. Whether such glacier advances, and also surges, may be triggered by excess sediment input related to catastrophic rock-slope failure remains uncertain. Vacco and his colleagues [24] presented numerical modelling results that show that thick supraglacial sediment cover fed by rock avalanches might cause glaciers to decelerate instead of advance. Moreover, their simulations suggested that any single moraine ridge would have to be associated with widespread, hummocky, glacial stagnation. In the case of the Waiho Loop, any such evidence may be buried beneath thick glaciofluvial outwash or lost through subsequent erosion. Menounous and co-workers [25] used a debris-advection module in an ice dynamics model, and concluded that rapid advance following instantaneous rockslide/rock avalanche loading was indeed plausible. Clearly, more research is necessary to elucidate the ice dynamic effects of rock-slope failures running onto glaciers. Sosio and her colleagues [26] provided a comprehensive catalogue of other historic ice–rock avalanches, and successfully reproduced several physical run-out characteristics through numerical modelling. Such models are partly dependent on reliable field data. For example, Hewitt [27] summarised some of these field data by relating impacts of historic supraglacial rock avalanches, surges, and supraglacial sediment budget on Bualtar Glacier in Pakistan. Yet even detailed short-term measurements based on remote sensing reveal changes in glacier surface velocities that may not be easily attributed to sudden loading by rockslide deposits [28]. Constraining the time and extent of historic large supraglacial rock-slides and avalanches that have escaped systematic coverage by remote sensing adds another uncertainty [15]. Deposits from such catastrophic slope failures are much thinner than those in ice-free settings (Figure 8.2), and the preservation potential, and hence the longevity of diagnostic geomorphic evidence, may be low.

Little is known about the fate and preservation of deposits from rock-slope failures in the proglacial environment. In this context, Cook and co-workers [29] provided an interesting report on the glacial reworking of rock-avalanche debris,

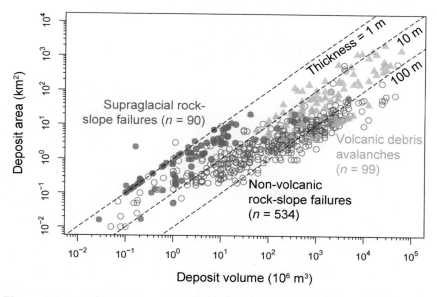

Figure 8.2 Relationship between deposit area and volume for three types of catastrophic rock-slope failures. Dashed lines show mean deposit thickness, which is lowest for supraglacial rock-slope failures running out onto broad valley glaciers.

which demonstrated a striking persistence in terms of overall shape after some 40 years of glacier overriding. The blocky carapace that characterises the surface of most rock avalanches seems to be capable of partly resisting glacier overriding to the degree that reworked rock-avalanche debris may resemble glacigenic deposits. Continued stagnation and formation of pore ice in deposits in the zone of alpine permafrost may result in a creeping mass that will eventually destroy some of the more unique sedimentary features of rock-avalanche deposits such as preserved lithological banding, and jigsaw clasts [30]. This process raises the important question of how many rock glaciers have largely formed from the debris from catastrophic rock-slope failures. Rock-avalanche deposits are hummocky, steep-fronted and steep-sided, show evidence of bulldozing and entrainment, and are composed of angular debris of local origin that is coarser at the top. This suite of features is not substantially different from those of either a talus-fed rock glacier with pore-space ice or a glacier-derived rock glacier with a core of foliated glacier ice. One method that offers the possibility of distinction is micro-sedimentological fingerprinting that is able to identify rock avalanche-generated fines from those produced by glacial and non-catastrophic mass movements [31]. Although this approach cannot provide conclusions about cause and effect, it might reveal if a proportion of a rock glacier is catastrophic landslide material. This important confusion potential feeds back to the debate about glacier advances and moraine formation related to supraglacial rock-slope failures, and may be summarised by two end-member opinions:

field glacial geologists should have no major difficulty in recognizing moraines associated with glacial advances that were triggered by landslides.

[24]

prominent moraines can be generated by rock-avalanche driven glacier advances; these cannot presently be distinguished from moraines reflecting climatically driven advances.

[22]

8.5 Rare and rapid mixtures

Any account on catastrophic mass wasting in high mountains would be incomplete without mentioning extremely large ice–rock avalanches, a phenomenon that has been rarely documented, let alone observed. The surprisingly mobile and far-reaching multi-phase mass movements at Nevado Huascarán in the Peruvian Andes is among the earliest studied and most prominent examples of such extreme events (Figures 8.3 and 8.4), long before research attention shifted to global warming as a potential cause or trigger. On May 31, 1970, an M_w = 7.8 earthquake offshore triggered a fall of rock and ice estimated to be 7.5×10^6 m^3

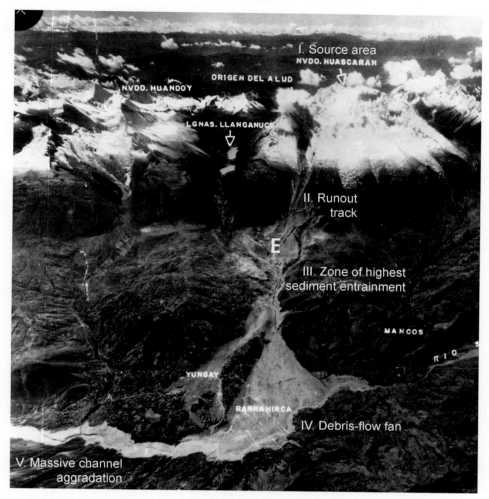

Figure 8.3 Oblique aerial view of the run-out track of the catastrophic ice–rock avalanche/debris flow that originated from the North Peak of Nevado Huascarán, Peru, following a strong offshore earthquake in 1970.
Source: after [32].

in volume from the west face of the North Peak (6654 m) of Nevado Huascarán [32]. On its rapid descent, the mass entrained an additional ~43 × 10^6 m^3 of snow and loose debris, transforming into a highly mobile debris flow that raced towards the Rio Santa, obliterating the town of Yungay, some 15 km from the source of the ice–rock fall, before carrying on as a diluted water and sediment pulse some 180 km to the Pacific Ocean. The Huascarán ice–rock avalanche and a similar, smaller event in 1962 (13–16 × 10^6 m^3) underpin one of the key problems when it comes to accurately reconstructing historic events, i.e. the accurate estimation of the volumes

Figure 8.4 Air photos showing the town of Yungay before and after the catastrophic debris flow generated by the 1970 Nevado Huascarán ice–rock
avalanche. The white line shows estimated extent of the built-up area.
Source: after [32].

of ice and rock involved. Evans and his colleagues [32] noted substantial inconsist-
ency between previous volume estimates, which differ by more than 150%, highlighting the difficulties of quantitatively assessing the hazard of future events. Of
even greater importance, published estimates of the loss of life from the two disasters
range from 6000 to 25 000 (Table 8.1), and depend heavily on the assumptions
made about the population density in Yungay at the time of the event [32].

On April 9, 2000, a massive rock–ice avalanche detached without warning from
the glacierised headwaters of Zhamu Creek in the Nyainqentanglha Mountains of
southeast Tibet. The landslide travelled some 10 km, and transformed into a highly
mobile debris flow that formed a 300×10^6 m^3 fan-shaped dam that blocked the
Yigong River, a tributary of the Yarlung-Brahmaputra (Figure 8.5) [33]. Chinese
authorities decided to create an artificial diversion channel through the dam, which
impounded a lake containing an estimated 3×10^9 m^3 (perhaps an overestimate).
However, the dam failed on June 10, 2000, causing a >30 m high flood wave that
wreaked damage downstream into India; the instantaneous peak discharge was
estimated at 120 000 m^3 s^{-1} some 17 km downstream of the dam [33]. Zhamu
Creek rapidly re-established its course through the valley-filling landslide debris,
thus delivering reworked sediment at extremely high rates (>100 000 m^3 km^{-2}
yr^{-1}) to a telescope fan that continues to divert the Yigong River. A similarly large
(~500×10^6 m^3) landslide previously blocked the Yigong River at this location in
1900, underscoring the potential for recurrence of such rare events at specific
locations, similar to what has been experienced at Nevado Huascarán. From a

Table 8.1 *Characteristics of three recent catastrophic ice-rock failures with excess run-out (see Figure 8.7).*

Article II.	Nevado Huascarán, Andes, 1970	Zhamu Creek, SE Tibet, 2000	Kolka-Karmadon, Caucasus, 2002
Source area (m asl)	5600–6400	5520	4200
Vertical drop height (m)	3200	3300	2900
Run-out (m)	15 000	10 000	19 500
Volume (m^3)	20–50 × 10^6	100–300 × 10^6	110 × 10^6
Velocity (m s^{-1})	50–85	10–20	50–65
Fatalities	6000–25 000(?)	109	125
Trigger	Earthquake (M_w = 7.9)	Rainfall, warming?	Glacial loading?
Precursor event	1962	1900	1902?

hazard assessment perspective, the remaining deposits of the 2000 ice–rock avalanche strongly resemble any other debris-flow fan that might have formed instead by multiple smaller events. This similarity curtails the unambiguous identification of similar events – at least from remote sensing reconnaissance – elsewhere in the region.

On September 20, 2002, a 2.7 km long section of Kolka Glacier in the Caucasus of southern Russia detached without warning, giving rise to an extremely rapid (65–80 m s^{-1}) ice–rock avalanche that buried 19 km of the Genaldon River beneath 110 × 10^6 m^3 of ice and debris, impressively demonstrating the excess run-out potential of such mixed-material events (Figure 8.6) [34]. The disconcerting aspects of this event are that (1) no conclusive trigger is evident – a problem frequently linked to such rare events – and (2) no comparable run-out had been documented historically for this type of mass movement. Several trigger mechanisms have been proposed, including a large supraglacial rock-slope failure that was believed to have sheared off the lower part of Kolka Glacier; a massive glacier surge driven by high water pressures within and below the glaciers, and catastrophic loss of effective stress to due to englacial drainage disruption by ice and debris loading [35].

Not surprisingly, research on the run-out dynamics of such rare and large mixed avalanches is an emerging field [36]. The scant previous work indicates that large ice avalanches can also be triggered by strong seismic shaking [37], thus complicating the search for unambiguous evidence of causes and triggers. The three historic case studies of ice–rock avalanches cited above rank among the most far-reaching and destructive non-volcanic catastrophic mass movements in high mountains since the beginning of the twentieth century. In particular, their vertical drop heights and very long run-out, together with large amounts of snow and ice and distinct debris-flow phases that dominated the final

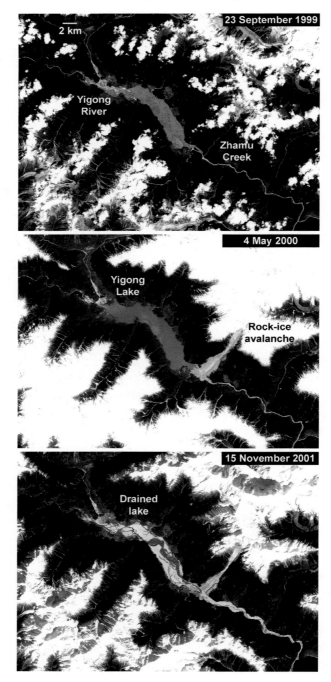

Figure 8.5 Multi-temporal ASTER satellite imagery of Yigong River near Zhamu Creek, which was impacted by the 2000 rock-ice avalanche (~300 × 10^6 m^3). The avalanche formed a large dam across the Yigong River, a tributary of the Yarlung Tsangpo-Brahmaputra. Yigong Lake developed behind the debris dam (middle image) on an aggraded reach of the valley floor (top image) that formed behind a large (~500 × 10^6 m^3) landslide dam in 1900 [33].
Source: images courtesy of NASA.

Figure 8.6 Pre- and post-event ASTER satellite imagery of Genaldon River, which was impacted by the highly mobile Kolka-Karmadon ice–rock avalanche in 2002.
Source: after [35].

emplacement phase, distinguish them from the larger record of catastrophic landslides throughout the world (Figure 8.7).

8.6 The forgotten mountains

Much of our knowledge about catastrophic mass wasting in high mountains is prone to spatial bias. Most data come from densely populated areas with long written or oral histories such as the European Alps. Many other mountain regions lack research at comparable detail. The result is either sparse data or, in some cases, incorrect interpretations, for example mistaken identification of rock-avalanche deposits as moraines [38]. High-latitude mountain belts are exemplars of understudied regions, although some show promising snapshots of activity. Contemporary warming in polar regions outpaces global trends, thus producing deglaciating landscapes prone to geomorphic change and catastrophic events. Yet

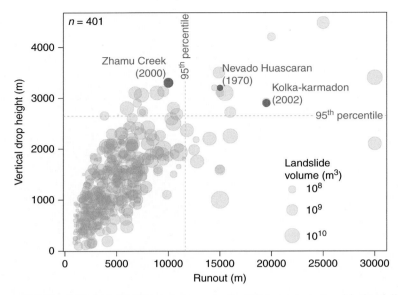

Figure 8.7 Vertical drop height and run-out of ~400 large catastrophic landslides throughout the world, including the examples discussed in the text. This sample is limited to landslides of individual volumes $>10^6$ m^3 and run-outs of >1 km.

our understanding of any future change in catastrophic mass wasting in these regions is hindered by our lack of understanding of contemporary rates and, even more so, past rates.

On November 21, 2000, a large (~90×10^6 m^3) rock avalanche fell from the southwest slope of Paatuut peak on the Nuussuaq Peninsula, West Greenland (Figure 8.8). The landslide entered Vaigat fjord at velocities of 140–200 km h^{-1}, triggering a tsunami that inundated the abandoned former mining town of Qullissat, 20 km away on Disko Island, destroying almost all buildings up to ~30 m above sea level [39]. Previous seismic surveys of Vaigat identified numerous submarine landslide deposits from both terrestrial and subaqueous failures, of which some were up to 200 m thick [39]. Surrounding the Paatuut deposits, along a 30 km stretch of the coast, are at least 19 other onshore rock-avalanche deposits (Figure 8.8). None of these shows evidence of glacial reworking, and all are likely to be younger than ~3000 years old based on their relation with sea-level markers [39]. The total volume of the deposits is $>330 \times 10^6$ m^3 (J. Benjamin, personal communication, 2013), roughly equivalent to an average of 3 mm yr^{-1} rockwall retreat. The geological setting of hard basalts overlying soft, deforming sedimentary rocks may make this cluster unique, or at least rare, but similar rates of rockwall retreat have been documented for the source areas of some 400 rock glaciers on Disko Island [40]. Many of the longer and larger glacier-derived rock glaciers extending down into Vaigat are bordered by areas of deep-seated

Figure 8.8 (a) Simplified geological map showing a dense cluster of catastrophic rock-avalanche deposits at Vaigat, West Greenland (after J. Benjamin, unpublished). The 2000 Paatuut rock avalanche is the easternmost deposit. (b) The 2000 Paatuut rock-avalanche deposit. A tsunami was generated by the failure of the frontal (toe) section of the deposit soon after emplacement; the backscarp is clearly visible at the shoreline (dashed line). The scar of a neighbouring deposit is visible on the left skyline; inset shows location in Greenland.

slumping. The Paatuut landslide, among others, has highlighted the lack of knowledge about the frequency of catastrophic rock-slope failures above flooded, glacially carved valleys such as fjords and lakes, especially in tectonically quiescent terrain. Although landslide-induced tsunamis or seiches have been indirectly inferred from sub-bottom profiling and bathymetric surveys, terrestrial remnants of the responsible landslides are generally poorly preserved, if not entirely absent, particularly where steep hillslopes offer little accommodation space for diagnostic deposits. The source scars may be easily confused with glacial cirques [41], such that the terrestrial record of former rock-slope failures is censured.

Compared to the European Alps, the mountains of Antarctica and the sub-Antarctic have a dearth of mass-movement studies. Most documented evidence comes from coastal areas and is limited to periglacial mass transport. Talus- and glacier-derived rock glaciers of the sub-Antarctic islands such as South Georgia, the South Shetlands, Marion Island, are reasonably well studied [42], and evidence of catastrophic mass movements in ice-free areas of Antarctica has been recognised, although mostly as sediment sources for periglacial landforms ([43] and references therein). Recent work has begun linking these classic periglacial landforms to a variety of catastrophic mass-wasting processes in Antarctica. For example, McGowan and co-workers [44] reinterpreted what was formerly believed to be Late Pleistocene shorelines in the McMurdo Dry Valleys as parts of much older slump deposits, thus undermining previous models of extensive lakes in this hyperarid area.

In general, the lack of observational, field, and time series data makes it difficult to distinguish between catastrophic and background rates of activity in this region. A notable exception is on South Georgia, which, although not strictly polar, is a useful proxy for a warming Antarctica, and in particular the Antarctic Peninsula. The island has high relief and is dominated by cryospheric processes. In 1976 a rock avalanche fell from Paulsen Peak (1877 m) and ran ~4 km down Lyell Glacier over a vertical distance of 1630 m [45]. The fine dust generated by the rock avalanche was distributed over a 110 km^2 area. The deposit volume ($2–3 \times 10^6$ m^3) was the equivalent to 93 years of 'normal' sediment delivery, based on calculations of supraglacial glacier cover. The question, then, is how frequent are these events in the sub-Antarctic and mainland Antarctica, and what is the relative past, present, and future importance of these high-magnitude events for debris cover and glacier dynamics?

The Antarctic mainland proper hosts abundant evidence of mass movements, dominated by talus cones, ice avalanches, and serac failures. Many talus cones and small rockfall deposits are found near Fossil Bluff (71 °S), situated on the coast of Alexander Island overlooking the George VI Ice Shelf. Several outlet glaciers that enter George VI Sound are bounded by bedrock slopes with evidence of large-scale mass movements. Two Step Cliffs is the best example, a 2.6 km wide, several-hundred-metre high concave scar backing a deep-seated bedrock failure that is protruding >1 km out into the ice (Figure 8.9a). It does not appear to have been overridden by ice, but rather failed against a thinning outlet glacier and the ice shelf itself [46]. There are no data to determine if the Two Step landslide is active,

Figure 8.9 (a) Oblique aerial photograph of rock-slope failure at Two Step Cliffs, Alexander Island, western Antarctic Peninsula, with the 2.6 km wide, displaced mass in the foreground. (b) Post-glacial rock-slope failure deposit at Threlkeld Knotts, English Lake District. Chain-dashed black line demarcates approximate front of deposit. Note the comparable morphologic setting and topography of the deposits in (a) and (b).

Source: (a) Image reproduced from the Polar Geospatial Centre, USGS-EROS Data Centre (image reference CA180031L0013). (b) After [47].

if much of the toe of the deposit has been removed [44], and if it has potential to transition into a more catastrophic failure if the ice-shelf were to thin as rapidly as it did after the Last Glacial Maximum. This deglacial debuttressing is often debated as either a causal trigger with a significant, but unpredictable, lag-time, or simply a common condition of many high-relief slopes [5]. We speculate that removal of the George IV Ice Shelf could well leave a failure-prone landscape very similar to Vaigat in Greenland, with remnant deposits with uncertain temporal links to the post-glacial history as found elsewhere [47] (Figure 8.9b).

8.7 Conclusions and scope for future research

High mountains are prone to various forms of catastrophic mobilisation of large volumes of snow, ice, debris, and bedrock. Causes or preparatory factors of such mass movement are topographic stress gradients owing to steep slopes and high-valley relief accentuated by glacial erosion, and loss of shear strength following pronounced physical weathering (frost shattering, freeze–thaw cycles, etc.) and permafrost degradation. Earthquakes, strong precipitation, and melt events, as well as slope undercutting by glacial, fluvial, or anthropogenic processes, are among the most frequently reported triggers of catastrophic mass wasting. However, for a substantial – and growing – number of reported sudden rock-slope failures, the type of trigger mechanism remains elusive or simply unobserved. This disconcerting observation has important repercussions for quantitative hazard and risk assessments in mountainous terrain, given that some of these slope failures are highly mobile, involve vertical drop heights of several kilometres, and entrain substantial amounts of snow and ice, and result in run-out of several, to several tens of kilometres.

One of the most pressing questions about high-mountain environments is the potential impact of climate change, specifically contemporary global warming, on mass wasting. Much recent research has focused on alpine permafrost, and whether its continued degradation is likely to cause significantly more frequent mass wasting in the future. This question is not an easy one to answer, as it requires careful consideration and disentangling of many potential triggers, lag times, and delayed response to deglaciation. Future research may establish with more confidence changes to the frequency or magnitude of large rock-slope failures in mountainous terrain [12,48]. Current data limitations make this task difficult, and we caution against drawing conclusions prematurely.

Much recent research has considered the role of supraglacial rock-slope failures as accelerators or decelerators of glacier motion. Two key implications, which are partly rooted in the possibility of confusing bedrock-landslide with glacial debris, are that (1) palaeoclimatic interpretations and chronologies based on dated moraines may be undermined by advances caused by emplacement of debris sheets on glaciers by rock avalanches; and (2) the hazard potential of such rare landslides may be commensurately over- or underestimated. Studies tracking the dispersion of landslide deposits on glaciers attest to the decisive local effects of the debris on the glacier mass balance. Future research will have to establish more conclusively whether landslide-generated moraines have distinctive characteristics that allow them to be discriminated from moraines that record past climatic events.

Little is known about landslide–glacier interactions that result in highly mobile ice–rock avalanches. These rare and highly destructive events have largely gone undocumented. A future challenge will be to improve our knowledge of these events using geomorphic and sedimentary archives that may be masked by land-forms and sediments produced by more gradual and episodic processes.

Finally, we argue that much of the current knowledge about catastrophic mass wasting in high mountains is subject to spatial bias: research tends to focus on easily accessible and readily monitored sites. Mountain ranges in the polar regions might be even more sensitive to future climate change, but are presently little studied. Large tsunamigenic landslides from Greenland's deglaciating coasts are but one example of a scenario that has been inadequately studied but should be considered in the future.

Acknowledgements

Many colleagues not mentioned in this brief overview have contributed tirelessly to furthering our knowledge on mass wasting in high mountains. We acknowledge and appreciate their work. We thank Philip Deline, John Clague, and Christian Huggel for their constructive comments during the editing phase. This contribution was funded in part by the Potsdam Research Cluster for Georisk Analysis, Environmental Change and Sustainability (PROGRESS), The Polish National Science Centre (2011/01/B/ST10/01553), and the Natural Environment Research Council (NE/I025840/1).

References

1. O Korup, JJ Clague, Natural hazards, extreme events, and mountain topography. *Quaternary Science Reviews*, **28** (2009), 977–990.
2. JT Weidinger, JM Schramm, F Nuschej, Ore mineralization causing slope failure in a high-altitude mountain crest: on the collapse of an 8000 m peak in Nepal. *Journal of Asian Earth Sciences*, **21** (2002), 295–306.
3. O Korup, DR Montgomery, K Hewitt, Glacier and landslide feedbacks to topographic relief in the Himalayan syntaxes. *Proceedings of the National Academy of Sciences of the United States of America*, **107** (2010), 5317–5322.
4. C Ballantyne, Paraglacial geomorphology. *Quaternary Science Reviews*, **21** (2002), 1935–2017.

5. ST McColl, Paraglacial rock-slope stability. *Geomorphology*, **153–154** (2012), 1–16.
6. SG Evans, JJ Clague, Recent climatic change and catastrophic geomorphic processes in mountain environments. *Geomorphology*, **10** (1994), 107–128.
7. S Gruber, M Hoelzle, W Haeberli, Permafrost thaw and destabilization of Alpine rock walls in the hot summer of 2003. *Geophysical Research Letters*, **31** (2004), doi:10.1029/2004GL020051.
8. SK Allen, SC Cox, IF Owens, Rock avalanches and other landslides in the central Southern Alps of New Zealand: a regional study considering possible climate change impacts. *Landslides*, **8** (2011), 33–48.
9. C Huggel, JJ Clague, O Korup, Is climate change responsible for changing landslide activity in high mountains? *Earth Surface Processes and Landforms*, **37** (2012), 77–91.
10. S Gruber, W Haeberli, Permafrost in steep bedrock slopes and its temperature-related destabilization following climate change. *Journal of Geophysical Research*, **112** (2007), F02S18.
11. M Krautblatter, D Funk, FK Günzel, Why permafrost rocks become unstable: a rock–ice–mechanical model in time and space. *Earth Surface Processes and Landforms*, **28** (2013), 876–887.
12. L Fischer, C Huggel, A Kääb, W Haeberli, Slope failures and erosion rates on a glacierized high-mountain face under climatic changes. *Earth Surface Processes and Landforms*, **38** (2013), 836–846.
13. L Fischer, RS Purves, C Huggel, J Noetzli, W Haeberli, On the influence of topographic, geological and cryospheric factors on rock avalanches and rockfalls in high-mountain areas. *Natural Hazards and Earth System Sciences*, **12** (2012), 241–254.
14. RH Guthrie, P Friele, K Allstadt, *et al.*, The 6 August 2010 Mount Meager rock slide–debris flow, Coast Mountains, British Columbia: characteristics, dynamics, and implications for hazard and risk assessment. *Natural Hazards and Earth System Sciences*, **12** (2012), 1277–1294.
15. M Uhlmann, O Korup, C Huggel, L Fischer, JS Kargel, Supra-glacial deposition and sediment flux of catastrophic rock-slope failure debris, south-central Alaska. *Earth Surface Processes and Landforms* **38** (2013), 675–682.
16. CF Larsen, RJ Motyka, JT Freymueller, KA Echelmeyer, ER Ivins, Rapid viscoelastic uplift in southeast Alaska caused by post-Little Ice Age glacial retreat. *Earth and Planetary Science Letters*, **237** (2005), 548–560.
17. C Waythomas, Formation and failure of volcanic debris dams in the Chakachatna River valley associated with eruptions of the Spurr volcanic complex, Alaska. *Geomorphology*, **39** (2001), 111–129.
18. SFL Watt, DM Pyle, JA Naranjo, TA Mather, Landslide and tsunami hazard at Yate volcano, Chile as an example of edifice destruction on strike-slip fault zones. *Bulletin of Volcanology*, **71** (2009), 559–574.
19. T Gorum, O Korup, CJ van Westen, M van der Meijde, C Xu, FD van der Meer, Why so few? Landslides triggered by the 2002 Denali earthquake, Alaska. *Quaternary Science Reviews*, **95** (2014), 80–94.
20. C Ballantyne, JO Stone, Timing and periodicity of paraglacial rock-slope failures in the Scottish Highlands. *Geomorphology*, **186** (2013), 150–161.
21. C Prager, C Zangerl, G Patzelt, R Brandner, Age distribution of fossil landslides in the Tyrol (Austria) and its surrounding areas. *Natural Hazards and Earth System Sciences*, **8** (2008), 377–407.
22. NV Reznichenko, TRH Davies, DJ Alexander, Effects of rock avalanches on glacier behaviour and moraine formation. *Geomorphology*, **132** (2011), 327–338.

23. J Shulmeister, TR Davies, DJA Evans, OM Hyatt, DS Tovar, Catastrophic landslides, glacier behaviour and moraine formation: a view from an active plate margin. *Quaternary Science Reviews*, **28** (2009), 1085–1096.

24. DA Vacco, RB Alley, D Pollard, Glacial advance and stagnation caused by rock avalanches. *Earth and Planetary Science Letters*, **294** (2010), 123–130.

25. B Menounos, JJ Clague, GKC Clarke, *et al.*, Did rock avalanche deposits modulate the Late Holocene advance of Tiedemann Glacier, southern Coast Mountains, British Columbia, Canada? *Earth and Planetary Science Letters*, **384** (2013), 154–164.

26. R Sosio, GB Crosta, JH Chen, O Hungr, Modelling rock avalanche propagation onto glaciers. *Quaternary Science Reviews*, **47** (2012), 23–40.

27. K Hewitt, Rock avalanches that travel onto glaciers and related developments, Karakoram Himalaya, Inner Asia. *Geomorphology*, **103** (2009), 66–79.

28. DH Shugar, BT Rabus, JJ Clague, DM Capps, The response of Black Rapids Glacier, Alaska, to the Denali earthquake rock avalanches. *Journal of Geophysical Research*, **117** (2012), F01006.

29. SJ Cook, PR Porter, CA Bendall, Geomorphological consequences of a glacier advance across a paraglacial rock avalanche deposit. *Geomorphology*, **189** (2013), 109–120.

30. JT Weidinger, O Korup, H Munack, *et al.*, Giant rockslides from the inside. *Earth and Planetary Science Letters*, **389** (2014), 62–73.

31. NV Reznichenko, TRH Davies, J Shulmeister, SH Larsen, A new technique for recognizing rock avalanche-sourced deposits in moraines and some palaeoclimatic implications. *Geology*, **40** (2012), 319–322.

32. SG Evans, NF Bishop, LF Smoll, PV Murillo, KB Delaney, A Oliver-Smith, A re-examination of the mechanism and human impact of catastrophic mass flows originating on Nevado Huascarán, Cordillera Blanca, Peru in 1962 and 1970. *Engineering Geology*, **108** (2009), 96–118.

33. Y Shang, Z Yang, L Li, D Liu, Q Liao, Y Wang, A super-large landslide in Tibet in 2000: background, occurrence, disaster, and origin. *Geomorphology*, **54** (2003), 225–243.

34. C Huggel, S Zgraggen-Oswald, W Haeberli, *et al.*, The 2002 rock/ice avalanche at Kolka/Karmadon, Russian Caucasus: assessment of extraordinary avalanche formation and mobility, and application of QuickBird satellite imagery. *Natural Hazards and Earth System Sciences*, **5** (2005), 173–187.

35. SG Evans, OV Tutubalina, VN Drobyshev, *et al.*, Catastrophic detachment and high-velocity long-runout flow of Kolka Glacier, Caucasus Mountains, Russia in 2002. *Geomorphology*, **105** (2009), 314–321.

36. D Schneider, P Bartelt, J Caplan-Auerbach, *et al.*, Insights into rock–ice avalanche dynamics by combined analysis of seismic recordings and a numerical avalanche model. *Journal of Geophysical Research*, **115** (2010), F04026, doi:10.1029/2010JF001734.

37. J van der Woerd, LA Owen, P Tapponier, *et al.*, Giant, ~M8 earthquake-triggered ice avalanches in the eastern Kunlun Shan, northern Tibet: characteristics, nature and dynamics. *Geological Society of America Bulletin*, **116** (2004), 394–306.

38. K Hewitt, Quaternary moraines vs catastrophic rock avalanches in the Karakoram Himalaya, Northern Pakistan. *Quaternary Research*, **51** (1999), 220–237.

39. SAS Pedersen, LM Larsen, T Dahl-Jensen, *et al.*, Tsunami-generating rock fall and landslide on the south coast of Nuussuaq, central West Greenland. *Geology of Greenland Survey Bulletin*, **191** (2002), 73–83.

40. O Humlum, The geomorphic significance of rock glaciers: estimates of rock glacier debris volumes and headwall recession rates in West Greenland. *Geomorphology*, **35** (2000), 41–67.

41. JM Turnbull, TRH Davies, A mass movement origin for cirques. *Earth Surface Processes and Landforms*, **31** (2006), 1129–1148.
42. W Nel, S Holness, KI Meiklejohn, Observations on rapid mass movements and screes on Sub-Antarctic Marion Island. *South African Journal of Earth Science*, **99** (2003), 177–181.
43. M Guglielmin, Advances in permafrost and periglacial research in Antarctica: a review. *Geomorphology*, **155–156** (2012), 1–6.
44. HA McGowan, DT Neil, JC Speirs, A reinterpretation of geomorphological evidence for Glacial Lake Victoria, McMurdo Dry Valleys, Antarctica. *Geomorphology*, **208** (2014), 200–206.
45. JE Gordon, RV Birnie, R Timmis, A major rockfall and debris slide on the Lyell Glacier, South Georgia. *Arctic and Alpine Research*, **10** (1978), 49–60.
46. MJ Bentley, JS Johnson, DA Hodgson, T Dunai, SPHT Freeman, CÓ Cofaigh, Rapid deglaciation of Marguerite Bay, western Antarctic Peninsula in the early Holocene. *Quaternary Science Reviews*, **30** (2011), 3338–3349.
47. TRH Davies, J Warburton, SA Dunning, A Bubeck, A large landslide event in a post-glacial landscape: rethinking glacial legacy. *Earth Surface Processes and Landform*, **38** (2013), 1261–1268.
48. K Hewitt, JJ Clague, JF Orwin, Legacies of catastrophic rock slope failures in mountain landscapes. *Earth-Science Reviews*, **87** (2008), 1–38.

9

Glacier- and permafrost-related slope instabilities

MICHAEL KRAUTBLATTER AND KERRY LEITH

9.1 Introduction

Glacier retreat and permafrost degradation are considered to be major hazards in alpine regions, as both induce slow rock slope deformation and rockfall activity that can endanger infrastructure and cause casualties (Figure 9.1). While the scientific observation of glaciers and changes in their extent dates back more than 150 years (e.g. Louis Agassiz in 1841, James D. Forbes in 1842), the observation of permafrost as a hidden subsurface phenomenon in alpine soils, debris, and rocks has only received serious attention since the late 1970s [1,2]. The time span of scientific observation of permafrost in mountains and of the respective instabilities is relatively short and focused on a handful of well-observed study sites. This explains why the retrospective correlation of instability and changing climate conditions is a difficult task. This chapter rather exploits our physical understanding of changes in stress and the physical strength of slopes over time in order to understand systemic patterns of glacier and permafrost-related slope stabilities.

9.1.1 Cryospheric rock slope systems

Permafrost is a thermally defined phenomenon referring to ground that remains below 0 °C for at least two consecutive years, irrespective of the presence of water or ice in the system [3]. Rock permafrost is not synonymous with perennially frozen rock, as rock often only freezes significantly below the datum freezing point T_0 (0 °C), due to the effects of solutes, pressure, pore diameter, and pore material [4]. Ice develops in pores, cavities, and joints (hereafter used as general term) such as fissures and (macro-)fractures (>0.1 mm aperture) [5]. The systemic difference between non-permafrost and permafrost rock walls is, thus, the potential perennial

The High-Mountain Cryosphere, ed. Christian Huggel, Mark Carey, John J. Clague and Andreas Kääb. Published by Cambridge University Press. © Cambridge University Press 2015.

(a)

Figure 9.1 (a) Massive rock slope failure subsequent to glacier retreat at the Eiger (CH) in 2006. Gullies in moraine deposits (top left) indicate the Little Ice Age height of the glacier. (b) Recent changes in thermal, hydrological, mechanical, and chemical conditions in permafrost-affected bedrock. Meltwater (waterfalls) provides massive heat input into fractured rocks, stress conditioned by glacier retreat and mechanical strength conditioned by abrupt thawing bedrock change, as well as weathering conditions (reddish zones) at sites where the protective ice cover from onfrozen glacierets has recently disappeared.

presence of ice (i.e. cryospheric), and its serious implications for the thermal, hydraulic, and mechanical properties of the rock wall system.

In cryospheric systems, glaciers and permafrost display a complex interplay and a high level of interconnectivity [6–8]. Principles of the thermal and mechanical interconnectivity of glaciers and permafrost are displayed in Figure 9.2, assuming that (1) active layer thickness decreases with altitude [9]; (2) cold glaciers and onfrozen glacierets are based on and favour the development of permafrost bedrock [10]; (3) warm glaciers conduct massive advective heat transfer with adjacent rocks [11,12]; and (4) the active layer is 'semiconductive' as thermal conduction in the frozen rock mass performs better than in unfrozen rock and across air-filled rock discontinuities [13].

(b)

Figure 9.1 (*cont.*)

Thus, glacier advance and retreat might not only change stress patterns across (formerly) glaciated valleys, but every glacial retreat and advance will cause dynamic changes in glacier–permafrost interconnectivity, which may promote instability. At the same time, permafrost is a steadily changing thermal system in which present-day near-surface heat fluxes govern future permafrost aggradation and degradation. The heat flux at the surface is controlled by short- and long-wave radiation, latent and sensible surface heat fluxes, and geothermal and transient heat fluxes. Variations in short-wave irradiance due to aspect, slope angle, and shading are the most important factors that account for the heterogeneity of mountain rock permafrost [14]. Variations in *potential* short-wave incoming radiation and albedo control the seasonal dynamics of permafrost [15], while inter-annual variations in the *real* incoming short-wave radiation, due to mean atmospheric conditions and reduced albedo (e.g. hot summer of 2003) control annual fluctuations in thaw depth [9]. Long-wave emission also shows strong seasonal fluctuations [15]. The geothermal heat flux from depth (~0.05 W m^{-2}) is several orders of magnitude lower than the external heat fluxes described above. However, transient effects on a Holocene timescale (Little Ice Age, Climatic Optima) as well as on a Quaternary

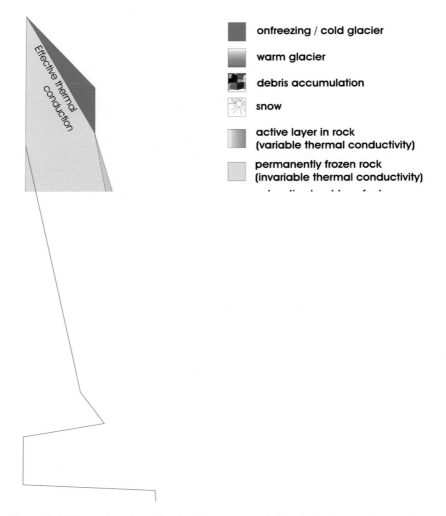

onfreezing / cold glacier

warm glacier

debris accumulation

snow

active layer in rock
(variable thermal conductivity)

permanently frozen rock
(invariable thermal conductivity)

Figure 9.2 Thermal and mechanical interconnectivity of glaciers and permafrost-affected rock walls.

timescale (Last Glacial Maximum) must be considered for adequate modelling of present-day thermal fluxes [16,17].

On a metre to decametre scale, the rock permafrost distribution is more complex. The highly structured topography of rock walls results in a multi-directional rather than a uni-directional heat flux [18]. In fractures, sensible and latent heat fluxes transfer energy in an advective manner [11]. Decadal memory effects occur with respect to both past climatic changes and snow patch/glacier fluctuations [19]. Rock permafrost in marginal glacier positions is often favoured due to the occurrence of polythermal glaciers and marginal onfreezing in alpine conditions [6].

Due to the complex interplay of topographic, hydrologic, climatic, and memory influences, and the lack of appropriate monitoring techniques, present knowledge of local permafrost distribution is limited.

For 2055, Nogués-Bravo *et al.* [20] anticipate a 3–4 °C warming for mid-latitude alpine environments and a 4–7 °C warming scenario, and for high-latitude mountain environments for the A1FI (global economic) scenario, and, respectively, a 2–3 °C and 3–5 °C warming for the B1 (global environmental) scenario. Recent studies have either tried to abductively reconstruct the response of rock slopes to warming in the last decades or century, or, deductively, tried to develop a mechanistic understanding of why rock walls must respond to climate change [21]. From a geomorphological point of view, rockfall inventories point towards an increasing frequency and magnitude of rockfalls from permafrost-affected rock walls [22,23]. From a geotechnical point of view, the mechanical properties of both ice and water-saturated rock are highly susceptible to temperature changes when close to the thawing point [24–26]. Permafrost-related rock–ice avalanches of Bergsturz size (release volumes >1 mio. m^3) were documented at Mt. Steller, Alaska ($5(\pm1) \times 10^7$ m^3) in 2005, at Dzhimarai-Khokh, Russian Caucasus (4×10^6 m^3) in 2002, at Mt. Steele, Yukon ($5.5 \times (\pm2.5) \times 10^7$ m^3) in 2007, at Harold Price, British Columbia (1.6×10^6 m^3) in 2002 and at the Brenva (2×10^6 m^3) and the Punta Thurwieser (2×10^6 m^3) in the Italian Alps in 1997 and 2004 [27]. Accordingly, enhanced activity of cliff falls (10^4–10^6 m^3), block falls (10^2–10^4 m^3), boulder falls (10^1–10^2 m^3) and debris falls (<10 m^3) was detected in several permafrost-affected rock faces [7,23,28]. However, the distinction of the climate change signal from non-climate determined long- and short-term dynamics of rock slope evolution [29] remains a difficult task, especially as most inventories cover only the last two decades.

9.1.2 Prerequisites for instability

The spatial and temporal development of rock slope instability results from either an increase in driving stresses, or a reduction in material strength. Structural controls predispose slope for a particular mode of failure and play a crucial role in determining the failure mechanism, the geometry of the failure plane (critical path), and volume. Short-term (i.e. $1 \times 10^{0-2}$ yr, hot summer 2003), mid-term (i.e. $1 \times 10^{0-2}$ yr, Little Ice Age), and long-term (i.e. $1 \times 10^{2+}$ yr, Glacial–Interglacial Cycles) climatic events can contribute to the development of rock slope instability, although the failure plane itself is predefined by geological structures. Focusing on physical changes in stress or strength provides a means of directly comparing rock slope system changes independent of spatial or temporal scale, and provides insight into climatically driven contributors that precondition and condition rock slopes for failure and, therefore, control patterns of instability in alpine regions.

Glacier retreat and permafrost degradation *precondition* rock slopes for failure by inducing changes in *in situ* stress within a rock mass. These changes may occur as a result of variations in topographic form by glacial and fluvial erosion, loading and unloading by ice, lakes, and sediments, or changing hydro- and cryostatic pressure and thermal stresses. *Conditioning* processes then operate in response to these changes, and reduce the cohesive and frictional strength of a rock mass, for example through reactivation and dilation of existing fractures, nucleation and propagation of new fractures, yielding of rock bridges or interlocking asperities, brittle–ductile block deformation, fatigue, strain softening, reduction of cohesive ice contacts, and increased weathering in association with these processes. Aside from weathering, these processes are all associated with the concept of progressive failure and internal rock mass deformation [30].

9.2 Glacially induced preconditioning and conditioning of alpine rock slopes

9.2.1 Preconditioning by long-term erosional processes

Slope instability is a common occurrence in regions experiencing a loss of glacial ice [31]. The removal of ice loading during deglaciation affects both the distribution and magnitude of near-surface crustal stresses [31–34]. In addition, the retreat of glacial ice also removes a degree of support to steep rock slopes [35], and initiates rock mass strength degradation through stress corrosion and progressive failure (Figures 9.3 and 9.4). While the progression of rock slope failure following deglaciation is becoming increasingly clear from the fossil record, our understanding of *in situ* stress changes during deglaciation, and the consequent onset of (for example) seismicity, does not yet allow a causative link to be drawn between seismicity and catastrophic rock slope failure. Morphological evidence for the effects of crustal stress changes is preserved in prominent post-glacial faults and uphill facing scarps in both the Scottish Highlands [36,37] and throughout the Alps [34,38,39]. Numerical investigations for both Alpine and cratonic regions suggest these are likely to form shortly after glacial unloading [33,34,40], although the delay time is uncertain as model results are dependent on poorly constrained crustal stress and strain conditions.

Deglaciation has been observed to induce instabilities on rock slopes in different ways, depending on the rock mass structure, strength, and valley geometry. Slopes oversteepened relative to their intrinsic rock mass strength may require only minor conditioning to become kinematically unstable. They rapidly readjust to a more

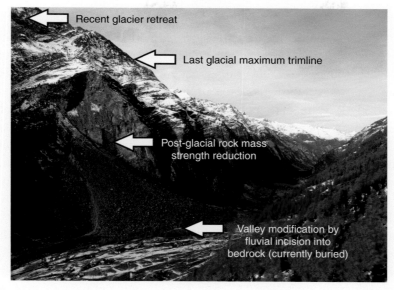

Figure 9.3 Preconditioning of the Randa rock slope failure by Quaternary glacial and interglacial erosion.

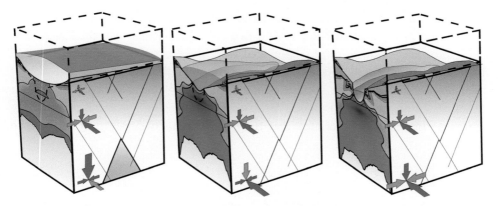

Figure 9.4 Impact of stress changes of bedrock fracturing during a period of strong glacial erosion. Colours indicate critical fracture stress (grey), micro-crack generation (purple), tensile fracture (yellow), propagation of existing fractures (orange), and extensional (or exfoliation) fracture (red).

natural angle through slope failure processes such as large deep-seated landsliding along pre-existing faults or weakness planes [36,41–45]. On the other hand, steep, high-strength slopes often develop slope–parallel exfoliation joints commonly associated with glacial stress release [46] which promote relatively small failures (e.g. rockfalls and block slides).

9.2.2 *Temporal and spatial patterns of post-glacial rock slope instability*

Large rock slope failures in formerly glaciated regions are commonly attributed to the effects of glacial erosion, stress redistribution following deglaciation, the progressive weakening of rock slopes, and/or the emergence of extraordinary environmental triggers throughout the Holocene [36,47,48].

In a steady-state model, rock instability is only controlled by geological/geotechnical properties of the rock slope and is not influenced by Lateglacial Glacier Retreat (Figure 9.5). This model would be implicitly used when constant Holocene recurrence rates for rock slope failure are inferred. As instabilities commonly exploit limited pre-existing tectonic structures or well-developed fracture sets, the exhaustion model postulates a rapid readjustment of oversteepened slopes and, subsequently, an approximately exponentially decrease in rock slope failure activity with time [36]. The peak and subsequent exponential decrease is attributed to simple kinematic exhaustion as the number of potential (i.e. unfailed) sites progressively reduces [49]. Instabilities that take place later in the post-glacial timeline are believed to require conditioning of the rock mass in order to become kinematically feasible. While these failures commonly also exploit pre-existing or well-developed fracture sets, they are controlled by conditioning processes within the slope.

A number of recent inventories point towards a bimodal (or multimodal) distribution of rock slope failures with two (or more) peaks over the Lateglacial and Holocene period. Strong post-glacial rock slope activity in both Alpine and Highland regions were concentrated in two periods, the first shortly (~1–2 kyr) after deglaciation, and the second at approximately 4–2 kyr BP [47,48,50]. For instance, Prager *et al.* [48] plotted 60 dated rock slides in Tyrol and surrounding areas in Germany, Switzerland, and Italy, and found two peaks of rock slope

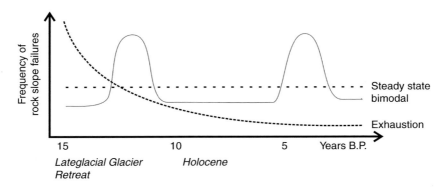

Figure 9.5 Three models explaining the frequency of rock slope failure subsequent to the Lateglacial Glacier Retreat.

failure from 10500–9400 cal BP and from 4200–3000 cal BP. Soldati *et al.* [52] found clusters of peak activity from 13000 to 9000 cal BP and from 6500–2300 cal BP in the Italian Dolomites. As the first peak occurred significantly after deglaciation, they postulated a relaxation time of a few thousand years due to brittle fracture propagation in rock masses. It seems likely that the initial period of activity was associated with preconditioning due to changes in glacial valley morphology, and subsequent conditioning through rock slope strength degradation [51]. The timing may coincide with either a significant climatic change or a delayed onset of post-glacial seismicity [47]. It has also been hypothesised that progressive weakening of rock slopes during the preceding ~10 kyr was essential in order to condition rock slopes which passed through the initial period of rock slope activity for failure during the second [47]. It is discussed whether the second peak in activity coincides with climatic changes, increased precipitation and possible increase in seismicity within the Central Alps or, partially, by warming of permafrost-affected bedrock [25,48,50,52,53]. Prager *et al.* [48] suggest 'striking environmental changes in the middle Holocene'. Other articles postulate more specifically that rockfalls within this timeframe originating from present-day permafrost rock walls could be a response to permafrost degradation subsequent to the Holocene Climatic Optimum [52,54,55].

9.3 Permafrost-induced preconditioning and conditioning of alpine rock slopes

9.3.1 Instability assessment of permafrost-affected slopes

Permafrost distribution and glacier and snow dynamics can significantly influence the stability of permafrost rock slopes, and will itself respond quickly to climatic fluctuations [56]. The influence of permafrost dynamics on rock slope stability is only considered in detail in a few studies. The presence of ice in the detachment zone of instabilities has often been reported [57]. A spatial relationship between permafrost degradation and rockfall was detected for the European Alps and the Southern Alps, New Zealand [23,58]. In some cases the detachment of rock–ice avalanches (Kolka-Karmadon, Caucasus, and Mt. Steller, Alaska) and rock slope failures (Monte Rosa, Italy) can be correlated with thermal disturbances caused by the interaction of permafrost and glacial ice, volcanic/geothermal effects, and climate change [10]. The sensitivity of permafrost to atmospheric warming and the subsequent enhanced activity of rockfall events were demonstrated in the European Alps on an annual timescale during the hot summer of 2003 [9], and on a decadal timescale in a 100-year record of rockfall activity on the permafrost-affected Les Drus Peak (France) [59].

9.3.2 Temporal and spatial controls of permafrost-affected rock slope failure

Insights into the influence of low temperatures and ice infillings on the properties of intact rock and rock masses are derived from cold regions geomorphology, glaciology, and cold regions geotechnics and engineering. From a mechanical point of view, the presence of permafrost can *increase shear stress* due to changing hydrostatic pressure and cryostatic pressure, i.e. by volumetric expansion or ice segregation (Figure 9.6). Thawing permafrost can also act to *decrease shear resistance* of rock masses as thawing alters the mechanical behaviour of intact rock, crack propagation, and frictional processes of rock–rock contacts, rock–ice contacts, and ice/frozen fill-material.

9.3.2.1 Increased shear stress

Hydrostatic pressure can be elevated due to perched groundwater above permafrost bedrock. Experiments indicate that the permeability of frozen fissured rock is one to three orders of magnitude lower than the permeability of identical thawed rock [60]. The ratio increases for highly weathered rocks close to the surface with greater ice contents. The combination of perched groundwater and deep-reaching unfrozen fracture systems causes significant problems for tunnelling structures by inundating water in permafrost rocks, e.g. at the Aiguille du Midi (France) and at the Jungfrau (CH) in 2003, as well as for the Kunlun Mountain tunnel of the QingHai–Tibetan railway track [61]. Hydrostatic pressures due to the sealing of rock surfaces by ice can play a vital part in the destabilisation of rock slopes, as

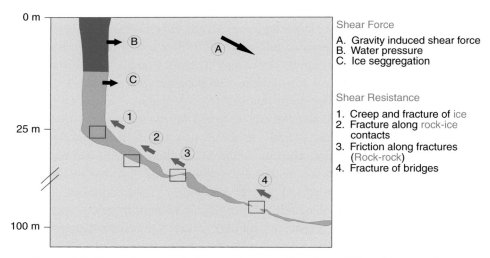

Figure 9.6 Shear forces and shear resistances for slope failure in permafrost-affected bedrock.
Source: modified from [25].

outlined by coupled hydro-mechanical modelling (Tschierva, Italy) and the observation of outflow of pressurised water (Kolka-Karmadon and Mt. Steller) [56].

Cryostatic pressure due to ice segregation requires a sub-zero temperature gradient (typically –0 °C to –6 °C) and water supply. These conditions frequently coincide at the base of the active layer above the permafrost table [62]. Heaving pressures of 20–30 MPa exceed even the tensile strength of strong rocks and can cause crack propagation [63]. For potential shear planes at greater depths (deca-metres), reduced temperature gradients and water availability, combined with increased normal load, would effectively squeeze out ice between protruding rock–rock contacts over time and exceed the shut-off pressure for ice segregation.

9.3.2.2 Reduced shear resistance

The shear resistance of ice-filled fractures responds to different mechanical processes acting individually, in succession, or in combination: (1) friction/fracture along rock–rock contacts; (2) friction/fracture along rock–ice contacts; (3) fracture/ deformation of cleft ice; and if present (4) deformation of frozen infill material (for a detailed mechanical overview, see [25]).

(1) For saturated intact rock, Mellor [26] could show an increase in uniaxial compressive strength from 20% to 50% and, respectively, an increase of the instantaneous tensile strength from 15% to 70% when freezing rock samples down to −10 °C, which was later confirmed by others [64–66]. The loss of strength correlates with rock porosity and water content and corresponds to changes in Poisson's ratio, Young's modulus, and joint stiffness [67]. The most likely explanation is that freezing of absorbed water in micro-cracks and fissures leads to an effective grouting of the rock. The effect is dependent on porosity and water content, and more pronounced for tensile strength than for compressive strength [67,68]. The abrupt decrease in strength, when thawing, influences crack propagation, subcritical crack propagation, and friction. The presence of ice within micro-cracks also raises the fracture toughness by 8% to 37% [69], depending on moisture content and temperature. Accordingly, subcritical fracture propagation is also heavily reduced in frozen rock and the frictional shear resistance of ice-free smoothened rock–rock surfaces decreases subsequent to thawing [25,65]. Thus, all components of rock mechanical strength reduce significantly between −10 °C and the thawing point.

(2) For moderate levels of normal load, the failure of ice-filled fractures occurs at the connection between ice and concrete, and not within the ice. Constant stress (acceleration allowed) experiments of concrete–ice samples indicate that the maximum strength of the rock–ice connection decreases linearly for temperatures between −3 °C and 0 °C [25].

(3) The behaviour of polycrystalline ice in fractures under constant load is dependent on the stress-strain conditions and the rate of loading. Ice demonstrates elastic and ductile creep behaviour (primary, secondary, and tertiary creep) without failure when slowly compressed. Exceeding certain thresholds for stress level, strain rate, or strain level (for uniaxial compressive strength this is 5–10 MPa, tensile: 1–2 MPa, 10^{-3} s^{-1} and 1%, respectively), ice deforms in a brittle and ductile-brittle manner until complete fracture occurs [70]. Constant strain (no acceleration) experiments of ice–concrete samples indicate that the shear strength of ice-filled fractures is a function of temperature and normal stress, i.e. that shear strength of the fracture declines with increasing temperature of the ice between –5 °C and 0 °C [71].

(4) The behaviour of fractures in permafrost bedrock with *frozen infill material* can probably be derived from studies on permafrost soils. Volumetric ice content and strain rate are key factors for the strength characteristics of frozen soils [72]. The strength increases as ice content decreases because of enhanced friction between solid particles. Ice is the bonding between particles and provides cohesion, resulting in a stiffer behaviour at the beginning of shearing at low confining stress compared to unfrozen samples. At high strain rates the resistance of frozen soil is similar to that of unfrozen soil. During strain relaxation, the ice-bonding heals itself due to refreezing and causes a strengthening of laboratory samples [73].

9.4 Present-day and anticipated response of para- and periglacial rock slope systems

We now refer to periglacial slopes as those affected by cryogenic weathering and permafrost, and para-glacial slopes as those preconditioned by former glaciations [74]. Environments that no longer contain ice and, thus, do not belong to the cryosphere can still show signs of glacial and periglacial preconditioning due to transient effects.

Gruber and Haeberli [57] postulate that 'the destabilisation of steep bedrock by permafrost degradation implies a portion of bedrock slopes in permafrost that is steeper than it would be when thawed'. The argument implicitly describes the existence of equilibrium and disequilibrium slopes and indicates a need to describe the sensitivity of slopes to climate change and to define the portion of slopes that reacts. We need to define the timescale of adaptation. Timescales of adaptation are therefore critical, and further research is required to describe the transience from one configuration to another in order to better predict the future response of rock slopes in alpine environments.

The slope angle α for an infinite (steady-state equilibrium) slope is conditioned by the cohesion of the rock mass, the density, the internal friction, the gravitational acceleration, and the depth of the sliding plane. Equilibrium ends if 'there were a change in any of the parameters involved'. For instance, in degrading permafrost rock, cohesion and total friction change abruptly close to the melting point [75]. Thus, thawing rock permafrost slopes are non-equilibrium slopes by definition. However, the sensitivity of a landscape to produce a sensible, recognisable, and persistent response includes two aspects [76]: the *propensity for change* and the *capacity* of the system *to absorb* the *change*. The propensity for change is given by driving forces and barriers to that change. The capacity of the system to absorb change is the ability to absorb and store energy, water, and materials. Once change has been initiated, the rate of change determines the relaxation time or the persistence of characteristics of the former state.

9.4.1 Reaction and relaxation time of peri/paraglacial rock slope systems

Exposed rock summits in high alpine conditions have persisted in these positions for tens to hundreds of thousands of years during several climatic fluctuations. As enormous 'driving forces' in terms of potential energy are available for unstable rock slopes, two explanations must be considered: (1) significant barriers to change exist and/or (2) the relaxation time spans a long timeframe. Apparently, significant barriers to change exist in a fractured rock wall of highly cohesive rock material where immediate adaptation can only occur along predefined planes of weakness. This shows that the understanding of relaxation time is a key concern in such environments.

The reaction time (lag) is the time taken to react to an impulse of change [77]. Then the relaxation (recovery, form adjusting) time persists until the new equilibrium slope is approached (Figure 9.7). The first 'sensible' impulse of change of a rock slope (i.e. reaction time) that develops towards a more gentle slope angle is given by the first rockfall that exceeds the limits of the mass wasting rates, which defined the steady-state equilibrium of the prior slope. In practice, these first rockslides are likely to occur along existing planes of weakness and to exploit predefined ice-filled fractures close to the surface by ice-mechanical weakening [25]. This is given by the speed of propagation of the thermal signal to the ice inside the rock fracture. On permafrost rock faces with significant predisposed critical ice-filled fractures near the rock surface the reaction time is in the range of weeks to a few months. For instance, the highest frequency of rockfalls in summer 2003, which was 5 °C above mean summer temperatures in Switzerland, was observed in July (e.g. Matterhorn) and August (e.g. Dent Blanche), earlier than expected from thermal modelling [9], and is complemented by the observation of ice or running water at the detachment zones [57].

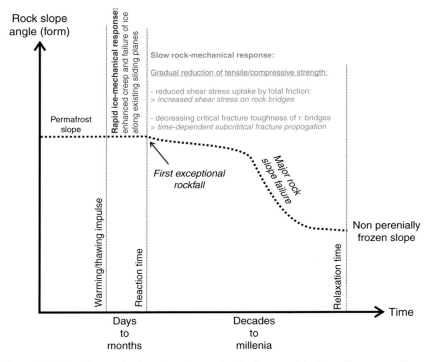

Figure 9.7 Reaction and relaxation time of rock slopes with degrading permafrost.

Total adjustment (i.e. relaxation time) is achieved as soon as the slope angle reflects the rock mass stability of unfrozen rock [77]. In terms of rock-/ice-mechanics, this long-term mechanical failure criterion responds to temperature-induced change in the compressive and tensile rock strength, and as well as fracture toughness. The onset of the rock bridge destruction is given by the time when the thermal impulse reaches the affected rock fracture. The reduction in rock strength upon thawing onsets an effect that causes brittle fracture propagation over a longer timeframe. Modelling examples show that in some cases the formation of a critical failure plane can take years to hundreds of years for larger rock failures [78].

9.4.2 The foreseeable future: transient rock slope systems

Plotting rockfall size versus time after impulse (Figure 9.8) has become a generally accepted conceptual model in permafrost research [79,80]. The underlying assumption is that a decrease of the speed of thermal propagation with depth and thus larger rockslides require an increasing response time (reaction + relaxation time) [11]. The response time of large rock slope failures to environmental change is often hundreds to thousands of years due to the slow formation of a sliding plane

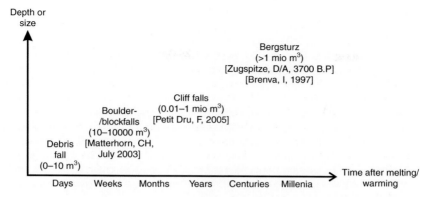

Figure 9.8 Time after warming versus magnitude of rock slope failure. CH, Switzerland; F, France, D/A, Germany/Austria, I, Italy.
Source: after [79].

by brittle fracture propagation. As evident from the fact that few large rock slides (>100 mio. m^3) occurred before 10 ka BP, 5000 years after ice had retreated from the flanks of the valleys in the Central Eastern Alps [48]. Any situation where the interval between form-changing events is smaller than the response time can be referred to as transient [77]. Even if the thermodynamic application of the term transience [17] is quite different from the geomorphologic definition, it may point towards the same direction: large rock masses inherit (thermal) signals that date back several thousand years and still influence present systems. In other words, the thermal signal caused by present-day climate change will influence rock slope stability in sensitive (i.e. preconditioned) deglaciated and permafrost-affected rock slopes over timescales of decades to hundreds of years and may condition both slow rock deformations and rapid rock slope failure.

References

1. D Barsch, H Fierz, W Haeberli. Shallow core drilling and bore-hole measurements in the permafrost of an active rock glacier near the Grubengletscher, Wallis, Swiss Alps. *Arctic and Alpine Research.* 1979;**11**(2):215–228.
2. H French, CE Thorn. The changing nature of periglacial geomorphology. *Géomorphologie: Relief, Processus, Environnement.* 2006;**3**:1–13.
3. NRC-Permafrost-Subcommitee. Glossary of permafrost and related ground-ice terms. *NRC Technical Memorandum.* 1988;**142**:1–156.
4. GSH Lock. *The Growth and Decay of Ice.* Cambridge: Cambridge University Press; 2005.
5. C Jaeger. *Rock Mechanics and Engineering.* Cambridge: Cambridge University Press; 2009.
6. B Etzelmüller, JO Hagen. Glacier–permafrost interactions in Arctic and alpine mountain environments with examples from southern Norway and Svalbard. In: C Harris, JB Murton (eds) *Cryospheric Systems: Glaciers and Permafrost.* London: Geological Society Special Publication; 2005. pp. 11–27.

7. L Fischer, C Huggel, A Kaab, W Haeberli. Slope failures and erosion rates on a glacierized high-mountain face under climatic changes. *Earth Surface Processes and Landforms*. 2013;**38**(8):836–846.

8. W Haeberli. Investigating glacier–permafrost relationships in high-mountain area: historical background, selected examples and research needs. In: C Harris, JB Murton (eds) *Cryospheric Systems: Glaciers and Permafrost*. London: Geological Society Special Publication; 2005. pp. 29–37.

9. S Gruber, M Hoelzle, W Haeberli. Permafrost thaw and destabilization of Alpine rock walls in the hot summer of 2003. *Geophysical Research Letters*. 2004;**31**(13):L15054.

10. C Huggel. Recent extreme slope failures in glacial environments: effects of thermal perturbation. *Quaternary Science Reviews*. 2009;**28**(11–12):1119–1130.

11. M Wegmann, GH Gudmundsson, W Haeberli. Permafrost changes in rock walls and the retreat of Alpine glaciers: a thermal modelling approach. *Permafrost and Periglac Process*. 1998;**9**:23–33.

12. BJ Moorman. Glacier–permafrost hydrological interconnectivity: Stagnation Glacier, Bylot Island, Canada. In: C Harris, JB Murton (eds) *Cryospheric Systems: Glaciers and Permafrost*. London: Geological Society Special Publication; 2005. pp. 63–74.

13. A Hasler, S Gruber, M Font, M Dubois. Advective heat transport in frozen rock clefts: conceptual model. *Permafrost and Periglacial Processes*. 2011; **22**(4):378–389.

14. S Gruber, M Hoelzle, W Haeberli. Rock-wall temperatures in the Alps: modelling their topographic distribution and regional differences. *Permafrost and Periglacial Processes*. 2004;**15**(3):299–307.

15. M Hoelzle, C Mittaz, B Etzelmüller, W Haeberli. Surface energy fluxes and distribution models of permafrost in European mountain areas: an overview of current developments. *Permafrost and Periglacial Processes*. 2001;**12**(1):53–68.

16. J Noetzli, S Gruber. Transient thermal effects in Alpine permafrost. *The Cryosphere*. 2009;**3**:85–99.

17. T Kohl. Transient thermal effects below complex topographies. *Tectonophysics*. 1999;**306**(3–4):311–324.

18. J Noetzli. Modeling transient three-dimensional temperature fields in mountain permafrost. PhD. University of Zurich; 2008.

19. IT Kukkonen, J Safanda. Numerical modelling of permafrost in bedrock in northern Fennoscandia during the Holocene. *Global Planetary Change*. 2001;**29**:259–274.

20. D Nogués-Bravo, MB Araújo, MP Errea, JP Martínez-Rica. Exposure of global mountain systems to climate warming during the 21st century. *Global Environmental Change*. 2007;**17**:420–428.

21. M Krautblatter, C Huggel, P Deline, A Hasler. Research perspectives on unstable high-alpine bedrock permafrost: measurement, modelling and process understanding. *Permafrost and Periglacial Processes*. 2012;**23**(1):80–88.

22. C Huggel, JJ Clague, O Korup. Is climate change responsible for changing landslide activity in high mountains? *Earth Surface Processes and Landforms*. 2012;**37**(1):77–91.

23. L Ravanel, P Deline. Climate influence on rockfalls in high-Alpine steep rockwalls: the north side of the Aiguilles de Chamonix (Mont Blanc massif) since the end of the Little Ice Age. *Holocene*. 2011;**21**:357–365.

24. MCR Davies, O Hamza, C Harris. The effect of rise in mean annual temperature on the stability of rock slopes containing ice-filled discontinuities. *Permafrost and Periglacial Processes*. 2001;**12**(1):137–144.

25. M Krautblatter, D Funk, F Guenzel. Why permafrost rocks become unstable: a rock–ice-mechanical model in time and space. *Earth Surface Processes and Landforms*. 2013;**38**(8):876–887.

26. M Mellor (ed.). *Mechanical Properties of Rocks at Low Temperatures. 2nd Int Conference on Permafrost*; 1973; Yakutsk, Russia.
27. SP Pudasaini, M Krautblatter. A two-phase mechanical model for rock-ice avalanches. *Journal of Geophysical Research: Earth Surface.* 2014;**119**(10):2272–2290.
28. O Sass. Spatial patterns of rockfall intensity in the northern Alps. *Zeitschrift für Geomorphologie.* 2005; **138**:51–65.
29. M Krautblatter, JR Moore. Rock slope instability and erosion: toward improved process understanding. *Earth Surface Processes and Landforms.* 2014;**39**(9): 1273–1278.
30. E Eberhardt, D Stead, JS Coggan. Numerical analysis of initiation and progressive failure in natural rock slopes: the 1991 Randa rockslide. *International Journal of Rock Mechanics and Mining Sciences.* 2004;**41**:69–87.
31. K Hewitt, JJ Clague, JF Orwin. Legacies of catastrophic rock slope failures in mountain landscapes. *Earth-Science Reviews.* 2008;**87**(1–2):1–38.
32. A Hampel, R Hetzel, G Maniatis. Response of faults to climate-driven changes in ice and water volumes on Earth's surface. *Philosophical Transactions of the Royal Society A: Mathematical, Physical and Engineering Sciences.* 2010;**368**(1919): 2501–2517.
33. K Leith, JR Moore, F Amann, S Loew. In situ stress control on the development of near-surface extensional fractures in alpine landscapes. *Journal of Geophysical Research, Solid Earth.* 2014;**119**:1–22.
34. M Ustaszewski, A Hampel, O Pfiffner. Composite faults in the Swiss Alps formed by the interplay of tectonics, gravitation and postglacial rebound: an integrated field and modelling study. *Swiss Journal of Geosciences.* 2008;**101**(1):223–235.
35. ST McColl, TRH Davies. Large ice-contact slope movements: glacial buttressing, deformation and erosion. *Earth Surface Processes and Landforms.* 2013;**38**(10): 1102–1115.
36. CK Ballantyne Paraglacial geomorphology. *Quaternary Science Reviews.* 2002;**21** (18–19):1935–2017.
37. IS Stewart, J Sauber, J Rose. Glacio-seismotectonics: ice sheets, crustal deformation and seismicity. *Quaternary Science Reviews.* 2000;**19**(14–15):1367–1389.
38. P Eckardt, HP Funk, T Labhart. Postglaziale Krustenbewegungen an der Rhein-Rhone-Linie. *Vermessung, Photogrammetrie, Kulturtechnik.* 1983;**2**: 43–56.
39. J-C Hippolyte, G Brocard, M Tardy, *et al.* The recent fault scarps of the Western Alps (France): tectonic surface ruptures or gravitational sackung scarps? A combined mapping, geomorphic, levelling, and 10Be dating approach. *Tectonophysics.* 2006; **418**(3–4):255–276.
40. J-C Hippolyte, D Bourlès, R Braucher, *et al.* Cosmogenic 10Be dating of a sackung and its faulted rock glaciers, in the Alps of Savoy (France). *Geomorphology.* 2009; **108**(3–4):312–320.
41. A von Poschinger, P Wassmer, M Maisch. The Flims rockslide: history of interpretation and new insights. Landslides from massive rock slope failure. *NATO Science Series.* 2006;**49**:329–356.
42. A Meigs, WC Krugh, K Davis, G Bank. Ultra-rapid landscape response and sediment yield following glacier retreat, Icy Bay, southern Alaska. *Geomorphology.* 2006; **78**(3–4):207–221.
43. P Augustinus. Rock mass strength and the stability of some glacial valley slopes. *Zeitschrift Fur Geomorphologie.* 1995;**39**(1):55–68.
44. E Cossart, R Braucher, M Fort, DL Bourles, J Carcaillet. Slope instability in relation to glacial debuttressing in alpine areas (Upper Durance catchment, southeastern France):

evidence from field data and 10Be cosmic ray exposure ages. *Geomorphology.* 2007;**95**(1–2):3–26.

45. MJ Selby. Controls on the stability and inclinations of hillslopes formed on hard rock. *Earth Surface Processes and Landforms.* 1982;**7**(5):449–467.
46. D Bahat, K Grossenbacher, K Karasaki. Mechanism of exfoliation joint formation in granitic rocks, Yosemite National Park. *Journal of Structural Geology.* 1999;**21**(1): 85–96.
47. C Ballantyne, GF Sandeman, JO Stone, P Wilson. Rock-slope failure following Late Pleistocene deglaciation on tectonically stable mountainous terrain. *Quaternary Science Reviews.* 2014;**86**(15):144–157.
48. C Prager, C Zangerl, G Patzelt, R Brandner. Age distribution of fossil landslides in the Tyrol (Austria) and its surrounding areas. *Natural Hazards and Earth Systems Science.* 2008;**8**(2):377–407.
49. DM Cruden, XQ Hu. Exhaustion and steady state models for predicting landslide hazards in the Canadian Rocky Mountains. *Geomorphology.* 1993;**8**(4):279–285.
50. M Strasser, K Monecke, M Schnellmann, FS Anselmetti. Lake sediments as natural seismographs: a compiled record of Late Quaternary earthquakes in Central Switzerland and its implication for Alpine deformation. *Sedimentology.* 2013;**60**(1):319–341.
51. ST McColl. Paraglacial rock-slope stability. *Geomorphology.* 2012;**153–154**:1–16.
52. M Soldati, A Corsini, A Pasuto. Landslides and climate change in the Italian Dolomites since the Late glacial. *Catena.* 2004;**55**(2):141–161.
53. K Nicolussi, M Kaufmann, G Patzelt, J van der, A Thurner. Holocene tree-line variability in the Kauner Valley, Central Eastern Alps, indicated by dendrochronological analysis of living trees and subfossil logs. *Vegetation History and Archaeobotany.* 2005;**14**(3):221–234.
54. H Jerz, Av Poschinger. Neuere Ergebnisse zum Bergsturz Eibsee-Grainau. *Geologica Bavarica.* 1995;**99**:383–398.
55. W Tinner, P Kaltenrieder, M Soom, *et al.* The postglacial rockfall in the Kander valley (Switzerland): age and effects on palaeo-environments. *Ecologae Geologicae Helvetiae.* 2005;**98**(1):83–95.
56. L Fischer, F Amann, JR Moore, C Huggel. Assessment of periglacial slope stability for the 1988 Tschierva rock avalanche (Piz Morteratsch, Switzerland). *Engineering Geology.* 2010;**116**(1–2):32–43.
57. S Gruber, W Haeberli. Permafrost in steep bedrock slopes and its temperature-related destabilization following climate change. *Journal of Geophysical Research – Earth Surface.* 2007;**112**(F2):F02S13.
58. SK Allen, S Gruber, IF Owens. Exploring steep bedrock permafrost and its relationship with recent slope failures in the Southern Alps of New Zealand. *Permafrost and Periglacial Processes.* 2009;**20**:345–356.
59. L Ravanel, P Deline. La face ouest des Drus (massif du Mont-Blanc). Évolution de l'instabilité d'une paroi rocheuse dans la haute montagne alpine depuis la fin du petit age glaciaire. *Geomorphologie.* 2008;**4**:261–272.
60. MI Pogrebiskiy, SN Chernyshev. Determination of the permeability of the frozen fissured rock massif in the vicinity of the Kolyma Hydroelectric Power Station. *Cold Regions Research and Engineering Laboratory.* 1977;**634**:1–13.
61. GZ Tang, XH Wang. Modeling the thaw boundary in broken rock zones in permafrost in the presence of surface water flows. *Tunnelling and Underground Space Technology.* 2006;**21**(6):684–689.
62. JB Murton, R Peterson, J-C Ozouf. Bedrock fracture by ice segregation in cold regions. *Science.* 2006;**314**:1127–1129.

63. B Hallet, JS Walder, CW Stubbs. Weathering by segregation ice growth in microcracks at sustained sub-zero temperatures: verification from an experimental study using acoustic emissions. *Permafrost and Periglacial Processes*. 1991;**2**:283–300.

64. L-O Dahlström. *Rock Mechanical Consequences of Refrigeration*. Göteborg: Chalmers University of Technology; 1992.

65. N Li, P Zhang, Y Chen, G Swoboda. Fatigue properties of cracked, saturated and frozen sandstone samples under cyclic loading. *International Journal of Rock Mechanics & Mining Sciences*. 2003;**40**:145–150.

66. RD Dwivedi, PK Singh, TN Singh, DP Singh. Compressive strength and tensile strength of rocks at sub-zero temperature. *Indian Journal of Engineerings and Materials Sciences*. 1998;**5**(1):43–48.

67. R Glamheden. *Thermo-Mechanical Behaviour of Refrigerated Caverns in Hard Rock*. Göteborg: Chalmers University of Technology; 2001.

68. Y Inada, K Yokota. Some studies of low-temperature rock strength. *International Journal of Rock Mechanics and Mining Sciences*. 1984;**21**(3):145–153.

69. RD Dwivedi, AK Soni, RK Goel, AK Dube. Fracture toughness of rocks under sub-zero temperature conditions. *International Journal of Rock Mechanics & Mining Sciences*. 2000;**37**:1267–1275.

70. T Sanderson. *Ice Mechanics and Risks to Offshore Structures*. Amsterdam: Springer; 1988.

71. MCR Davies, O Hamza, BW Lumsden, C Harris. Laboratory measurements of the shear strength of ice-filled rock joints. *Annals of Glaciology*. 2000;**31**:463–467.

72. L Arenson, S Springman, DC Sego. The rheology of frozen soils. *Applied Rheology*. 2007;**17**:1–14.

73. LU Arenson, S Springman. Triaxial constant stress and constant strain rate test on ice-rich permafrost samples. *Canadian Geotechnical Journal*. 2005;**42**:412–430.

74. CK Ballantyne, JO Stone. Timing and periodicity of paraglacial rock-slope failures in the Scottish Highlands. *Geomorphology*. 2013;**186**:150–161.

75. N Hovius, D Lague, S Dadson. Processes, rates and patterns of mountain belt erosion. In: PN Owens, O Slaymaker (eds) *Mountain Geomorphology*. London: Arnold; 2004.

76. D Brunsden, JB Thornes. Landscape sensitivity and change. *Transactions of the British Institute of Geographers*. 1979;**4**(4):463–484.

77. D Brunsden. Relaxation time. In: A Goudie (ed.) *Encyclopedia of Geomorphology*. London: Routledge; 2004. pp. 838–840.

78. J Kemeny. The time-dependent reduction of sliding cohesion due to rock bridges along discontinuities: a fracture mechanics approach. *Rock Mechanics and Rock Engineering*. 2003;**36**(1):27–38.

79. W Haeberli, M Wegmann, D Vonder Mühll. Slope stability problems related to glacier shrinkage and permafrost degradation in the Alps. *Eclogae Geologicae Helvetiae*. 1997;**90**(3):407–414.

80. C Harris, LU Arenson, HH Christiansen, *et al.* Permafrost and climate in Europe: monitoring and modelling thermal, geomorphological and geotechnical responses. *Earth-Science Reviews*. 2009;**92**(3–4):117–171.

10

Erosion and sediment flux in mountain watersheds

DIETER RICKENMANN AND MATTHIAS JAKOB

10.1 Introduction

Mountain hydro-geomorphic processes are responsible for significant loss of life and property in mountainous areas. Even in highly developed countries with significant capital and scientific investment in understanding and mitigating such hazards, losses from extreme regional events can still amount to hundreds of millions of dollars and lead to major business interruptions and human suffering. This realization continues to motivate and justify efforts to improve understanding of such processes and reduce associated risk. This chapter provides a brief summary of issues associated with erosion and sediment flux in mountain watersheds and attempts to highlight recent advances, while focusing on the principal problems and solutions associated with steep stream morphodynamics and their associated hazards and risks.

Viewed at the scale of river watersheds and over geologic time, sediment is produced, transferred, and deposited, moving from headwaters to alluvial plains [1]. Zones of sediment production are characterized by the entire spectrum of mass movement processes, ranging from high-frequency (daily), micro-scale processes such as rain splash erosion, needle ice lift of soil particles and downslope movement, and solute transport, to mega-scale, low-frequency events such as rock avalanches. The persistence of landforms generated by these processes is proportional to size: micro-scale landforms are ephemeral and very difficult to quantify, whereas rock avalanche deposits can be discerned for many thousands of years.

Every watershed is unique in terms of type, intensity, and rate of mass movement and fluvial processes and of the hazard and risk profile associated with such processes. Climate fluctuations can strongly influence rates of mass movement and

The High-Mountain Cryosphere, ed. Christian Huggel, Mark Carey, John J. Clague and Andreas Kääb. Published by Cambridge University Press. © Cambridge University Press 2015.

Figure 10.1 The Swiss village Baltschieder with coarse fluvial sediments deposited during the flood of October 2000.
Source: Swiss Federal Office for the Environment.

fluvial processes in space and time. Continental-scale glaciation is a particularly powerful landscape denudation agent, sculpting mountains and producing glacial landforms, whose legacy of sediment-delivery continues today.

Sediment transfer in steep channels may primarily occur as fluvial sediment transport (bedload and suspended load) (Figure 10.1), debris floods (Figure 10.2) or hyperconcentrated flows, and debris flows [2,3] (Figure 10.3). Analysis of more than 280 bedload transporting flood events in the Swiss Erlenbach stream [4] demonstrate that maximum volumetric bedload concentrations estimated at one-minute monitoring intervals rarely exceed 10%. In contrast, debris flows are often characterized by volumetric sediment concentrations in excess of 40% [5], at least in their frontal surge waves. Debris floods involve high concentrations primarily of coarse solid particles, and hyperconcentrated flows are characterized by elevated concentrations primarily of finer solid particles [2,6]. Therefore, debris floods typically occur in steep alpine streams with coarse channel beds. Because direct measurements of flow and sediment transport events in steep streams are rare, the processes governing the transition from fluvial bedload transport to debris flows are still poorly understood. Pierson and Costa [7] and Hungr *et al.* [2] point out that the limiting sediment concentration between different flow types may be highly variable and depend, for example, on sediment composition and mean flow velocity. Debris flows typically travel in surges, and the frontal part commonly

Figure 10.2 Debris flood at Cougar Creek, Canmore, Alberta, Canada, in June 2013. The channel aggraded over 5 m in this fan reach, and eroded its bank over 10 m in two days.
Source: Town of Canmore.

Figure 10.3 Fresh debris-flow deposits on a fan in the Lötschen valley, Switzerland, in October 2011. Note the strong sediment connectivity between the tributary and the valley river during the event.
Source: Jules Seiler, Geoplan AG, Steg, Switzerland.

has a higher sediment concentration and includes coarser particles than the rear part of the surge where the flow behavior transitions to debris flood and flood flow conditions.

In the European Alps, the term "torrent" refers to a steep channel and is usually associated with mountain watersheds in which debris flows can occur and travel to a fan in the main river valley. Based on several studies from the French, Italian, Swiss, Canadian, and Austrian Alps, torrent watersheds typically have channel gradients steeper than 0.1 and surface areas smaller than 15 km^2. In the case of 127 debris flow events investigated in the northeastern Italian Alps [8], all channels had a mean gradient steeper than 0.1, and 96% of the watersheds had areas less than 15 km^2. Similar limits were documented by Rickenmann and Koschni [9] for debris-flow watersheds during large floods in Switzerland in 2005. In such watersheds, both debris flow and fluvial sediment transport may be important sediment transfer processes. Larger watersheds (typically exceeding 15 km^2) also may have tributaries that are prone to debris flows, while the main channel may not be steep enough to carry debris flows.

Takahashi [10] demonstrated that debris flows can initiate on slopes with gradients steeper than about 0.27. Montgomery and Buffington [11] report that alluvial channels can be found in steep watersheds with gradients up to 0.3 and surface areas as small as about 1 km^2. A processed-based classification using a slope–area relation suggests that the transition between debris-flow dominated and fluvial channels occurs at a gradient range of 0.10 to 0.03 for watershed areas of about 0.1 to 1 km^2 [12].

Examination of sediment transfer from the alpine periglacial zones to lower regions in the watershed indicates that two aspects must be considered in the context of their hazard and landform formation: (1) the frequency–magnitude relation of sediment transporting events; and (2) the connectivity between smaller watersheds and larger watersheds. Kirchner *et al.* [13] found that erosion rates on millennial timescales for 32 Idaho mountain watersheds, ranging from small experimental watersheds (0.2 km^2) to large river watersheds (35 000 km^2), are on average 17 times higher than stream sediment fluxes measured over the past 10–84 years. These observations have important implications for risk perception and hazard assessment. They suggest that sediment delivery from mountainous terrain is highly episodic and that it may be dominated by rare large-magnitude events that have a lower chance of occurring in a relatively short observation period. In mountain watersheds where both debris flows and fluvial sediment transport occur, low-frequency, high-magnitude debris flows or debris floods may dominate from a hazard or risk perspective [14–16]. The importance of debris fans or alluvial fans in moderating downstream sediment transfer over longer time periods has long been recognized, and Harvey [17] provides a recent review on this subject.

The objective of this chapter is to discuss erosion and sediment flux in mountain watersheds from a hazard perspective, largely focusing on debris flows and fluvial sediment transport. We focus on extreme and high-magnitude events that are typically most hazardous, may dominate the long-term sediment budget, and which may be particularly affected by climate change.

10.2 Debris flows and debris floods in high-mountain watersheds

Any structural mitigation against debris flows, floods, or landslides requires a sound and defensible design basis. Such design can be based on arbitrarily chosen or legislated event return periods, which is typically referred to as the "design event" [e.g. 18]. Alternatively, a risk-based approach can be applied in which the hazard scenarios are systematically combined with a formal consequence assessment that may include loss of life, economic and environmental losses, or in the case of corporations, losses in reputation and market share [19–21]. The latter approach is advantageous as it recognizes a much wider spectrum of potential hazard scenarios and informs a gamut of consequences for individual and combined scenarios.

10.2.1 Frequency–magnitude relations

The fundamental input to any debris-flow hazard or risk assessment is the establishment of a sound relationship between the magnitude and the frequency of the hazard, including estimates of the unavoidable error involved in this analysis. The fundamental paradigm is that larger magnitude geophysical events are less frequent than smaller magnitude ones. Frequency is expressed as the annual probability of occurrence, and the return period is defined as the inverse of frequency. The encounter probability is the chance that specific elements at risk will be affected by a hazardous process. In the context of this discussion, magnitude can be expressed as the total volume or peak discharge of a debris flow, or the integrated combined water and sediment hydrograph of debris floods and its peak flow. Volume and peak discharge estimates directly feed into numerical modeling. Because numerical simulations are becoming the standard basis for hazard intensity maps, defensible frequency–magnitude (F–M) curves typically are the foundation for mitigation and land use and zoning decisions.

Debris-flood and debris-flow volumes and discharges have been monitored in the field through instrumentation and retention basins. But this is largely the exception rather than the rule. In most instances, the practitioner has to estimate the F–M relationship for an unmitigated site with few or no direct observations. Moreover, data typically have large uncertainties. One can learn from adjacent monitored watersheds where such F–M information exists. However, we caution against the idea that F–M relationships can be extrapolated from one watershed to another, even adjacent ones. Our caution stems from the fact that individual watersheds have peculiarities in their hydrology, engineering geology, sediment delivery processes, or stream channel connectivity to hillslope processes that would introduce significant error if extrapolated to another watershed. It is thus of paramount importance that the study team examines hillslope and fluvial processes while applying the full range of field and analytical methods to decipher process frequency and magnitude.

Many peer-reviewed papers have described methods for determining debris-flow frequency, volumes, and peak flows. One of the most promising methods is dendrochronology, as it yields precise years of occurrence and sometimes can be used to pinpoint the season of disturbance [e.g., 22]. However, this and other methods rarely allow the establishment of a continuous record for more than approximately 200 years. This time window is useful in the case of jurisdictions in which the hazard scenario spectrum falls within this range (i.e., Austria with a 150-year return period, or Switzerland with a 300-year return period); it is inadequate for those countries or provinces that explicitly demand more conservative return periods in their hazard assessments (i.e. up to 10 000-year and 3000-year return periods in British Columbia and in Alberta, respectively). Radiocarbon dating is often applied where natural exposures or trenched sediments contain organic materials for dating [23]. If neither trees nor datable organics are available, lichenometry [24] might be applied but only if measurements can be tied to a nearby lichen growth curve to yield reliable ages.

Estimates of debris-flow magnitudes for a larger spectrum of return periods remains challenging: careful mapping of dated deposits and estimates of their respective thickness can yield magnitudes. Where thickness measurements are not available, empirical relations of debris-flow deposition area to volume can be applied [25,26]. However, the spatial extent and thus the magnitude of the buried deposits may be difficult to determine unless a comprehensive test trenching program is carried out. In channels underlain by bedrock, it may be helpful to estimate channel yield rates in m^3/m through estimation of debris stored in the channel and summing all channel reaches for which the yield rate or erodibility index has been estimated [8,15,27,28]. This method is particularly well suited for supply-limited watersheds where sediment point sources such as debris avalanches and debris slides are absent, or where their volumes can be estimated and added to the summed channel debris volumes. This method, however, can only provide estimates of debris-flow volumes at the time of the investigation. Channel recharge rates would have to be known to estimate how the total debris-flow volume increases with time in absence of debris flows. Another method is empirical

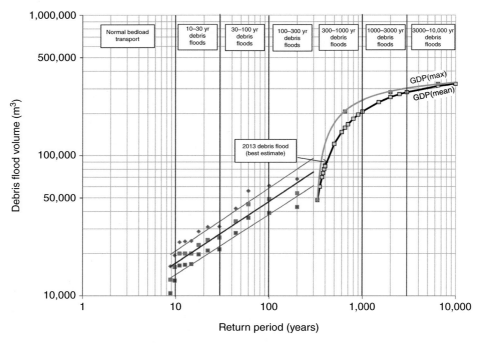

Figure 10.4 Relationship between return period and debris flood volume for debris floods at Cougar Creek, Canmore, Alberta. The curved lines indicate the mean (black) and maximum (gray) volume estimate of the general Pareto distribution (GPD).

estimation of debris-flow volumes from morphometric watershed characteristics [e.g., 29], but it is not suitable for establishing F–M relationships and should thus only be applied for regional-scale reconnaissance-level studies.

Several statistical procedures can be used to construct debris-flow F–M relationships [30]. Magnitudes are either extrapolated to return periods that lie outside the observation range or interpolated to specific return periods mandated by regulations. Such relationships may be based on single processes or multiple processes, as illustrated in Figure 10.4, where debris floods from exceedence of critical runoff thresholds and from dam outbreak floods are considered.

10.2.2 Climate change and debris-flow activity

It is now indisputable that global warming over at least the past 60 years can largely be attributed to human greenhouse gas emissions [31]. Climate change can alter debris-flow frequencies in two ways: In case of more soil moisture, particularly during the debris-flow season, antecedent moisture conditions and thus phreatic surfaces could be higher or can reach critical levels more frequently,

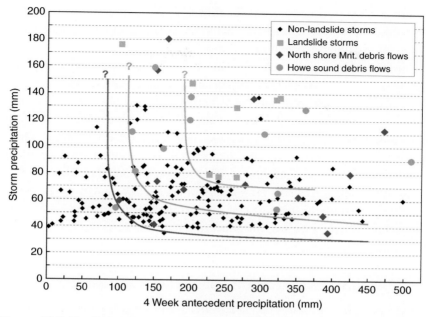

Figure 10.5 Envelopes for 24-hour storm rainfall and antecedent rainfall for landslides along Howe Sound, British Columbia, Canada.

which pre-disposes soils to fail during high-intensity rainfall events. Similarly, debris-flow frequency will increase when high-intensity rainfall events become more frequent or when system-intrinsic debris-flow mobilization thresholds are exceeded more often. Jakob and Lambert [32] show in the case of the British Columbia Coast Mountains that both rainfall amounts and intensities are likely to increase, which will likely lead to higher frequencies of debris flows (Figure 10.5). This seemingly straightforward relationship, however, is challenged by the geo-morphic intricacies of specific debris-flow systems: supply-limited watersheds will not necessarily experience higher debris-flow frequencies unless channel recharge rates increase in concert with changes in rainfall amounts and intensities; supply-unlimited watersheds will respond more directly to increases in hydroclimatic extremes. Stoffel *et al.* [33] used data from eight Swiss debris-flow-prone and supply-unlimited watersheds to examine possible impacts of climate change. They concluded that small increases in debris-flow frequency can be expected by the middle of the twenty-first century, accompanied perhaps by an increase in debris-flow volumes. In contrast, conditions favorable for debris-flow initiation might decrease in the second half of the twenty-first century. This analysis underscores the complexity of mountain watersheds responding to climate change.

Changes in debris-flow magnitude due to climate change are equally complex. If debris-flow frequency increases in supply-limited watersheds, shorter recharge times will separate successive debris flows, which implies lower volumes. In supply-unlimited watersheds and in alluvial channels, more rainfall and runoff should result in higher debris-flow volumes.

Additional complications are introduced by higher-order effects of climate change. Glacier recession might expose large amounts of unconsolidated debris that is then subject to erosion and possible transport by debris flows. Degradation of mountain permafrost would lead to the disappearance of ground ice at the southern and lower limits of permafrost, with an increase in the thickness of the active layer. Both processes will likely increase debris volumes available for debris-flow initiation or entrainment [34–36]. Warming can also lead to the formation of thermokarst lakes on glaciers or rock glaciers that might eventually drain suddenly. Similarly, new glacial lakes might form between lateral and end moraines and receding glaciers, introducing a new potential hazard.

10.3 Fluvial sediment transport in steep channels

Fluvial sediment transport in steep headwater watersheds is strongly influenced by hillslope processes. In smaller watersheds the channel bed is dominated by colluvial rather than alluvial processes [11]. In addition, many steep channels are armored by large and relatively immobile clasts, often imbricated in jammed structures that remain stable except in extreme events. As a result, fluvial bedload transport is controlled not only by the applied hydraulic forces (transport capacity limited conditions) but also by sediment availability or sediment supply [3]. Therefore, sediment transport for a given discharge level may differ, and transport rates predicted by conventional equations may strongly overestimate observed sediment transport [37–38].

10.3.1 Bedload transport calculations in steep streams

Steep natural channels are characterized by very low relative flow depth, and macro-roughness conditions possibly are important also during normal flood events [39]. Flow conditions can change between super- and subcritical along the channel, with hydraulic jumps resulting in spill resistance that substantially increases total flow resistance compared to deeper flows in channels with lower bed slopes. Observed bedload transport in natural steep streams can be several orders of magnitude smaller than predicted by conventional, flume-based transport equations [9,37,38,40,41]. The following may be responsible for the considerable overestimation of bedload transport in steep natural streams: (1) increased flow resistance [38,42]; and (2) development of a strong armor layer, including imbrication, clustering, and particle wedging, all of which increase the structural stability of the bed and limit sediment transport [40,43–45].

Several studies emphasize the importance of accounting for increased total flow resistance in steep streams when making bedload transport calculations [e.g., 9,38,42,46]. Bedload transport equations for steep channels were typically developed and calibrated with flume experiments, where grain flow resistance dominated and macro-roughness bed elements were absent. Concepts similar to the grain/form resistance partitioning in lowland rivers have been applied in steep streams to quantify the additional flow resistance that reduces the flow energy available for bedload transport. Earlier approaches to correct for overestimation of bedload transport in steeper streams [47,48] have been improved in more recent studies to separate base-level resistance from macro-roughness flow resistance. These recent studies are based on a significantly larger number of field observations of flow resistance [49] and have been successfully tested with observations of bedload transport in steep streams [41,50].

In the case of large flows, some macro-roughness elements become submerged, resulting in a decrease in total flow resistance, or a substantial part of the surface bed material may start to move, thus destroying some of the macro-roughness elements. In such situations, flow resistance partitioning becomes less important, and in steep channels there may be a transition to debris-flood transport conditions [50]. The severe floods in August 2005 in Switzerland were associated with many debris flows in headwater watersheds and fluvial bedload transport in mountain rivers. Rickenmann and Koschni [9] analyzed bedload volumes transported by both fluvial processes and debris flows. Most of the debris-flow data are in line with the trend defined by the fluvial transport events, indicating a continuous transition to debris-flood and debris-flow conditions (Figure 10.6). The debris-flow events with the largest sediment volumes were associated with large sediment inputs by landslides.

10.3.2 Limitations of sediment availability

It is generally assumed that fluvial sediment transport in steep channels is limited by sediment availability or sediment supply rather than by the hydraulic transport capacity [11]. However, it is difficult to quantitatively estimate sediment availability along steep streams unless they are underlain by bedrock. Even in that case, very detailed and time-consuming fieldwork is required to estimate reasonable sediment volumes [51]. In addition, more than just knowledge of the amount of stored material is required to understand the relations between sediment storage and sediment delivery. Over time, sediment storage will increase unless interrupted by a debris flow, thus channel recharge rates must be estimated. Such estimates can

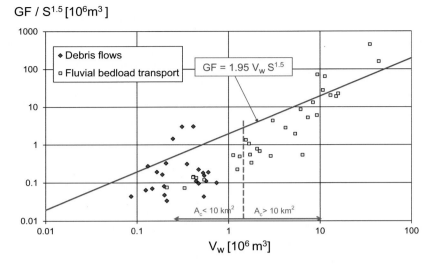

Figure 10.6 Observations of bedload volumes (GF) transported by debris flows and channel runoff during the August 2005 floods in Switzerland. The bedload volumes are normalized by a channel gradient factor ($S^{1.5}$) and shown against the integrated runoff volume (V_w). An integrated version of the bedload transport equation of Rickenmann [38] is also shown for comparison, accounting for 30% of pore volume. The debris-flow data plotting above the equation line refer to events where a large landslide supplied much sediment. Watershed area (A_c) is a rough indicator of dominant process activity.

be made by comparing sediment volumes in supply-limited channels as a function of time since the last debris flow [52].

Mueller and Pitlick [53] related sediment-supply proxies to the characteristics of over 80 drainage watersheds ranging in area from 1.4 to 35 000 km^2 in the northern Rocky Mountains of the United States. They correlated landscape controls on sediment supply to observed sedimentology and channel patterns. The relative sediment supply correlated poorly with relief, mean watershed slope, and drainage density. Instead, contemporary sediment supply to stream channels in the northern Rocky Mountains appeared to be controlled primarily by watershed lithology. According to this study for the northern Rocky Mountains, sediment supply rates were smaller in watersheds in granite and other hard rocks than in watersheds dominated by sedimentary and volcanic rocks. Recking [54] studied the effect of sediment supply on bedload transport in 13 streams with channel slopes steeper than 0.05 in the United States and European Alps. He found larger transport rates in channels identified as being connected to an active sediment source. Recking [54] attributed this to a higher percentage of fine sediments, increase in bed instability due to scouring, and increase in transport efficiency of coarse materials.

10.4 Considerations for sediment-related hazard assessment

Any sediment-related geohazard assessment must be preceded by identification of the existing hazards acting in the watershed, including the channel and the fan. Different hazards act on different temporal and spatial scales. In steep headwater watersheds with relatively steep fans and Melton numbers larger than about 0.3 [26,55], sediment transfer along a channel is typically dominated by debris flows. For larger events this process is likely to be more important in terms of total sediment delivery than fluvial sediment transport [8,26,55].

A field-based distinction between debris flows, debris floods, and fluvial sediment transport can be difficult after an event, as later flow phases are likely to alter the previous ones [56]. In steep headwater channels, landslides from slope or stream-bank failure may contribute substantially to the overall sediment supply and may trigger or transition into a debris flow in the principal channel. Steep headwater watersheds commonly have a high drainage density and short hillslope lengths, thus promoting the conveyance of landslides to the main channel [57].

A field investigator may assume that a creek is prone to only fluvial sediment transport if limited natural exposures show only well-developed stratification, distinct laminae or some degree of imbrication. It is unwise to assume, however, that fluvial processes have shaped the entire fan. At depth, and with the help of excavators, one might encounter matrix-supported diamicton from previous debris-flow events that have been covered by fluvial deposits. It is thus important to examine sedimentary evidence from test trenching as well as geomorphic evidence, such as old landslide dams or landslide scars within the watershed. In such cases, modeling of landslide dam outbreaks may reveal that the hazard is under-estimated. Process cascades of evolving hydro-geomorphic events and other long-run-out landslides should be considered in watershed-scale geohazard assessments.

Once the suite of hydro-geomorphic processes acting in a given watershed has been identified, their frequency must be estimated. In most instances this analysis will require a multidisciplinary approach employing several dating methods, each of which spans a different timeframe. Stereo air photographs typically span several decades; historical accounts and maps may be helpful to identify debris-flow and flood events over the past century or two [15,58]; dendrochronology can lengthen the record to two centuries or more [59]; and radiocarbon dating [23], [10]Be [13], or sometimes well-calibrated lichenometry [24] can extend the record to millennia.

Ideally the hazard analysis should be based both on evaluating the date and characteristics of past events, and on predicting and modeling future events. While dating of past events may provide insight into their frequency, magnitudes are much more difficult to estimate. Obliteration of the evidence of past events by more recent ones, erosion of older deposits, poorly developed stratigraphy in natural exposures and test trenches are serious obstacles to the practitioner tasked with a hazard or risk assessment. The yield rate concept explained in Section 10.2.1 is useful, but in absence of knowledge of channel debris recharge rates or point sources, it does not provide sufficient information on potential magnitudes. Comparative repeat LiDAR survey of watersheds – if available – can shed light on recharge rates, although channel recharge may be episodic and linked to specific weather events.

Work summarized herein suggests alternative methods in determining the volume of sediment that can be mobilized in the steep headwaters of watersheds. Linking channel slope and total rainfall volume to sediment movement for fluvial transport processes is promising because it lends itself to a frequency–magnitude analysis. A precipitation time series can be analyzed with extreme value statistical methods, thus coupling sediment movement rates directly to return periods of a threshold-exceeding rainfall event. For debris-flow volume prediction, total rainfall volumes alone are not suitable because debris flows are triggered by a combination of rainfall intensities and duration and antecedent moisture conditions, and because, unlike many flood-prone streams, steep watersheds may be sediment supply-limited.

Channel reaches on alluvial fans are important components of steep mountain watersheds. Some channels have configurations that are stable in specific configurations related to the dimensionless shear stress. Deepening the channels and confining them will change the fluvial regime, rendering them more prone to abrupt widening through bank erosion during high discharge events. Understanding the thresholds at which sudden channel widening, deep scour, or abrupt fan aggradation can occur is vital in predicting the system's response to extreme hydroclimatic events.

An increasing number of numerical codes for modeling flow and depositional behavior of sediment transporting flows are becoming available. However, it can be difficult to determine the most appropriate one for a specific application. Many practitioners lack the mathematical background or time to fully comprehend the rheological foundation and limitations of specific models. Blind application of these models will lead to significant error in the intensity parameters on which hazard and risk maps are based. It is thus crucial to calibrate or compare the models to known events [60]. In the absence of well-documented events, the best approach is to allow for a large variety of model input parameters and provide a range of outcomes. For engineering and zoning applications, the more conservative outcome (higher run-out distance, deeper flow depths, and higher velocities) might be favored.

One significant obstacle to the successful application of the methods and approaches outlined in this chapter is the cost of their full implementation. In many nations, geohazard and risk assessments are completed by consultants. Requests for proposals are posted by regional governments and companies must provide competitive proposals. The cheapest proposals may be chosen without due consideration of the issues highlighted herein. Decision-makers should be aware that the more reliable the study, the higher the potential savings in risk management.

Affluent societies make large capital expenditures on protection measures to reduce geohazard risk in mountainous environments, yet two contrary tendencies have developed. First, with increasing human populations, pressure to build in or at the piedmont of mountain ranges has increased in many parts of the world. These pressures are particularly acute where mountains are also viewed as preferred recreational arenas. In numerous mountain valleys, high flood hazard and the environmental importance of floodplains have directed development onto alluvial fans with potential for rare but highly destructive events. More and higher-density developments on such fans will increase risk to residents and the infrastructure that is associated with such developments. Second, there is increasing observational evidence and theory that climate change will lead to more, and likely more severe, hydroclimatic events, including landslides, debris floods, and debris flows. For supply-unlimited alluvial channels, a higher frequency of events feeding into debris retention structures will imply higher capital investment in excavating, transporting, and stockpiling accumulated debris.

10.5 Conclusions and outlook

Alluvial fans continue to be favored sites for urban development and for transportation, energy, and communication infrastructure in mountainous areas. They are favored for development, in part, because of increasing pressures on land that can be safely developed, and because of flood risks on river plains. Alluvial fans, however, are dominantly formed by extreme events accompanied by recurrent channel avulsion. It is therefore imperative that they are recognized as sites of high risk for development. Understanding, characterizing, and simulating the hazard on alluvial fans and their upstream channel networks require a thorough knowledge of coupled geomorphic processes that define erosion and sediment flux in mountain streams. Knowledge is required of mass movement processes on open slopes and channelized hillsides, sediment storage and mobilization in the feeder channels at all temporal scales, and sediment transport to and beyond the alluvial fan. Finally, an understanding of the morphodynamics of the main valley stream, which itself interacts with the growth and erosion of fans, is essential.

Many breakthroughs have been made in understanding mountain stream sediment flux over the past several decades. Sediment movement rates and steep creek morphodynamics and debris-flow mechanics are now better understood, and sediment-laden flows can be modeled with reasonable confidence. Nonetheless, the authors of this chapter feel that there are still subjects that are under-researched. For example, multi-process frequency–magnitude relations and process chains are still poorly understood, and the application of extreme value statistics to incomplete data sets also needs to be improved. The prediction of sediment supply to the channel by hillslope processes as well as deposition and re-entrainment through bed scour and lateral erosion along the channel during extreme events remains challenging. Finally, a disconnect, or at least a significant time lag exists between scholarly achievements in the subjects addressed in this chapter and their application by practitioners.

Acknowledgments

We thank John Clague and Brett Eaton for very helpful and constructive review comments.

References

 1. JM Buffington, DR Montgomery, Geomorphic classification of rivers. In *Treatise on Geomorphology, Volume 9, Fluvial Geomorphology*, ed. J Shroder, E Wohl (San Diego, CA: Academic Press, 2013), pp. 730–767.
 2. O Hungr, S Evans, MJ Bovis, JN Hutchinson, A review of the classification of landslides of the flow type. *Environmental and Engineering Geoscience*, **7** (2001), 221–238.
 3. M Church, Steep headwater channels. In *Treatise on Geomorphology, Volume 9, Fluvial Geomorphology*, ed. J Shroder, E Wohl (San Diego, CA: Academic Press), pp. 528–549.
 4. D Rickenmann, BW McArdell, Continuous measurement of sediment transport in the Erlenbach stream using piezoelectric bedload impact sensors. *Earth Surface Processes and Landforms*, **32** (2007), 1362–1378.
 5. F Lavigne, H Suwa, Contrasts between debris flows, hyperconcentrated flows and stream flows at a channel of Mount Semeru, East Java. *Geomorphology*, **61** (2004), 41–58.
 6. TC Pierson, Hyperconcentrated flow-transitional process between water flow and debris flow. In *Debris-flow Hazards and Related Phenomena*, ed. M Jakob, O Hungr (Chichester: Springer Praxis Publishing, 2005), pp. 159–202.
 7. TC Pierson, JE Costa, A rheologic classification of subaerial sediment–water flows. *Geological Society of America, Reviews in Engineering Geology*, **7** (1987), 1–12.
 8. L Marchi, V D'Agostino, Estimation of debris-flow magnitude in the Eastern Italian Alps. *Earth Surface Processes and Landforms*, **29**:2 (2004), 207–220.
 9. D Rickenmann, A Koschni, Sediment loads due to fluvial transport and debris flows during the 2005 flood events in Switzerland. *Hydrological Processes*, **24** (2010), 993–1007.
10. T Takahashi, *Debris Flow* (Rotterdam: Balkema, 1991).
11. DR Montgomery, JM Buffington, Channel-reach morphology in mountain drainage basins. *Geological Society of America Bulletin*, **109** (1997), 596–611.

12. J Stock, WE Dietrich, Valley incision by debris flows: evidence of a topographic signature. *Water Resources Research*, **39** (2003), 1089. doi: 10.1029/2001WR001057.

13. JW Kirchner, RC Finkel, CS Riebe, *et al.*, Mountain erosion over 10 yr, 10 k.y., and 10 m.y. time scales. *Geology*, **29**:7 (2001), 591–594.

14. D Rickenmann, Empirical relationships for debris flows. *Natural Hazards*, **19** (1999), 47–77.

15. M Jakob, Debris flow hazard analysis. In *Debris-flow Hazards and Related Phenomena*, ed. M Jakob, O Hungr (Chichester: Springer Praxis Publishing, 2005), pp. 411–443.

16. M Jakob, A size classification for debris flows. *Engineering Geology*, **79** (2005), 151–161.

17. AM Harvey, The coupling status of alluvial fans and debris cones: a review and synthesis. *Earth Surface Processes and Landforms*, **37** (2012), 64–76.

18. A Petrascheck, H Kienholz, Hazard assessment and mapping of mountain risks in Switzerland. In *3rd International Conference on Debris-Flow Hazards Mitigation*, ed. D Rickenmann, CL Chen (Rotterdam: Millpress, 2003), pp. 25–38.

19. P Greminger, Managing the risks of natural hazards. In *3rd International Conference on Debris-Flow Hazards Mitigation*, ed. D Rickenmann, CL Chen (Rotterdam: Millpress, 2003), pp. 39–56.

20. R Fell, KKS Ho, S Lacasse, E Leroi, A framework for landslide risk assessment and management. In *Landslide Risk Management*, ed. O Hungr, R Fell (London: Taylor & Francis, 2005), pp. 3–25.

21. M Porter, N Morgenstern, *Landslide Risk Evaluation. Canadian Technical Guidelines and Best Practices Related to Landslides. A National Initiative for Loss Reduction.* Geological Survey of Canada, Open File 7312, 2013, doi:10.4095/292234.

22. M Schneuwly-Bollschweiler, C Corona, M Stoffel, How to improve dating quality and reduce noise in tree-ring based debris-flow reconstructions. *Quaternary Geochronology*, **18** (2013), 110–118.

23. R Chiverrell, M Jakob, Radiocarbon dating: alluvial fan/debris cone evolution and hazards. In *Dating Torrential Processes on Fans and Cones*, ed. M Schneuwly-Bollschweiler, M Stoffel, F Rudolf-Miklau. (New York: Springer, 2013), pp. 265–282.

24. JL Innes, Lichenometric dating of debris-flow deposits in the Scottish Highlands. *Earth Surface Processes and Landforms*, **8** (2006), 579–588.

25. JP Griswold, RM Iverson, *Mobility Statistics and Automated Hazard Mapping for Debris Flows and Rock Avalanches.* U.S. Geological Survey, Scientific Investigations Report 2007–5276, 2008.

26. C Scheidl, D Rickenmann, Empirical prediction of debris-flow mobility and deposition on fans. *Earth Surface Processes and Landforms*, **35** (2010), 157–173.

27. O Hungr, GC Morgan, R Kellerhals, Quantitative analysis of debris torrent hazards for design of remedial measures. *Canadian Geotechnical Journal*, **21** (1984), 663–677.

28. H Kienholz, E Frick, E Gertsch, Assessment tools for mountain torrents: SEDEX© and bed load assessment matrix. In *Proceedings of the International Interpraevent Symposium, Taipei, Taiwan* (2010), pp. 245–256.

29. M Jakob, Morphometric and geotechnical controls of debris flow frequency and magnitude in southwestern British Columbia. PhD thesis, University of British Columbia (1996).

30. M Jakob, The fallacy of frequency: Statistical techniques for debris-flow frequency–magnitude analyses. In *Proceedings of the 11th International Symposium on Landslides*, Banff ISL, June 3–8, 2012.

31. IPCC, *The Scientific Basis* (Cambridge: Cambridge University Press, 2014).

32. M Jakob, S Lambert, Climate change effects on landslides along the south-west coast of British Columbia. *Geomorphology*, **107** (2009), 275–284.

33. M Stoffel, T Mendlik, M Schneuwly-Bollschweiler, A Gobiet, Possible impacts of climate change on debris-flow activity in the Swiss Alps. *Climate Change*, **122** (2014), 141–155.

34. M Zimmermann, W Haeberli, Climatic change and debris flow activity in high-mountain areas: a case study in the Swiss Alps. *Catena Supplement*, **22** (1992), 59–72.

35. C Huggel, JJ Clague, O Korup, Is climate change responsible for changing landslide activity in high mountains? *Earth Surface Processes and Landforms*, **37** (2012), 77–91.

36. LU Arenson, M Jakob, *Periglacial geohazard risks and ground temperature increases*. IAEG XII Congress – Engineering Geology for Society and Territory, Torino, Italy, September 15–19, 2014.

37. V D'Agostino, MA Lenzi, Bedload transport in the instrumented watershed of the Rio Cordon: Part II. Analysis of the bedload rate. *Catena*, **36** (1999), 191–204.

38. D Rickenmann, Comparison of bed load transport in torrents and gravel bed streams. *Water Resources Research*, **37** (2001), 3295–3305.

39. F Comiti, L Mao, Recent advances in the dynamics of steep channels. In *Gravel-Bed Rivers: Processes, Tools, Environments*, ed. M Church, PM Biron, AG Roy. (Chichester: John Wiley & Sons, 2012), pp. 353–377.

40. JJ Barry, JM Buffington, JG King, A general power equation for predicting bedload transport rates in gravel bed rivers. *Water Resources Research*, **40** (2004), doi:10.1029/2004WR003190.

41. M Nitsche, D Rickenmann, JM Turowski, A Badoux, JW Kirchner, Evaluation of bedload transport predictions using flow resistance equations to account for macro-roughness in steep mountain streams. *Water Resources Research*, **47** (2011), W08513, doi:10.1029/2011WR010645.

42. A Zimmermann, Flow resistance in steep streams: an experimental study. *Water Resources Research*, **46** (2010), doi:10.1029/2009WR007913.

43. MA Hassan, M Church, TE Lisle, Sediment transport and channel morphology of small, forested streams. *Journal of the American Water Resources Association*, **41** (2005), 853–876.

44. JC Bathurst, Effect of coarse surface layer on bed-load transport. *Journal of Hydraulic Engineering*, **133** (2007), 1192–1205.

45. MP Lamb, WE Dietrich, JG Venditti, Is the critical Shields stress for incipient sediment motion dependent on channel-bed slope? *Journal of Geophysical Research*, **113** (2008), F02008, doi:10.1029/2007JF000831.

46. EM Yager, JW Kirchner, WE Dietrich, Calculating bedload transport in steep, boulder-bed channels. *Water Resources Research*, **43** (2007), doi:10.1029/2006WR005432.

47. SM Palt, Sedimenttransportprozesse im Himalaya-Karakorum und ihre Bedeutung für Wasserkraftanlagen. Institut für Wasserwirtschaft und Kulturtechnik, Universität Karlsruhe (TH), Mitteilungen 209 (2001) [in German].

48. M Chiari, K Friedl, D Rickenmann, A one dimensional bedload transport model for steep slopes. *Journal of Hydraulic Research*, **48** (2010), 152–160.

49. D Rickenmann, A Recking, Evaluation of flow resistance equations using a large field data base. *Water Resources Research*, **47** (2011), W07538, doi:10.1029/2010WR009793.

50. D Rickenmann, Alluvial steep channels: flow resistance, bedload transport and transition to debris flows. In *Gravel-Bed Rivers: Processes, Tools, Environments*, ed. M Church, PM Biron, AG Roy. (Chichester: John Wiley & Sons, 2012), pp. 386–397.

51. ME Oden, *Debris recharge rates in torrented gullies on the Queen Charlotte Islands.* MSc thesis, University of British Columbia (1994).

52. M Jakob, M Bovis, M Oden, The significance of channel recharge rates for estimating debris-flow magnitude and frequency. *Earth Surface Processes and Landforms,* **30** (2005), 755–766.

53. ER Mueller, J Pitlick, Sediment supply and channel morphology in mountain river systems: 1. Relative importance of lithology, topography, and climate. *Journal of Geophysical Research Earth Surfaces,* **118** (2013), doi:10.1002/2013JF002843.

54. A Recking, Influence of sediment supply on mountain streams bedload transport. *Geomorphology,* **175–176** (2012), 139–150.

55. DJ Wilford, ME Sakals, JL Innes, RC Sidle, WA Bergerud, Recognition of debris flow, debris flood and flood hazard through watershed morphometrics. *Landslides,* **1** (2004), 61–66.

56. JE Costa, Rheologic, geomorphic, and sedimentologic differentiation of water floods, hyperconcentrated flows, and debris flows. In *Flood Geomorphology,* ed. VR Baker, RC Kochel, PC Patton (New York: John Wiley & Sons, 1988), pp. 113–122.

57. O Korup, Landslides in the fluvial system. In *Treatise on Geomorphology, Volume 9, Fluvial Geomorphology,* ed. J Shroder, E Wohl (San Diego, CA: Academic Press, 2013), pp. 244–259.

58. M Zimmermann, P Mani, H Romang, Magnitude–frequency aspects of Alpine debris flows. *Eclogae Geologicae Helvetiae,* **90** (1997), 415–420.

59. M Stoffel, M Bollschweiler, Tree-ring analysis in natural hazards research: an overview. *Natural Hazards and Earth Systems Science,* **8** (2008), 187–202.

60. D Rickenmann, D Laigle, BW McArdell, J Hübl, Comparison of 2D debris-flow simulation models with field events. *Computational Geosciences,* **10** (2006), 241–264.

11

Glaciers as water resources

BRYAN G. MARK, MICHEL BARAER, ALFONSO FERNANDEZ,
WALTER IMMERZEEL, R. DAN MOORE, AND ROLF WEINGARTNER

11.1 Introduction

Glaciers are perennial ice features that temporarily store freshwater at the higher altitudes and latitudes of Earth. Although comprising only a small fraction of Earth's cryosphere, the net mass loss of mountain glaciers worldwide has become one of the most widely recognized indicators of the reality of human-induced climate change. And since freshwater is essential for human society, there is a critical link between the fate of glaciers and sustainability of water resources downstream. However, it is not accurate to assume retreating glaciers comprise a generalizable water shortage to society. Many variables relating to location, watershed scale, and timing (of both seasonal availability and demand) impact glacier water resources, and present challenges for models used to project future scenarios. Assessing the relative magnitude of glacier influence on water resources requires not only careful assessments of climate control and hydrologic response, but also a critical review of the social factors that ultimately hold sway over water allocation and access [1].

The key conditions necessary for glacier formation are possible at nearly all latitudes (Figure 11.1), while regional differences in climate and proximity to human society impact the supply and exploitation of glacier water. In general, mountain topography constrains the nature of glacier water resources, from modifying climate required to sustain glaciers, to influencing all surface and subsurface flow and reservoirs. Glaciers are effective "water towers" [2], accumulating seasonal snow and ice above the snowline, while also releasing freshwater as the ice flows downward and melts. The gravitational potential energy yields secondary resource gains as water flowing from mountain regions is often utilized for hydroelectrical power generation. Nevertheless, human demands for water vary regionally and impact risks to water resources.

The High-Mountain Cryosphere, ed. Christian Huggel, Mark Carey, John J. Clague and Andreas Kääb. Published by Cambridge University Press. © Cambridge University Press 2015.

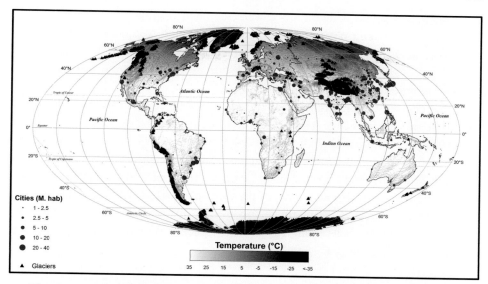

Figure 11.1 World distribution of glaciers (Randolph glacier inventory) and location of cities with more than one million inhabitants (United Nations, Department of Economics and Social Affairs), scaled by size. The background on land areas is the global temperature distribution derived from CRU 3.10 and ERA Interim reanalysis (Map projection is Mollweide, datum WGS84).
Sources: www.glims.org/RGI/; http://esa.un.org/unpd; www.cru.uea.ac.uk; www.ecmwf. int; Julien Nicolas is gratefully acknowledged for preprocessing ERA data.

With ongoing global climate change, glacier water resources are vulnerable to a high degree of climate sensitivity. Studies of warming climate scenarios mostly feature glacier recession that will alter the volume and variability of stream flow. The hydrological response to glacier retreat follows a characteristic sequence for regions undergoing net glacier mass loss. Under conditions of continuous retreat, glaciers generate a temporary increase in glacial-melt-derived stream runoff as they lose mass. As the remaining glacier volume diminishes, the annual runoff hydrograph reaches a maximum called "peak water" and is then followed by a persistent annual decrease [3].

However, evaluating glaciers as water resources is more complex than projecting glacier-melt contributions to stream flow. Stream discharge is certainly a key dimension of how glacier changes impact water resources, but not the only one important to society. For example, glacier melting is linked to altered water quality, natural hazard risk, food supply, energy production, degraded tourism, and impacts on the cultural landscape. Furthermore, glacier growth and decay over long timescales preconditions proglacial valley landforms and sediments to be subterranean water storage units, recharged over millennia by glacier melt. Sustainability of future water resources depends as much on human demand as on glacier supply, and thus it is a social concern since water demand responds to dynamic market forces that can extend beyond topographically defined watershed boundaries. In particular, activities such as mineral extraction and agricultural irrigation that rely on the availability of glacier water resources are responsive to often-distant socio-economic forces.

In this chapter we highlight different key mountain ranges to demonstrate how mountain glaciers influence water resources, and how those resources are exposed to different environmental changes and risks. This chapter is not meant to provide an exhaustive review of all mountain ranges. Rather, we survey selected regions to highlight how risk might be evaluated given a variety of climatic, environmental, and social factors. To that end, each of the four subsections within the next section focuses on a glacierized mountain region from a different continent, wherein we explore how glaciers are changing and the specific regional issues relating to glaciers as water resources.

11.2 Regional review of glaciers as water resources

11.2.1 The Andes

Glaciers span more latitude and thus climatic environments along this mountain range than any other on Earth. Yet in this region work has shown that risks to water supply are large since glaciers are receding rapidly, while demand is growing. The Andes maintain an average elevation of nearly 4000 m, which forms a natural barrier to Southern Hemisphere prevailing winds: the Easterly trade winds towards the north and the Westerlies to the south (Figure 11.2; see also Chapter 2 in this book). The resulting orographic precipitation causes the Andes to capture a large amount of the moisture carried by the prevailing winds, and creates rain shadow areas downwind. These areas, situated to the west of the Andes in the north and to the east in the south suffer from a deficit of precipitation compared to the continental average. Being so surrounded by arid regions gives most glaciers of the Andes an important role in supplying populations, economies, and ecosystems with freshwater.

In the last few decades there has been increasing scientific concern about changes in Andean glaciers given a strong regional recession since the Little Ice Age with predicted ongoing recession and impact to water resources [4]. The few *in situ* glacier mass balance monitoring programs throughout the Andes have been essential for developing a quantitative understanding of climate change effects on mountain glaciers and impacts to water. However, less than 3% of the more than 6000 inventoried Andean glaciers have been observed at least by remote-sensing techniques, and an even smaller portion of them has been studied through local models [5].

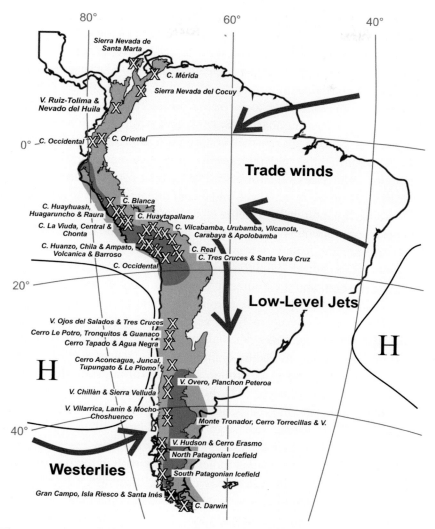

Figure 11.2 Illustrative representation of the Andes dominant winds (arrows) and rain shadow zones (transparent gray areas). Surfaces identified by an "H" illustrate the characteristic subtropical anticyclone positions. Main glacierized cordilleras (C.), volcanoes (V.) and isolated peaks are named and marked by white crosses. Due to the high number of scattered glacierized peaks and volcanoes south of the 26th parallel, not all glacierized areas are named.

Vulnerability to glacier retreat-related hydrological changes in the Andes is widespread, but projections of outcomes vary along the vast latitudinal extension of the range. In Colombia to the north, the glaciers from the Ruiz-Tolima Volcanic Massif that feed high-altitude water bodies and permanent water-reservoir habitats are shrinking rapidly, threatening water demand satisfaction for the lowland populations [6]. Three capital cities along the Andes – Quito (Ecuador), Lima (Peru), and La Paz (Bolivia) – depend to various degrees on the runoff from glaciers for water and energy supplies [7]. However, the relative significance of glaciers to these municipal water supplies changes greatly relative to glacier proximity; Lima has much less water originating from glaciers than La Paz. Moreover, consideration of future demographics and other social factors are likely to play a much greater role than diminishing glaciers in impacting water supplies to population centers [8].

In Chile and Argentina, glacier-melt contribution to stream flow is forecasted to diminish, with substantial and widespread socio-economic impacts, especially during extremely dry years [9]. However, glacier-melt contribution to stream flow has not been investigated in detail yet, and recent research seems to point out that it varies according to latitudinal gradients of temperature, precipitation, and topography. Glacier contribution is probably more important in the Central area (~30 °S to ~37 °S), a zone that displays a transition from dry to humid conditions, driven by the seasonal migration of the Subtropical Anticyclone, the Low Level Jet from the Amazonian region, and the Southern Westerlies – all of them modified by the Andean elevation gradient [10]. The current population, estimated to be over ten million and steadily increasing in Chile and Argentina combined, concurs with historically high and rising water demand for drinking, agriculture, industry, and mining [9]. Recent studies in Pacific-draining valleys, where 80% of the population resides, have attempted to determine general hydrological vulnerabilities in light of expected climate changes [11]. In the northern tip of this zone, Gascoin *et al.* [12] showed that during the period 2003/2004 to 2007/2008, glacier melt contributed 3–23% to annual discharge, a figure larger than what would be expected from current glacier coverage. These results disagree with previous, regional studies (29 °S to 32 °S) that indicated snow melt is more important than glaciers to discharge [13]. In watersheds closer to Santiago, Chile's capital, it seems that during wet years glacier-melt input is less important than snowmelt, but during abnormally dry years, the former may contribute as much as 90% to stream flow, despite the fact that glacierized area is 11.5% of the watershed [14].

Concerns about future water supplies are particularly pronounced on Peru's arid Pacific slope, where upstream glacier recession has been accompanied by rapid and water-intensive economic development [15]. This is the case in the Santa River watershed that drains the western slopes of the Cordillera Blanca, where both a glacier retreat-related water resources decline and a simultaneous increase of the coastal demand for water have been observed (Figure 11.3). At a point situated just downstream of the major glacierized valleys of the Cordillera Blanca, the Santa River has been recently diagnosed as having passed peak water [3]. This means that, in the highly probable scenario of continuing glacier retreat, the river is predicted to exhibit a steady decline in yearly discharge, as well as in dry season

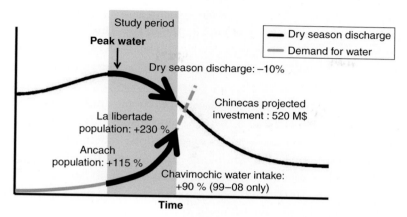

Figure 11.3 Conceptual representation of the projected water deficit at the Rio Santa outlet. The dark gray line represents the glacier's retreat-related decrease of the Rio Santa dry season flows; the light gray solid line represents the evolution of the demand for water at the coast. The light gray dotted line illustrates the projected evolution in demand. The shadowed area highlights the study focus period. The text on the figure provides explanations for the shape of the curves.

discharge, and an increase in inter-seasonal flows variability. So far, around 10% of the dry season average discharge would have been lost due to the recent glacier retreat and an additional 30% should disappear as the glaciers retreat further. At the same time, pronounced growth along the coast in population and economic activities is directly related to the diversion of the Santa River for intensive irrigation-based agriculture, municipal supply and hydropower generation [1]. The Chavimochic project in La Libertad has so far developed a mother canal to divert the water along 267 km to supply three valleys and the town of Trujillo with freshwater. The Chinecas project in Ancash uses a similar canal-based diversion development model with the objective of increasing the pre-project agricultural revenues of the covered area by 1000%. Both projects are still expanding. The existing development plans of both organizations will probably ensure that it will not be necessary to wait for the glaciers to lose their hydrological influence before the Rio Santa runs dry seasonally.

11.2.2 The Swiss Alps

Swiss glaciers have lost approximately 50% of their volume since 1850. The summer temperatures have risen about 1 °C since 1980, while the snow accumulation on glaciers has hardly changed [16]. This has resulted in a strong decrease in glacier thickness, currently proceeding at about 1 m per year on average. As the area extent of Swiss glaciers is approximately 1000 km^2 today, the annual volume loss of ice is about 1 km^3. Today's remaining estimated ice volume in Switzerland is 55 ± 15 km^3; it was around 100 km^3 in 1850 [16]. This shrinkage will likely accelerate in the future. By 2100, only 20–30% of current volume will remain [16]. This climate-induced change will particularly affect mountain societies, but it will also have an impact downstream in dry years.

Currently, net ice mass loss from glaciers contributes only about 2.5% of the total stream runoff of Switzerland, which is estimated to be 1000 mm/yr. This glacier runoff contribution corresponds exactly to the fractional area covered by glaciers. Thus, the impact of shrinking glaciers on the macro scale, i.e., Switzerland, is rather small. However, this does not apply to years with major summer droughts, e.g. 2003. Such years are likely to occur more frequently in the future, and in these years the contribution of ice melt to stream flow is substantial [17]. In the dry year 2003, the ice-fed Alpine rivers exhibited about 1.5 times larger summer runoff than average [17], which had a positive effect on the water supply in the densely populated Swiss Plateau.

In contrast to the macro-scale, the effects of consistently negative glacier mass balance glaciers are crucial to runoff at the meso-scale of heavily glaciated catchments. This watershed-scale influence is exemplified by the Massa River (195 km^2, mean altitude: 2945 m asl, area covered by glaciers: 65%) that is fed by the meltwater of Aletsch glacier. The mean annual discharge has increased in the Massa River since the 1970s (Figure 11.4) as a result of very intense glacier mass loss. Discharge will likely reach a maximum ("peak water") around 2050, and by 2100 it will still be larger than it was in 1900, despite the fact that the Aletsch glacier will be very small. It is important to note, however, that the main sources of stream discharge and their glacier-fed proportion will alter (Figure 11.4): up until 2050, approximately 50% of projected river discharge will originate from net glacier mass loss via ice melt; thereafter, snowmelt and liquid precipitation runoff will become more important. Total annual precipitation will not change significantly throughout this century, while both intra- and inter-annual variability of discharge will increase because the rain-fed component will gain in importance.

What are the implications of these changes for the seasonal runoff pattern in the future? To answer this question, glacier-fed catchments (mean altitude > 1900 m asl, Figure 11.5) can be compared with selected catchments in Switzerland which are either snow- or rain-fed, depending on their mean altitude. Köplin et al. [18] calculated for each catchment and each month a change coefficient of runoff for the end of this century (2070–2099). There is an obvious relationship between the mean catchment elevation and the monthly variable change coefficient. The projected future runoff will generally increase in the winter due to a higher proportion of liquid precipitation with warmer temperatures. This is also valid for catchments with a higher mean elevation. By the end of the twenty-first century, it

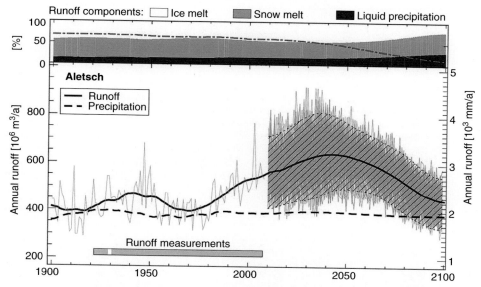

Figure 11.4 Evolution of runoff, precipitation and sources of runoff from 1900 to 2100.
Source: Farinotti *et al.* in [16].

appears that only the highest catchments will still have higher-than-average runoff from summer glacier ice melt. Overall, the strong decrease in summer runoff from glacier-fed catchments is compensated by increases from November to April (or even June), resulting in a minor change in total annual runoff volume.

From an economic point of view, Alpine waters are most important for hydropower production given the high specific water yield of glacier-fed catchments and steep topographic gradients. Hydroelectric power plants generate up to 60% of electricity in Switzerland; the contribution of plants located in glaciated Alpine catchments is substantial. Based on several case studies in the Swiss Alps, Weingartner *et al.* [19] investigated the effects of climate change on hydro power production using a complex model chain (time horizon 2015). Different ensembles of climate and hydrologic models under the A1B emission scenario were applied to assess what the impact of glacier change would be on the hydro power production, assuming that current market conditions of electricity generation will not change [16]. The results showed only a minor change in production. While this suggests hydroelectric production will not be affected dramatically in the near future, the changes will be quite large for the more distant future due to the combination of shrinking glaciers and drier summers.

The combination of decreasing runoff originating from ice melt and decreasing summer rainfall will have further impacts for future water resources as water access shifts:

B. G. Mark et al.

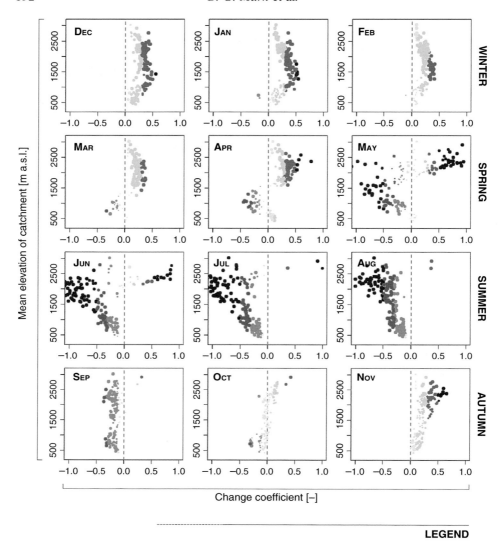

Figure 11.5 Change coefficient and number of significant scenarios (shown by symbol size) as a function of mean altitude of a catchment for the far future. The darker the tone of the symbol, the more significant the change.
Source: taken from [18].

- More frequent dry spells will lead to an increased need for irrigation [20], particularly in the already dry inner-Alpine valleys where agriculture is highly dependent on irrigation [21] that originates traditionally from meltwater of glaciers.
- People downstream who currently profit from the abundant mountain waters [2] may be affected when summer runoff in Alpine rivers will decrease significantly.

Two hydrological realities thus present themselves as Swiss glaciers recede: on one hand, a seasonal redistribution of runoff can be expected; on the other hand – remarkably – only a minor change in annual runoff can be expected. The latter situation opens ample opportunities for societies to adapt, among the choices for which the transformation of existing hydrological storage reservoirs that are exclusively used for hydro power production to multifunctional storage is the most obvious.

11.2.3 The Himalayas

Rivers fed from the Tibetan Plateau and adjacent mountain ranges sustain livelihoods for billions of people. In this area, here referred to as the Greater Himalayas, snow and glacial melt are important hydrologic processes, and climate change is expected to seriously affect melt characteristics and related runoff. The greater Himalayas contain the largest area of glaciers outside the poles and a total area of 123 604 km^2 is estimated for the Central Asia and South Asia combined with a total estimated ice volume of 12 807 km^3. Without central Asian glaciers the total area estimate is 55 634 km^2 with a total ice volume of 6327 km^3 [22].

There is some uncertainty about the cryospheric response to climate change in Asia, caused by lack of data and a limited number of studies. Most Himalayan glaciers are seemingly losing mass at rates similar to glaciers elsewhere, except for emerging indications of stability or mass gain in the western part of the greater Himalayas, e.g., in the Pamir and Karakoram ranges, during the last decade [23].

Regional differences in glacier response may also be explained by climatic variation. The central Himalayan region is dominated by the East Asian monsoon climate, in which most of the precipitation is concentrated during summer (June to September), while winters are rather dry. High elevation areas are significantly more arid because wet air masses are orographically forced out at elevations lower than 4000 m asl [24]. From east to west, the monsoon influence decreases and mid-latitude Westerlies become more important in the western part of the Greater Himalayas (Pamir and Karakoram ranges). Precipitation from Westerlies is highest in winter when low pressure systems reach the western margin of the Greater Himalayas. This supply of moisture can allegedly reach higher elevations than the summer monsoon, which might be related to the higher tropospheric extent of the westerly airflow.

Whether glaciers are important in the overall hydrology of a river basin does not depend solely on the extent and ice volume of the glaciers. Several other factors play an important role; e.g. the basin hypsometry, the relative glacier area, the downstream climate, and the downstream demand for water. One measure to quantify this is the normalized melt index (NMI) [22], which has been used to quantify the importance of meltwater from the upstream areas of five major Asian river basins (Indus, Ganges, Brahmaputra, Yangtze, and Yellow rivers) to the overall basin hydrology. NMI is defined as the volumetric discharge of upstream snow and glacier divided by the downstream natural discharge. Results show that glacier and snowmelt play a dominant role in the overall hydrology only for the Indus, where NMI is 151%; for the Ganges this is only 10%. The much smaller NMI is attributed to comparatively large downstream areas, limited upstream precipitation, smaller glaciers, and/or wet monsoon-dominated downstream climates. In the Indus and Ganges basin about 40% of the meltwater originates from glaciers, while in the other basins the glacial melt contribution is much less. Although at the basin scale glacier melt may only be relevant for a few basins, such as the Indus and the Amu and Syr Darya in the west, locally at the sub-basin level the importance of glacier melt varies considerably.

Figure 11.6 illustrates the great variability in runoff generated and the role that glacier melt has therein. These modeling results for the period 1998–2007 [25] show that along the monsoon-dominated Himalayan arc most runoff is generated from intense seasonal precipitation, and that the role of glacier melt is relatively limited. However, in the upper Indus considerable amounts of runoff are generated too, but here glaciers are the main supplier of water. Hence, it is concluded that the role of glaciers at the scale of a large river basin is generally not very large, but locally glaciers play a crucial role in the water supply.

Glaciers will continue to shrink in the future according to the latest climate models. It is estimated based on a climate model ensemble and a glacier mass balance model that the annual mass loss between the future periods 2003–2022 and 2081–2100 will be -0.4 ± 0.4 m w.e y^{-1}, -0.6 ± 0.4 m w.e y^{-1}, -0.2 ± 0.4 m w.e y^{-1} (meters water equivalent per year) for Central Asia, South Asia (West) and South Asia (East), respectively [26], and high-resolution local studies confirm these findings [27]. Although the glaciers will shrink further, it is not expected that river flow will decrease in the Greater Himalayas for two major reasons. The glacier melt is the product of the glacier extent and the melt rate per unit area. The decrease in area is compensated for by an increased melt rate, and the melt peak for the Greater Himalayas is expected to occur after 2050 [27]. Most climate models project an increase in precipitation of around 5% on average annually in 2050 relative to 1961–1990, but rapidly increasing afterwards, with an increasingly large spread between the climate models. By the time the melt peak is reached, the precipitation takes over and more than compensates for the reduction in glacier meltwater. Overall, no decreases in average water availability can thus be expected.

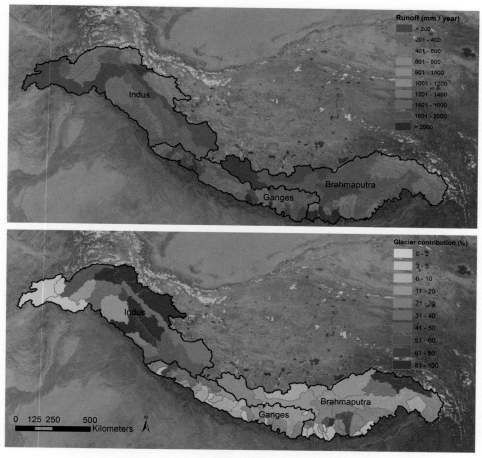

Figure 11.6 Total specific runoff per sub-basin for 1998–2007 (top) and the glacier contribution to total runoff (bottom).

11.2.4 Western North America

Glaciers are distributed throughout the mountain regions of western North America. The concentration of ice masses generally increases from south to north, and is highest in regions without major population centers, including northwest British Columbia (BC), southeast Alaska, and southwest Yukon (Figure 11.7). Many of the large ice masses in southeast Alaska terminate as tidewater glaciers and thus are not currently sources of water supply for human use or freshwater habitat. Our focus here is on land-terminating glaciers. In the conterminous United States, glaciers covered only about 688 km^2 at the end of the twentieth century, of which about 65% was located in Washington [28].

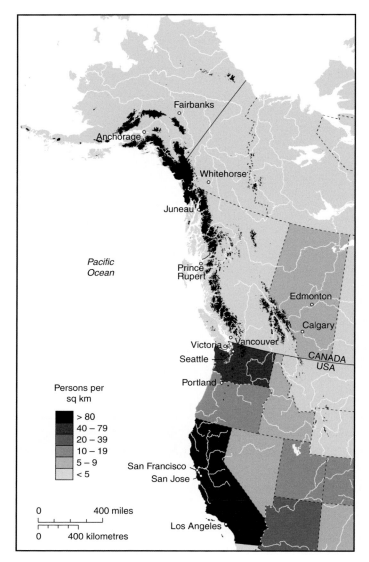

Figure 11.7 Western North America, showing locations of glaciers (black shading) and human population densities by state and province.

Glacier runoff is an important source of inflow to hydroelectric facilities that serve urban populations. In addition, many run-of-river hydroelectric projects in BC are developed on glacier-fed rivers. These facilities lack storage and are thus sensitive to seasonal changes in stream flow. Further, a range of resource extraction projects, primarily mining and oil and gas operations, are underway or planned for northwest BC. For these projects, glaciers must be considered in terms of being both a water resource and a potential hazard – for example, in relation to the siting and design of tailings ponds and river crossings associated with linear infrastructure such as roads and pipelines. Of particular concern in some areas is the potential for increased frequency of outburst floods and landslides [29].

Glaciers are important controls on freshwater habitat throughout the region, particularly as an important source of late-summer stream flow in rivers that support populations of salmon and other valued fish species [30]. In addition, glacier cover is associated with lower summer stream temperature, which is a critical variable that controls fish habitat suitability and bioenergetics [31].

Western North America generally experiences wet, cool winters and warmer, drier summers. In the south, the rainy season typically begins in October or November and extends to March or April. However, the rainy season begins earlier in the autumn and ends later in the spring in northern BC and southeast Alaska, and substantial amounts of rain can fall throughout summer. The accumulation season typically begins in October and ends in April or May. However, at higher elevations, particularly in Alaska and Yukon, air temperatures may remain low enough to suppress melting all year round.

Within this hydroclimatic context, the contributions of glacier melt to stream flow are most distinct between July and September, and are generally least obscured by seasonal snowmelt or rainfall in August. Stahl and Moore [32] showed that, in BC, the relation between August stream flow and air temperature shifts from negative to positive when glacier cover exceeds about 3% of the catchment area, suggesting that even a few percent of glacier cover can provide significant contributions to late-summer flow in dry weather.

A number of studies have quantified glacier-melt contributions to stream flow on both seasonal and annual timescales. Some studies applied water balance analyses, where glacier mass balance measurements at a single glacier were extrapolated to gauged catchments [33,34]. Nolin *et al.* [35] employed isotopic hydrograph separation. Others have used catchment-scale simulation models that incorporate glacier accumulation and melt processes, although problems with equifinality mean that these estimates will be subject to high uncertainty unless model calibration is constrained using measurements of glacier mass balance or volume loss [36]. A key result from these studies is that long-term averages mask the important contributions of glaciers in specific years or months. For example, Jost *et al.* [36] found that ice-melt contributions to annual stream flow in the Mica basin in the Columbia Mountains of BC, which has 5% glacier cover, varied from 3% to 9% and averaged 6%. In August and September, however, glacier ice melt contributed up to 25% and 35% of total stream flow, respectively, during summers following winters with low snow accumulation. Naz *et al.* [37] found that glacier-melt contributions to August stream flow for Bow River, located on the east side of the Canadian Rockies, varied from 20% to 50%. The magnitude of glacier-melt contributions to stream flow generally increases with catchment glacier cover [34].

Most glaciers in western North America retreated from their Little Ice Age maximum extents through most of the twentieth century, consistent with the dominant warming trend over that period. A period of slower recession, and in some cases advance, occurred between 1950 and 1980. This period coincided with a dominantly negative phase of the Pacific decadal oscillation (PDO), which is associated with cooler and wetter weather throughout much of BC and the US Pacific Northwest. Notwithstanding this pause in sustained glacier recession, the total glacier area throughout western North America decreased over the last half of the twentieth century [38,39]. Since 1980, nearly all glaciers have been retreating. Between 1985 and 2005, glacier area declined approximately 11% in BC and 25% in Alberta [40]. Within BC, the smallest relative decline occurred in the north and central coast regions (7–8%) and the greatest in the northern interior (24%).

There is general agreement that, following the onset of a warming trend accompanied by a shift to negative mass balance, there should be an initial increase in glacier-melt contributions followed by an ultimate shift to declining flows as glaciers recede [41]. Negative trends in summer stream flow or its glacier-melt component have been reported for the east slope of the Canadian Rocky Mountains [42], the Washington Cascades [43], and southern BC [32]. It thus appears that glaciers in these regions have passed the point of "peak water" and are on the declining phase. In contrast, increasing stream flow has been reported for glacier-fed streams in northwest BC [32,44] and the Yukon River basin [45].

Considering that glaciers are currently retreating throughout the region, and that projected climate scenarios indicate continued warming that would favor continued negative mass balance conditions, glaciers should continue to retreat over the twenty-first century. A small number of studies have made quantitative projections by using downscaled climate scenarios from General Circulation Models to drive models simulating both glacier response and catchment hydrology, focused on the southern Coast Mountains of BC [46], the Mica basin in the Columbia Mountains [36], and the east slope of the Rocky Mountains [47]. Projected declines in glacier cover, relative to current conditions, ranged from about 50% [46] to over 80% [36]. Marshall *et al.* [47] projected volume losses in excess of 80% for some scenarios. In all studies, the ultimate result was a decline in glacier melt contributions to stream flow, particularly in August and September.

11.3 Summary and future research priorities

Mountain glaciers are mostly receding in response to global-scale climate forcing, and this is transforming downstream hydrology. However, risks to water resources have to be analyzed regionally to account for the combination of biophysical and human factors.

- The Andes face high potential hazards and risks related not only to ongoing glacier mass loss but also to water access, a reality more constrained by socio-economic and political realities than by physical or climatic ones. Most of the rapid economic growth in Andean countries during the last decades has relied upon natural resources exploitation. An increasing global demand for primary products such as exported agricultural crops and minerals, which require large amounts of water, puts additional stress on water resources that may outpace the effects of projected climate changes on scenarios of water availability. In many Andean locations there remain uncertainties about the actual glacier contribution to the local hydrology relative to other components such as groundwater and seasonal snow.

- The persistent shrinking of Swiss glaciers has had, and will continue to have, significant impacts, particularly on the meso scale. Continued glacier mass loss is very likely since it is primarily driven by temperature. From a hydrological perspective, it is important to distinguish between the near and distant future: In the near future, heavily glaciated alpine catchments will produce more runoff as they can more or less profit from the negative mass balance of glaciers; in the more distant future (>50 years), summer drought will become a problem due to the fact that the glaciers will be too small to supply the river sufficiently and thus to compensate for the deficit in summer precipitation. Fortunately there is still time to react, and abundant capacity to adapt, but this should not preclude action now.

- The challenge for the future of Himalayan water resources lies in adaptation to intra-annual shifts in the hydrograph and the ability to deal with extremes. As the buffering capacity of the glaciers will diminish, the hydrological system will accelerate, and the risk for society is likewise related to the resilience of managing this altered hydrologic regime in combination with a strong rise in future water demand.

- Glaciers in western North America play a key role by maintaining stream flow in late summer and autumn, particularly during extended periods of dry weather or following winters with low snow accumulation. Available evidence points to a continued retreat of glaciers throughout the region. In the conterminous United States, southern BC, and Alberta, the current trend to decreasing stream flow in late summer and early autumn should continue until the glaciers establish equilibrium with the climate or ultimately disappear. A major question for northwest BC, Yukon, and southeast Alaska is when the current trend for increasing stream flow will shift to a decreasing trend.

Several themes are important for future consideration. (1) Ongoing glacier recession will reduce capacity of watersheds to buffer contrasts in precipitation that are likely to become more extreme. (2) With glaciers diminishing, the changes in hydrological timing are potentially more important to water resources than overall volume change. (3) Despite the obvious and dramatic loss of glacier water storage by melting, the contribution to water resources is potentially obscured by other components such as groundwater, seasonal snowmelt (especially within mid-latitudes), and rain seasonality. Regionally, the fate of seasonal snow cover is more crucial to water supply globally than glaciers. (4) There is a complex interplay between water demand (growing cities, hydropower, globalized agriculture) and less available water. Although the total loss from stream flow due to glacier loss may be less than some earlier estimates, changes are occurring faster than estimated, and the problems may arise in seasonal distribution of supply vis-à-vis peak usage. Civil water resource reservoir capacity and other infrastructure may not be adapted to these changes.

Future research should be directed at consolidating observations, establishing consistent monitoring, and analyzing glacier interactions with components of the water cycle, notably snow and groundwater. Social, economic, and demographic changes are likely to have equal influence on glacier water resources, and attention needs to be drawn to evaluating risks and adaptation options with rigorous, data-based scenario evaluations of how management decisions impact all end users. Integrated, long-term (decades to centuries) glacier hydroclimate modeling studies have to be combined with field sampling and measurements. Likewise, more high-elevation measurements of glacier mass balance and hydroclimatic variables should be instituted. Observations must be distributed globally to account for regional differences, but common protocols and data sharing should be standardized.

References

1. M Carey, M Baraer, BG Mark *et al.*, Toward hydro-social modeling: merging human variables and the social sciences with climate-glacier runoff models (Santa River, Peru). *Journal of Hydrology*, **518**(A) (2014), 60–70. http://dx.doi.org/10.1016/j.jhydrol.2013.11.006.
2. D Viviroli, HH Dürr, B Messerli, M Meybeck, R Weingartner, Mountains of the world, water towers for humanity: typology, mapping, and global significance. *Water Resources Research*, **43**:7 (2007), W07447. 10.1029/2006WR005653.
3. M Baraer, BG Mark, JM McKenzie, *et al.*, Glacier recession and water resources in Peru's Cordillera Blanca. *Journal of Glaciology*, **58**:207 (2012), 134–150. 10.3189/2012JoG11J186.
4. M Vuille, B Francou, P Wagnon, *et al.*, Climate change and tropical Andean glaciers: past, present and future. *Earth-Science Reviews*, **89**:3–4 (2008), 79–96. 10.1016/j.earscirev.2008.04.002.

5. F Pellicciotti, S Ragettli, M Carenzo, J McPhee, Changes of glaciers in the Andes of Chile and priorities for future work. *Science of the Total Environment*, **493**: (2013), 1197–1210.

6. D Ruiz, HA Moreno, ME Gutiérrez, PA Zapata, Changing climate and endangered high mountain ecosystems in Colombia. *Science of the Total Environment*, **398**:1 (2008), 122–132.

7. P Chevallier, B Pouyaud, W Suarez, T Condom, Climate change threats to environment in the tropical Andes: glaciers and water resources. *Regional Environmental Change*, **11** (2011), S179–S187. 10.1007/s10113-010-0177-6.

8. W Buytaert, B De Bievre, Water for cities: the impact of climate change and demographic growth in the tropical Andes. *Water Resources Research*, **48** (2012). 10.1029/2011wr011755.

9. MH Masiokas, R Villalba, BH Luckman, C Le Quesne, JC Aravena, Snowpack variations in the central Andes of Argentina and Chile, 1951–2005: large-scale atmospheric influences and implications for water resources in the region. *Journal of Climate*, **19**:24 (2006), 6334–6352.

10. R Garreaud, The Andes climate and weather. *Advances in Geosciences*, **22**:22 (2009), 3–11.

11. S Vicuña, RD Garreaud, J McPhee, Climate change impacts on the hydrology of a snowmelt driven basin in semiarid Chile. *Climatic Change*, **105**:3–4 (2011), 469–488.

12. S Gascoin, C Kinnard, R Ponce, S Macdonell, S Lhermitte, A Rabatel, Glacier contribution to streamflow in two headwaters of the Huasco River, Dry Andes of Chile. *The Cryosphere*, **5** (2011), 1099–1113.

13. V Favier, M Falvey, A Rabatel, E Praderio, D Lopez, Interpreting discrepancies between discharge and precipitation in high-altitude area of Chile's Norte Chico region (26–32 degrees S). *Water Resources Research*, **45** (2009). 10.1029/2008wr006802.

14. N Ohlanders, M Rodriguez, J McPhee, Stable water isotope variation in a Central Andean watershed dominated by glacier and snowmelt. *Hydrology and Earth System Sciences*, **17**:3 (2013), 1035–1050. 10.5194/hess-17-1035-2013.

15. J Bury, BG Mark, M Carey, *et al.*, New geographies of water and climate change in Peru: coupled natural and social transformations in the Santa River Watershed. *Annals of the Association of American Geographers*, **103**:2 (2013), 363–374. 10.1080/00045608.2013.754665.

16. FOEN, *Effects of Climate Change on Water Resources and Waters: Synthesis Report on the "Climate Change and Hydrology in Switzerland"* (FOEN, 2012).

17. BUWAL, BWG, MeteoSchweiz, Auswirkungen des Hitzesommers 2003 auf die Gewässer (Schriften reihe Umwelt, 2004).

18. N Köplin, O Rößler, B Schädler, R Weingartner, Robust estimates of climate-induced hydrological change in a temperate mountainous region. *Climatic Change*, **122** (2013), 171–184. 10.1007/s10584-013-1015-x.

19. R Weingartner, B Schädler, P Hänggi, Climate change versus Swiss hydro: what happens next? *International Water Power & Dam Construction*, **64**:4 (2012), 38–42.

20. E Reynard, M Bonriposi, O Graefe, *et al.*, Interdisciplinary assessment of complex regional water systems and their future evolution: how socioeconomic drivers can matter more than climate. *Wiley's Interdisciplinary Reviews: Water*, **1**:4(2014), 413–426. 10.1002/wat2.1032.

21. J Fuhrer, M Beniston, A Fischlin, *et al.*, Climate risks and their impact on agriculture and forests in Switzerland. In *Climate Variability, Predictability and Climate Risks*, ed. H Wanner, M Grosjean, R Rothlisberger, E Xoplaki (Springer, 2006), pp. 79–102.

22. WW Immerzeel, LP van Beek, MF Bierkens, Climate change will affect the Asian water towers. *Science*, **328**:5984 (2010), 1382–1385.
23. T Bolch, A Kulkarni, A Kääb, *et al.*, The state and fate of Himalayan glaciers. *Science*, **336**:6079 (2012), 310–314. 10.1126/science.1215828.
24. W Immerzeel, L Petersen, S Ragettli, F Pellicciotti, The importance of observed gradients of air temperature and precipitation for modeling runoff from a glacierized watershed in the Nepalese Himalayas. *Water Resources Research*, **50**:3(2014), 2212–2226.
25. A Lutz, WW Immerzeel, AB Shrestha, MF Bierkens, Increase in High Asia's future runoff confirmed at the large scale. *Geophysical Research Letters* (in prep).
26. V Radić, A Bliss, A Beedlow, R Hock, E Miles, J Cogley, Regional and global projections of 21st century glacier mass changes in response to climate scenarios from global climate models. *Climate Dynamics*, (2013). doi: 10.1007/s00382-013-1719-7.
27. W Immerzeel, F Pellicciotti, M Bierkens, Rising river flows throughout the twenty-first century in two Himalayan glacierized watersheds. *Nature Geoscience*, **6**:9 (2013), 742–745.
28. A Fountain, M Hoffman, K Jackson, H Basagic, T Nylen, D Percy, Digital outlines and topography of the glaciers of the American West, U.S. Geological Survey Open-File Report, (2007).
29. RD Moore, SW Fleming, B Menounos, *et al.*, Glacier change in western North America: influences on hydrology, geomorphic hazards and water quality. *Hydrological Processes*, **23**:1 (2009), 42–61. 10.1002/hyp.7162.
30. SW Fleming, Comparative analysis of glacial and nival streamflow regimes with implications for lotic habitat quantity and fish species richness. *River Research and Applications*, **21**:4 (2005), 363–379.
31. R Moore, M Nelitz, E Parkinson, Empirical modelling of maximum weekly average stream temperature in British Columbia, Canada, to support assessment of fish habitat suitability. *Canadian Water Resources Journal*, **38** (2013), 135–147. 10.1080/07011784.2013.794992.
32. K Stahl, R Moore, Influence of watershed glacier coverage on summer streamflow in British Columbia, Canada. *Water Resources Research*, **42**:6 (2006). 10.1029/2006WR005022
33. C Hopkinson, GJ Young, The effect of glacier wastage on the flow of the Bow River at Banff, Alberta, 1951–1993. *Hydrological Processes*, **12**:10-11 (1998), 1745–1762.
34. LE Comeau, A Pietroniro, MN Demuth, Glacier contribution to the North and South Saskatchewan rivers. *Hydrological Processes*, **23**:18 (2009), 2640–2653.
35. AW Nolin, J Phillippe, A Jefferson, SL Lewis, Present-day and future contributions of glacier runoff to summertime flows in a Pacific Northwest watershed: implications for water resources. *Water Resources Research*, **46** (2010). 10.1029/2009wr008968.
36. G Jost, R Moore, B Menounos, R Wheate, Quantifying the contribution of glacier runoff to streamflow in the upper Columbia River Basin, Canada. *Hydrology and Earth System Sciences*, **16**:3 (2012), 849–860.
37. B Naz, C Frans, G Clarke, P Burns, D Lettenmaier, Modeling the effect of glacier recession on streamflow response using a coupled glacio-hydrological model. *Hydrology and Earth System Sciences*, **18** (2014), 787–802. 10.5194/hess-18-787-2014.
38. N Barrand, M Sharp, Sustained rapid shrinkage of Yukon glaciers since the 1957–1958 International Geophysical Year. *Geophysical Research Letters*, **37**:7 (2010), L07501.
39. E Berthier, E Schiefer, GK Clarke, B Menounos, F Rémy, Contribution of Alaskan glaciers to sea-level rise derived from satellite imagery. *Nature Geoscience*, **3**:2 (2010), 92–95.

40. T Bolch, B Menounos, R Wheate, Landsat-based inventory of glaciers in western Canada, 1985–2005. *Remote Sensing of Environment,* **114**:1 (2010), 127–137.
41. RD Moore, SW Fleming, B Menounos, *et al.*, Glacier change in western North America: influences on hydrology, geomorphic hazards and water quality. *Hydrological Processes,* **23** (2009). 10.1002/hyp.7162.
42. M Demuth, V Pinard, A Pietroniro, *et al.*, Recent and past-century variations in the glacier resources of the Canadian Rocky Mountains: Nelson River system. *Terra Glacialis,* **11**:248 (2008), 27–52.
43. M Pelto, Skykomish River, Washington: impact of ongoing glacier retreat on streamflow. *Hydrological Processes,* **25**:21 (2011), 3356–3363.
44. SJ Déry, K Stahl, R Moore, P Whitfield, B Menounos, JE Burford, Detection of runoff timing changes in pluvial, nival, and glacial rivers of western Canada. *Water Resources Research,* **45**:4 (2009).
45. TP Brabets, MA Walvoord, Trends in streamflow in the Yukon River Basin from 1944 to 2005 and the influence of the Pacific Decadal Oscillation. *Journal of Hydrology,* **371**:1 (2009), 108–119.
46. K Stahl, R Moore, J Shea, D Hutchinson, A Cannon, Coupled modelling of glacier and streamflow response to future climate scenarios. *Water Resources Research,* **44**:2 (2008), W02422.
47. SJ Marshall, EC White, MN Demuth, *et al.*, Glacier water resources on the eastern slopes of the Canadian Rocky Mountains. *Canadian Water Resources Journal,* **36**:2 (2011), 109–134.

12

Glacier floods

DUNCAN QUINCEY AND JONATHAN CARRIVICK

12.1 Introduction

Glacier floods, also known as jökulhlaups, glacier outbursts, glacial lake outburst floods (GLOFs), aluviónes and debacles, refer to the sudden release of water from a glacier hydrological system or from glacial lakes impounded by moraine sediments and/or ice [1]. The biggest floods ever to have occurred on the Earth's surface (reaching peak discharges up to 17–18×10^6 m^3 s^{-1}) were from glaciers and affected thermohaline circulation and caused widespread and intense landscape change [2]. Holocene and modern events may not have had such a significant global impact, but they have posed a hazard to communities living downstream, and to infrastructure and land in the flood path (peak discharges reaching up to $\sim 0.1 \times 10^6$ m^3 s^{-1}) [1]. In the last decade alone, glacier floods have threatened communities in Peru, Bolivia, Nepal, India, Pakistan, Iceland, Greenland, New Zealand, Switzerland, Russia and North America (Figure 12.1), destroying hydroelectric installations, road and rail infrastructure, farmland, housing and, in some cases, causing loss of human life.

This chapter aims to review current knowledge regarding glacier floods, building on recent subject-specific contributions on proglacial lakes [3], jökulhlaup mechanisms [4,5], geomorphological impacts [6] and outburst floods [7]. It will bring together understanding of five components: (1) the processes leading to the development and outburst of glacial lakes; (2) the form that different types of floods take (i.e. their character and dynamics) and their geomorphic impact; (3) flood modelling and associated challenges; (4) hazard identification and assessment procedures; and (5) hazard management (i.e. phenomena prediction, lake remediation, reducing exposure). We will focus primarily on advances made over the previous decades and in doing so will highlight remaining knowledge gaps.

The High-Mountain Cryosphere, ed. Christian Huggel, Mark Carey, John J. Clague and Andreas Kääb. Published by Cambridge University Press. © Cambridge University Press 2015.

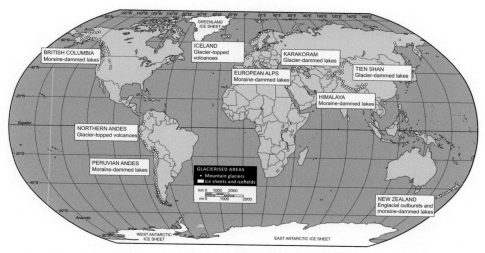

Figure 12.1 High-mountain regions of the world at threat from glacier floods.

12.2 Lake development and outburst processes

Glacial lakes can form on top of the glacier (supraglacial), in front of the glacier (moraine-dammed), marginal to the glacier (often ice-dammed), within the glacier (englacial) or at the glacier–bedrock interface (subglacial) (Figure 12.2). Lake volumes can range from some tens through to many millions of cubic metres, and peak discharges can range from just above normal proglacial river flow (<10 cumecs) through to $>10^6$ cumecs, although events of this larger size are rare.

Supraglacial lakes typically begin as a series of ponds that subsequently coalesce into a larger lake, dammed either by the stagnating glacier tongue or by the end-moraine, which can also be ice-cored. Supraglacial ponds tend to grow particularly rapidly, with ablation rates on the surrounding ice cliffs being typically one or two orders of magnitude greater than sub-debris melt rates [8]. They are often ephemeral, draining periodically as they connect (at their base) to the englacial drainage system, but can develop into supraglacial and then base-level lakes, where their surface elevation is at or below the elevation at which water leaves the glacier. One of the best cited examples of such evolution is the development of Imja Tsho (Nepal), which began as a series of disconnected supraglacial ponds in the 1950s that subsequently coalesced during the 1970s, and then expanded to the current lake area of ~1.25 km². In such cases, lakes will persist until the moraine dam fails, and it is therefore base-level lakes that are of greatest concern in high-mountain regions.

Moraine-dammed lakes may fail through the gradual degradation of a permafrost core, saturation and seepage through the moraine sediments, piping

Figure 12.2 Glacial lake types from around the world: (a) Imja Tsho (Nepal), a supraglacial lake that has grown from a series of small supraglacial ponds in the 1960s to its >2 km length today; (b) the Griessee, which occupies the Little Ice Age extent of the Griesgletscher, and is now dammed for hydro-electric power production; (c) an ephemeral ice-dammed lake trapped between Thompson and White glaciers on Axel Heiberg Island, Canadian Arctic; and (d) supraglacial ponds on Khumbu Glacier that currently occupy an area of suppressed topography and are likely to expand further in coming years.
Sources: (a) Photograph by M. J. Hambrey, www.glaciers-online.net; (b) Photograph by J. Alean, www.glaciers-online.net; (c) Photograph by J. Alean, www.glaciers-online.net; (d) Photograph by D. J. Quincey.

and headcut retreat, or overtopping and distal scour following the influx of landslide or avalanche material into the lake. The processes occurring during overtopping are perhaps the best understood. Overtopping most often occurs across the full width of the dam crest in a uniform sheet (e.g. [9]) but may also occur as a focused incision if there is an existing outlet (e.g. [10]). Once the boundary shear stress of the overtopping water exceeds the cohesive strength of the dam material, downcutting will occur, resulting in runaway incision as underlying, often less consolidated sediments become exposed and eroded. Incision will only cease when drainage is complete or when a layer of increased cohesion (or even bedrock) is reached. Processes leading to the internal failure of the dam structure are less well understood, but are thought to begin with the saturation and removal of fine sediments within the dam material, leading to piping through the moraine structure. The local physical strength of the dam material is thus compromised

[11], and breach formation can then follow as overlying material slumps and collapses into the sediment voids.

Ice-dammed lakes occur where glaciers advance across drainage routes or where ice-avalanche deposits block river drainage, resulting in much larger lakes than their moraine-impounded counterparts and threatening populations and infrastructure for many hundreds of kilometres downstream [12]. Ice-dammed lakes are therefore most common in areas where glaciers are in positive mass-balance, or in areas where surging glaciers cause large frontal advances. The Karakoram Himalaya has historically seen some of the biggest floods, with 69 destructive outburst floods recorded on the upper Indus in the last 200 years alone, nearly all prior to the 1940s [13]. However, recent mass-gain associated with anomalous climatic conditions in the region [14,15] has led to renewed fears of ice-dammed lake formation, and satellite imagery has captured the cyclical formation and drainage of a number of lakes in the last decade (e.g. [16]). The coupling between ice- and lake-dynamics is particularly interesting in this context, with recent work suggesting that submarine melting may significantly enhance glacier thinning rates [17] and thus reduce potential maximal glacial lake volumes with each drainage–formation cycle, at least in the short-term [16].

Ice-dammed lakes drain either through subglacial tunnel flow (e.g. [18]) or by mechanical failure of the ice dam. Subglacial tunnel flow occurs when the gravitational force preventing the glacier from floating is exceeded by the buoyant force of the lake water. As a rule of thumb, the glacier may become buoyant when the water level reaches ~90% of the dam height. Once initiated, subglacial tunnel growth can evolve over hours to days as a result of thermal erosion, and once peak flow is reached discharge will typically decrease to previous levels over minutes to hours. Mechanical failure of the ice dam can occur where existing lines of weakness fail under the pressure exerted by the lake water, or are exploited by percolation of relatively warm water through sediments within the ice. In such cases peak discharge can be reached within minutes.

Englacial floods tend to be small, often caused by the release of water from a blocked conduit or moulin [19] and are not location-specific. They are very difficult to predict given there is usually no surface expression of their presence. They are most common on active glaciers where the englacial hydrological system is under constant modification and steep surface topography leads to slumping of sediment and boulders by glaciofluvial undercutting of moraine slopes [20].

12.3 Flood form, character and geomorphological impact

Regardless of its source, the flood hydrograph is largely dependent on the initial lake hypsometry and volume, dam geometry, the nature of the failure mechanism, the downstream topography and sediment availability [21].

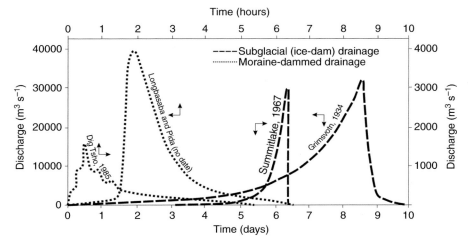

Figure 12.3 Typical hydrographs of floods sourced from subglacial and moraine-impounded sources. Arrows point to relevant axes in each case. Longbasaba and Pida hydrographs are simulated and have not burst to date.
Source: data derived from [25,75,76,77].

Discharge from moraine-dammed lakes is largely controlled by the rate and extent of the breach growth [21], and overall flood volumes of $10–30 \times 10^6$ m^3 and discharges of several thousand cumecs are not exceptional. Floods originating from moraine-impounded lakes are characterised by an exponential increase in discharge associated with rapid and often self-perpetuating incision [22], a short period of maximum discharge, and a more gentle, often intermittent, decreasing phase as drainage is regulated by the (in)stability of the remaining moraine sediments (Figure 12.3). Moraine-impounded lakes tend to drain rapidly because of their low width-to-height geometry and non-cohesive composition [23], and with little warning, making them particularly hazardous (e.g. [24,25]), but they tend to be one-off events because they completely destroy the moraine dam.

Floods from subglacial ice-dammed lakes are thought to almost always occur through ice-walled channels at the ice-cap's bed. The Grimsvötn jökulhlaup hydrograph shown in Figure 12.3 is typical of subglacial floods from marginal and subglacial lakes in magnitude and shape [26]; discharge increases approximately exponentially over one or two weeks and decreases relatively rapidly following a peak of up to several thousand cubic metres per second. Theory suggests that this shape results from the way in which the size of an ice-walled subglacial channel, assumed to convey the floods, evolves over time through a competition between enlargement through melt caused by the flowing water and closure due to the flow of ice [23].

The form and character of floods from glacier-dammed lakes, at least based on available Karakoram evidence, is more akin to those sourced from moraine-impounded lakes [12]. Again, there is a paucity of observations close to the ice dams to robustly constrain the hydrograph, but those that do exist suggest a very steep rise to peak and a gradual recessional limb. Because of the sheer volume of water that can be stored in a glacier-dammed lake, sudden and complete outbursts can impact many hundreds of kilometres downstream. For example, the 18 km long lake that had formed in 1929 behind the Chong Khumdan glacier in the Upper Shyok, Karakoram, released an estimated 1.5×10^9 m^3 of water into the lower valley, creating a flood that still registered on the hydrograph recorded at a gauging station some 1452 km downstream [27].

Although glacier floods are initially 100% water by volume and are sediment supply-limited, downstream they are potent agents of rapid landscape change, causing erosion of bedrock and entrainment and redistribution of sediment. For example, the 1985 flood from Dig Tsho was reported to have mobilised sediment 20–50 m above the main river bed and induced terrace failures of $>10^5$ m^3 volume. Further downstream, aggradation was up to 10 m in places, including transportation of boulders >1 m in diameter. Much of the sediment transported in a flood event may be sourced from the moraine dam itself (Figure 12.4); the Nostetuko lake outburst in British Columbia mobilised more than 1×10^6 m^3 of morainic

Figure 12.4 Moraine-dam failure adjacent to the Ama Dablam Glacier, Everest-region, Nepal. Floods from moraine-dammed lakes are rarely 100% water because of the large volumes of sediment available from the moraine structure itself, as evidenced by the debris-fan in the image. Further debris can be entrained as sediments are mobilised along the flood route, and it is this debris that can cause most destruction.

sediment and deposited it directly downstream as a large fan on top of a former meadow [28]. Many outburst floods mobilise sufficient glacial and colluvial debris from downstream areas to become debris-flows, increasing their volume and discharge and becoming particularly erosive as they move through steep channels, exerting up to six times as much shear stress on channel beds as an equivalent water-only flood [29].

12.4 Flood modelling and associated challenges

12.4.1 Flood magnitude

Knowledge of the source lake and accurate and detailed simulation of the glacier lake drainage trigger and drainage mechanism is central to producing an initial flood hydrograph and thus to an assessment of the downstream hazards posed by a glacier flood (Figure 12.5). Predicting the absolute timing of floods and even the probability of occurrence in any given time period is virtually impossible because of rapid changes in the nature of glacial systems, the low frequency of events and the high complexity of the processes involved [30]. Therefore, predictions focus on the magnitude of events, and on the hydraulic evolution of events. In order of increasing complexity, analysis of magnitude can be based on (1) empirical relationships derived from historical events; (2) analytical models; or (3) mechanistic models (Table 12.1).

For ice-dammed lakes, discharge–volume (Q–V) data compiled for glacier floods have been presented by a number of authors [31–33]. Only floods thought to have drained subglacially were included. The power laws derived between discharge and volume can generally be physically explained [33]. However, the scatter of the data is significant: (1) multiple flood events from a given lake commonly span up to an order-of-magnitude range in Q and V; and (2) if a Q–V relation was constructed for each lake the resulting best-fit would generally differ significantly from that derived for all lake systems. Furthermore, many glacier lakes emptied, while other floods terminated with the lake still partially filled.

For moraine-dammed lakes, examples of empirical equations used for dam-break modelling are provided in Table 12.2. Thornton *et al.* [34] have undertaken comprehensive investigations of uncertainty estimates surrounding the use of such equations. The limitations and assumptions behind empirical relationships include (1) that they have been derived almost exclusively from physical scale, laboratory-based experiments (e.g. [35,36,37]); (2) such experiments cannot typically simulate the effects of complex basin bathymetry on wave propagation and run-up; and (3) the interaction of a displacement wave, or a series of seiche waves, with the

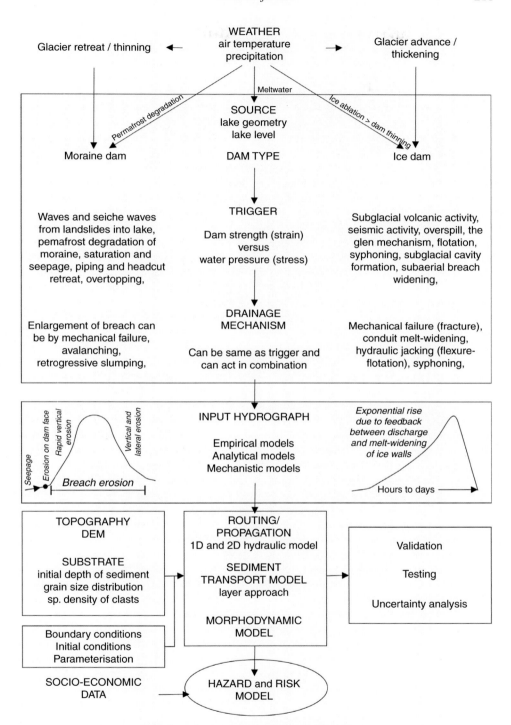

Figure 12.5 Suggested workflow for modelling glacier floods.

Table 12.1 *Overview of types of models available for dam-breach simulation, highlighting required input data, output data and key advantages and limitations of each approach* [78].

Model type	Required input data	Output data	Advantages	Limitations
Empirical regression equations (see Table 12.2).	One or more of: reservoir/lake depth; dam height; dam width; lake surface area; lake volume; potential energy of stored water.	Peak discharge (Q_p) and time to peak (T_p).	Expeditious implementation. Requires only simple input data. May be suitable for 'first-pass' rapid assessment of potential Q_p or T_p.	No physical basis. Derivation from case study data which may differ significantly from intended application often overlooked with serious implications for validity of results. Full hydrograph not produced
Analytical / parametric numerical models.	Time required for full breach development; breach cross-sectional form; erodibility coefficient.	Full breach outflow hydrograph.	Rapid implementation. Full outflow hydrograph produced. Codes/models readily available to public.	Limited physical basis. Sediment transport/breach development assumed to be steady-state – typically an incorrect assumption given the mechanics of breach expansion. Time to full breach development often unknown.
Physically based numerical models	Lake hypsometry; cross-sectional and elevation of downstream valley floor; dam geometry; material properties of dam.	Full breach outflow hydrograph. Possible to extract sediment evacuation rates.	Process-based. Consider sediment transport equations, flow hydraulics and soil mechanics in solution. 'Free-form' breach development permitted. Full outflow hydrograph obtainable.	High volume of data input required; many input parameters and initial conditions often only quantifiable by detailed field investigation. Often proprietary; not widely available.

Table 12.2 *Factors for consideration in determining the level of hazard posed by a glacial lake.*

	Factor	Influence on glacier floods and their assessment
Glaciers	Type of margin	Hanging margins particularly prone to avalanches; retreating margins associated with lake formation and paraglacial slope reworking; advancing margins inundate land, block valleys and are prone to avalanches.
	Relationship to lake	Avalanches can cause displacement waves in lakes. Floating termini can collapse catastrophically following fall in lake level.
	Surface gradient	Low average gradients are associated with stagnation and lake development. Steeper glaciers are prone to avalanching (>25° for warm-based; >45° for cold-based).
	Surface structures	Influence drainage patterns and lake growth processes. Can indicate potential break-off points for avalanches and calving blocks.
	Flow velocities	Low-velocity glaciers prone to lake development; sudden increases in velocity can be a precursor to avalanche or surge activity.
	Glacier hydrology	Nature of conduit system influences drainage and thus ponding potential. Sudden meltwater inputs into lakes can lead to outbursts.
	Existence of debris cover	Low-angle debris-covered valley glaciers prone to lake development. Debris buffers climate signal – exposed ice can melt very rapidly.
	Mass balance	Negative balance indicates glacier recession and potential for lake development; positive balance indicates thickening snout and potential avalanche activity. Long-term trends affect water resource planning.
Dam structure	Dam geometry	Tall, narrow dams more prone to failure (defined by height:width ratio).
	Freeboard	Height difference between crest of dam and lake provides protection against overtopping by waves.
	Geology	Influences strength, porosity, permeability and factor of safety to slope failures.

Table 12.2 (*cont.*)

	Factor	Influence on glacier floods and their assessment
	Presence of buried ice (moraine dam)	Melting ice reduces dam height and/or width. Percolating meltwaters reduce dam stability, massive ice impedes drainage and can control lake level.
	Hydrogeology	Seepage can reduce internal angle of friction and/or lead to piping (removal of fines) causing instability and possibly outbursts.
	Previous instability	Can indicate a tendency towards further instability where detailed observations or measurements are not available.
Lake system	Lake basin and surrounding geometry	Lakes with steep shores and narrow basin geometry are particularly susceptible to impact waves. Needed for calculation of potential discharge volume and design of remedial works.
	Surface area	Required as input into the calculation of potential discharge volume.
	Drainage outlet	Point of lowest freeboard. Focal point of erosion during 'normal' water flow and overtopping events. No outlet means potential for increase in lake volume.
	Rate of expansion	Important factor governing the level of risk, and in determining the time at which to take action.
Local environment	Ice avalanche sources	Ice avalanches into lakes can trigger outburst floods and also represent direct hazards. Failure detectable by glacier bed morphology (ramp- or cliff-type), gradient, ice temperature, bed conditions (material types, meltwater).
	Rockfall sources	As triggers of lake outburst or direct hazards. Rock mass strength, discontinuity patterns and slope angle/aspect with respect to bedding and presence of permafrost control stability. Juxtaposed hazard types can act as triggers (e.g. avalanche to lake outburst to debris flow).
	Interplay between hazards	Downstream lakes can act as buffers or exacerbate floods/debris flows. Flow/flood material may cause secondary blockages.
	Existing remedial measures	May no longer be appropriate as hazards develop in response to climate change. Inappropriate measures can exacerbate hazard.

Table 12.2 (*cont.*)

	Factor	Influence on glacier floods and their assessment
Regional environment	Tectonic activity	High-mountains especially prone to earthquake-triggered mass movement due to seismic amplification. Can destabilise moraine dams and damage remedial structures.
	Seasonal variations	Seasonal rains saturate materials, fill lakes and increase discharge rates (e.g. strong correlation between Asian monsoon seasons and lake outbursts. Entrained winter snow masses can increase volume/run-out distance of ice avalanche.
	Extreme weather events	Unseasonably warm and wet conditions can trigger glacier outbursts; effects of regional phenomena (e.g. El Niño can exacerbate hazards).

moraine-dam structure is of most interest from a hazard perspective, and this approach provides no such information.

Analytical and parametric models of dam breaches represent simplified physical processes; specifically of the breach development phase [38]. They typically assume that the rate of breach growth is purely time-dependent and that flow over the dam can be represented by a weir equation. Early flood routing models developed by the United States National Weather Service (NWS), including DAMBRK [39] and FLDWAV [40], utilise such parametric approaches to compute an upstream hydrograph. Glacier flood modelling studies that have adopted semi-physical breach models include [41], who used DAMBRK to reconstruct the 1981 GLOF from Zhangzanbo Lake, Nepal, and [42], who used an enhanced version (BOSS-DAMBRK) to estimate patterns of flood inundation in the Sun Koshi basin, Nepal. Osti and Egashira [43] used the NWS Simplified Dam Break (SMPDBK) model combined with an empirical predictor of failure duration to reconstruct the 1998 GLOF from Sabai Tsho glacial lake in the Khumbu Himal, Nepal.

Physically based mechanistic numerical models can simulate moraine breach development, flow hydraulics and sediment transport. Most notably they include geotechnical structural models that simulate breach widening (e.g. [22,44]). Physically based numerical models of moraine dam breaches have not seen widespread use in glacier flood applications, due partly to the high computational requirement for these models but also due partly to the input data requirements for parameterisation and specification of initial conditions, although usage of advanced models is becoming more common (e.g. [22]).

12.4.2 *Proglacial flood hydraulic routing/propagation*

Frontal waves of glacier floods travel slower than the main body as channel roughness is stage dependant and channels are usually virtually dry before the flood. Initially glacier outburst floods accelerate due to gravity, but quickly converge to an inertial regime. As channel slopes decrease and sediment transport proceeds, bed friction exerts more of an effect and the flow regime becomes more viscous [45]. A glacier flood hydrograph evolves through time and down channel with topography expansions/contractions, rapid changes in valley floor gradient [46] and strong contrasts in channel roughness and channel substrate. Multiple peaks in outburst flood hydrographs are not uncommon.

One-dimensional models are based on the one-dimensional St-Venant or Shallow Water equations, usually derived by integrating the Navier–Stokes equations over the cross-sectional surface of the flow. The shallow water assumption is that horizontal motion is far more significant than any vertical acceleration. 1D models are limited to use where the direction of water movement is aligned to the centre line of the river channel. Furthermore, the cross-sectional averaged velocity has poor physical meaning in a situation where large variations in velocity magnitude exist across the floodplain. Perhaps most importantly, 1D models require that a flow path is pre-defined. As well as predefined cross-sections, initial boundary conditions for 1D models include either an input hydrograph at the uppermost cross-section, or, if unavailable, downstream water profiles calculated from palaeo-stage indicators (PSIs). The most popular 1D models used in the glacier floods literature are HEC-RAS (e.g. [47,48]), NWS DAMBRK (e.g. [41]) and FLDWAV (e.g. [24]).

2D models are based on a simplified version of the 2D shallow water equations where some terms are neglected, resulting in the kinematic and diffusive wave representations. 2D models usually have a much finer regular discretisation of the physical space than 1D models. They require an input hydrograph and there is no predefined flow path or routeway. 2D models can be implemented on, among others, regular grids. This not only makes 2D models quite efficient computationally, but also visually realistic and also highly profitable in providing a wealth of spatio-temporal information. Advantages of using 2D models over 1D models include their ability to simulate multi-directional and multi-channel flows, super-elevation of flow around channel bends, hydraulic jumps (i.e. in-channel transitions between supercritical and subcritical flow regimes) and turbulent eddying [49,50]. Despite their representation of the flow conditions intrinsic to glacier floods and of more transient and complex hydraulics, applications of 2D models in the glacier floods hazard literature are rare (e.g. [22]), mainly because the fine spatial and temporal resolution field data required for parameterisation and validation do not exist.

The very nature of outburst floods makes them extremely challenging to model, and while advances are being made in the computational sciences, engineering and geoscience disciplines, arguably much more could be made of the synergies in interests between these [45]. Ongoing challenges in the mechanistic modelling of outburst floods include assessing accuracy, prediction of shocks (discontinuities), treatment of source terms and wetting and drying, sediment transport and morphodynamics.

12.4.3 Sediment transport

Sediment entrainment has a major influence on hydraulics of outburst floods, and flood routeways are subject to rapid and dramatic changes in geometry. Research to improve understanding of outburst floods must consider the rapid exchange of sediment with a bed because entrained sediment (1) affects the mass and momentum energy of a flow; (2) results in erosion and deposition which further perturbs flow hydraulics; and (3) constitutes the major hazard associated with outburst floods. Transient bulking (i.e. gain of mass due to sediment entrainment) and dilution of an outburst flood are common. Carrivick *et al.* [51,52] have discussed that the mass and energy transfers due to sediment entrainment and deposition and subsequent channel geometry modifications can be both intense and rapid; channel depth can change by over 100% in just a few minutes. The transport of sediment in outburst floods is arguably what constitutes the main hazard; boulder impact on structures; submergence of property and land by sediment, scour around bridge piers, and excavation of pre-existing natural and man-made material, for example. Ultimately, volume and peak discharge can increase by an order of magnitude with full transition to debris-flow.

12.5 Identifying hazards and hazard assessment procedures

Systematic inventories of glacial lakes have become increasingly common over the last decade (e.g. [53]), coinciding with increased availability of wide-swath remotely sensed (satellite) imagery and a growing awareness of the potential hazard posed by lakes in high-mountain areas. For instance, inventories for parts of Nepal [54], India [53], Bhutan [55] and the wider Hindu-Kush Himalaya [56] now exist from satellite-based assessments.

In the context of atmospheric warming, glacial lakes are expected to grow in number and size, and the prediction of future lake growth is an emerging field. So far, research has focused on identifying glacier characteristics that may indicate lake development susceptibility on a timescale of the order of a decade. Reynolds [57] found a slope gradient of 2° to be the critical threshold for supraglacial lake formation on debris-covered glaciers in the Himalayas. Quincey *et al.* [58] combined this slope threshold with information on glacier velocity, concluding that shallow-angled debris-covered glacier parts with low flow velocities are most likely sites with a potential for supraglacial lake formation. Frey *et al.* [59] presented a multi-level strategy to anticipate the formation of future glacier lakes at different scales where local overdeepenings in the glacier bed were estimated by analysing the current glacier surface characteristics based on digital elevation models (DEMs), digital glacier outlines and satellite imagery. With an ever-increasing archive of satellite imagery it is becoming easier to evaluate the indicators of future lake development, and it will soon be possible to validate predictions of lake extent with real data captured over many decades.

Although glacial lake volumes may exceed several millions of cubic metres, most glacial lakes are not unstable and have low probability of catastrophic failure. Objective, accurate and repeatable methods for assessing hazard, and their effective communication and dissemination, are therefore crucial if credibility with local stakeholders and populations is to be assured [53,60]. Several different assessment procedures exist [30,53,61,62], but commonly focus on quantifying (on an ordinal scale) lake-dam characteristics such as lake volume, dam composition, dam geometry, freeboard, potential triggers and potential for lake impact [63]. Once assessed, lakes can be ranked based on their probability of outburst, but as with the challenge of predicting lake development, there remains an insufficient (temporal) record of previous outburst events to be able to evaluate the efficacy of any of these published schemes. Furthermore, glaciers are such dynamic systems that statistical and empirical approaches based on historical data are commonly unrepresentative of the modern situation.

12.6 Managing glacier floods

The risk posed by a glacier flood is commonly expressed as the product of hazard (i.e. the probability of the occurrence of an event), vulnerability (i.e. the susceptibility to experiencing loss from an event) and exposure (the number/type of assets/people that may be lost). Vulnerability in high-mountain areas can be extremely dynamic, and in comparison to lowland areas is exacerbated by physical and socio-economic characteristics such as extreme topography, inaccessibility, isolation, poor infrastructure, resource degradation, political tension, poverty and health problems. Much work has been done (usually on a case-by-case basis) to reduce exposure in affected areas, mainly through education and adaptation strategies and the movement of people out of (or the protection of) areas of potential inundation. Conversely, a rapidly changing climate that is leading to larger numbers and sizes of glacial lakes [64], and rapid population growth that is leading to higher numbers

Figure 12.6 Examples of remediation: (a) concrete channel at Arhueyacocha, Cordillera Blanca, Peru; (b) concrete spillway and dam-strengthening works at Laguna Llaca, Cordillera Blanca, Peru; (c) pumps in operation to drain a supraglacial lake on Belvedere Glacier, Italian Alps; (d) the flood-damaged tunnel constructed in 1978 through the moraine dam at Laguna Safuna, Cordillera Blanca, Peru.
Source: (b) Photograph by M. J. Hambrey, www.glaciers-online.net.

of people living in hazardous zones, serve to keep both vulnerability and exposure at an elevated level.

Reducing the probability of an event occurring usually focuses on decreasing lake volume and/or increasing the integrity of the dam structure (Figure 12.6). Previously successful approaches have included the excavation of trenches or the installation of tunnels to reduce lake level [65], the construction of concrete spillways to increase dam stability and maintain or increase freeboard [66], the use of siphons and pumps to reduce lake volume [67] and the construction of flood channels and barriers downstream to divert or restrict water and debris from impacting on habitation.

The use of tunnelling to reduce lake level has seen particular prominence in the Cordillera Blanca, Peru, having been used at Laguna Paron, at Hualcán [65] and at Laguna Safuna [68], among other lake sites. Concrete spillways and open channels have been used in many glacierised environments to both lower lake levels and increase the dam integrity. Perhaps the most famous example can be found at Tsho Rolpa, located about 110 km northeast of Kathmandu in the Rolwaling Valley in

Nepal. Construction work began in 1998 on an open (concrete-lined) channel through the western end-moraine, which was completed in 2000 and successfully lowered the lake by three metres [69]. Siphoning and pumping has been used in several remediation efforts in the European Alps, most notably at Belvedere Glacier [70], Grubengletscher [67] and to drain Lac de Rochemelon close to the Franco-Italian border [71].

Once an outburst event has been initiated, focus turns to the exposure of downstream populations. Relatively little can be done to halt or alter the progress of a glacier outburst downstream and mitigation efforts are of a more passive nature. The use of downstream channels and barriers to contain or divert water-based outburst events is relatively uncommon, but both rigid and deformable barriers for containing debris-flow sediments have been shown to be effective in real events as well as recent experiments [72]. Flood modelling can be used to identify areas where such engineering solutions will be most effective, as well as identifying areas most likely to be inundated by future outburst events. Most recently, the coupling of several different physically based numerical models to simulate lake hazards threatening Carhuaz in Peru provided timings of chain processes to aid the design of early warning systems [73], and is now being used to inform structural interventions at the lake. In a number of cases early warning systems have been successfully used to alert populations of an approaching flood [16] but usually require regular maintenance and are often subject to vandalism and theft. At a more basic level, awareness-raising among local communities, through school education and government and religious institutions, can be equally valuable.

For many high-mountain regions, the consequences of a glacier flood are devastating, both socially and economically, because dwellers have few resources to fall back on. Many people inhabit the foothills, and may even farm the land directly beneath moraine dams. Individual families are therefore vulnerable to losing their lives and all property (buildings, land, livestock), and whole villages can be deprived of their agricultural subsistence base through the destruction of cultivable land and forest [25]. On a broader scale, many mountain communities depend on footpaths and roads to trade staple goods and, often, to attract tourism. If supplies become limited, and access routes (e.g. main bridges) are not able to function, drastic and rapid price rises can also ensue.

Vulnerability to outburst floods is therefore already high in many mountain regions, and will only increase with changing physical and socio-economic conditions. Population dynamics, new economic growth (increasing access and desire for mountain tourism) and a changing climate (often most rapid at high elevations) means that traditional and balanced adaptation mechanisms may be losing efficacy [74]. On a more positive note, however, traditional problems relating to (1) lack of understanding of glaciers and their dynamics, (2) lack of experience in hazard assessment, (3) lack of local awareness and preparedness, (4) poor linkages between stakeholders and (5) poor coordination between national and local authorities, are being addressed by an increasing number of local charities and international NGOs working in high-mountain regions. As glacial hazards become better integrated within national disaster management plans, vulnerability to glacier outburst floods in high-mountain regions should thus be further reduced.

12.7 Summary

Glacier floods currently pose a hazard to many communities living in high-mountain regions through their potential destruction of hydroelectric installations, road and rail infrastructure, farmland and housing and life loss.

Many glacial lakes are associated with the recession of debris-covered ice, but some of the most destructive floods have historically emanated from ice-dammed lakes formed by glacier advances, although these are locally specific, being confined to those areas where glaciers are expanding, or where glacier surges are common. The exact timing of an outburst event is difficult, if not impossible, to predict; triggers can be associated with continuous processes, e.g. filling due to melt, ice-cored moraine degradation, piping; or episodic events, e.g. intense rainfall, landslide or avalanche impact, earthquake destabilisation. Floods of $>10^6$ m^3 and discharges of several thousand cumecs are not uncommon, but are largely dependent on the size of the water reservoir and the nature of the failure mechanism. Knowledge of the trigger, failure mechanism and source lake geometry are all required if routing/propagation models are to be properly parameterised.

There remain many gaps in understanding; predictions of outburst and future lake formation are poorly constrained and the spatio-temporal complexity of their flow is difficult to replicate in a modelling environment, mainly because physically based models require significant computational time, and parameterisation data are difficult to acquire. Direct observations and measurements associated with glacier floods are few and there remains no definitive source compiling all previous events (in contrast to those from man-made dam failures, for example). Such historical data should be treated with caution when being used for model development and parameterisation, however, because future dynamic systems may differ significantly from those in the past. Recent studies have shown that glacial lakes are increasing in both number and size [64] and the likelihood of future outbursts is therefore similarly growing. Careful monitoring of such hazards will therefore become critical in future years.

References

1. SD Richardson, JM Reynolds, An overview of glacial hazards in the Himalayas. *Quaternary International*, **65–66** (2000), 31–47.
2. VR Baker, Global late Quaternary fluvial paleohydrology: with special emphasis on paleofloods and megafloods. In J Shroder, E Wohl (eds), *Treatise on Geomorphology*, Academic Press, San Diego, CA, vol. **9** (2013), pp. 511–527.
3. JL Carrivick, FS Tweed, Proglacial lakes: character, behaviour and geological importance. *Quaternary Science Reviews*, **78** (2013), 34–52.
4. H Björnsson, Understanding jökulhlaups: from tale to theory. *Journal of Glaciology*, **56** (2010), 1002–1010.
5. MJ Roberts, Jökulhlaups: a reassessment of floodwater flow through glaciers. *Reviews of Geophysics*, **43** (1) (2005), RG1002.
6. AJ Russell, MJ Roberts, H Fay, *et al.*, Icelandic jökulhlaup impacts: implications for ice-sheet hydrology, sediment transfer and geomorphology. *Geomorphology*, **75** (2006), 33–64.
7. JE O'Connor, JJ Clague, JS Walder, V Manville, RA Beebee, Outburst floods. In J Shroder, E. Wohl (eds), *Treatise on Geomorphology*. Academic Press, San Diego, CA, Vol. **9** (2013), pp. 475–510.
8. DI Benn, T Bolch, K Hands, *et al.*, Response of debris-covered glaciers in the Mount Everest region to recent warming, and implications for outburst flood hazards. *Earth Science Reviews*, **114** (2012), 156–174.
9. JA Kershaw, JJ Clague, SG Evans, Geomorphic and sedimentological signature of a two-phase outburst flood from moraine-dammed Queen Bess Lake, British Columbia, Canada. *Earth Surface Processes and Landforms*, **30** (2005), 1–25.
10. JE Costa, RL Schuster, The formation and failure of natural dams. *GSA Bulletin*, **100** (1988), 1054–1068.
11. O Korup, F Tweed, Ice, moraine, and landslide dams in mountainous terrain. *Quaternary Science Reviews*, **26** (2007), 3406–3422.
12. K Hewitt, J Liu, Ice-dammed lakes and outburst floods, Karakoram Himalaya: historical perspectives on emerging threats. *Physical Geography*, **31** (6) (2010), 528–551.
13. K Hewitt, Natural dams and outburst floods of the Karakoram Himalaya. *Hydrological Aspects of Alpine and High Mountain Areas*, **138** (1982), 259–269.
14. J Gardelle, E Berthier, Y Arnaud, Slight mass gain of Karakoram glaciers in the early twenty-first century. *Nature Geoscience*, **5** (2012), 322–325.
15. A Kääb, E Berthier, C Nuth, J Gardelle, Y Arnaud, Contrasting patterns of early twenty-first-century glacier mass change in the Himalayas. *Nature*, **488** (2012), 495–498.
16. C Haemmig, M Huss, H Keusen, *et al.*, Hazard assessment of glacial lake outburst floods from Kyagar glacier, Karakoram mountains, China. *Annals of Glaciology*, **55** (2014), 34–44.
17. BL Trüssel, RJ Motyka, M Truffer, CF Larsen, Rapid thinning of lake-calving Yakutat Glacier and the collapse of the Yakutat Icefield, southeast Alaska, USA. *Journal of Glaciology*, **59** (2013), 149–161.
18. P Klingbjer, Recurring jökulhlaups in Sälka, Northern Sweden. *Geografiska Annaler: Series A, Physical Geography*, **86** (2004), 169–179.
19. J Gulley, DI Benn, L Screaton, J Martin, Mechanisms of englacial conduit formation and implications for subglacial recharge. *Quaternary Science Reviews*, **28** (2009), 1984–1999.
20. SD Richardson, DJ Quincey, Glacier outburst floods from Ghulkin Glacier, upper Hunza Valley, Pakistan. *Geophysical Research Abstracts*, **11** (2009), 12871.

21. JJ Clague, SG Evans, Formation and failure of natural dams in the Canadian Cordillera. *Geological Survey of Canada Bulletin*, **464** (1994), 1–35.
22. R Worni, M Stoffel, C Huggel, C Volz, A Casteller, B Luckman, Analysis and dynamic modeling of a moraine failure and glacier lake outburst flood at Ventisquero Negro, Patagonian Andes (Argentina). *Journal of Hydrology*, **444–445** (2012), 134–145.
23. JJ Clague, SG Evans, A review of catastrophic drainage of moraine-dammed lakes in British Columbia. *Quaternary Science Reviews*, **19** (2000), 1763–1783.
24. B Bajracharya, AB Shrestha, L Rajbhandari, Glacial lake outburst floods in the Sagarmatha region. *Mountain Research and Development*, **27** (2007), 336–344.
25. D Vuichard, M Zimmermann, The 1985 catastrophic drainage of a moraine-dammed lake, Khumbu Himal, Nepal: cause and consequences. *Mountain Research and Development*, **7** (1987), 91–110.
26. MJ Roberts, Jökulhlaups: a reassessment of floodwater flow through glaciers. *Reviews of Geophysics*, **43**(1) (2005), RG1002.
27. JP Gunn, Hydraulic observations on the Shyok flood of 1929, Lahore, Pakistan. Government of Punjab, Irrigation Branch Paper #32 (1930).
28. SG Evans, JJ Clague, Recent climatic change and catastrophic geomorphic processes in mountain environments. *Geomorphology*, **10** (1994), 107–128.
29. JE Costa, Physical geomorphology of debris flows. In J Costa, P Fleisher (eds) *Developments and Applications of Geomophology*. Berlin: Springer (1984), pp. 268–317.
30. C Huggel, W Haeberli, A Kääb, D Bieri, SD Richardson, Assessment procedures for glacial hazards in the Swiss Alps. *Canadian Geotechnical Journal*, **41** (2004), 1068–1083.
31. JJ Clague, WH Mathews, The magnitude of jökulhlaups. *Journal of Glaciology*, **12** (66) (1973), 501–504.
32. JS Walder, JE Costa, Outburst floods from glacier-dammed lakes: the effect of mode of lake drainage on flood magnitude. *Earth Surface Processes and Landforms*, **21** (1996), 701–723.
33. F Ng, H Björnsson, On the Clague–Mathews relation for jökulhlaups. *Journal of Glaciology*, **49** (165) (2003), 161–172.
34. CI Thornton, MW Pierce, SR Abt, Enhanced predictions for peak outflow from breached embankment dams. *Journal of Hydrologic Engineering*, **16** (2011), 81–88.
35. B Ataie-Ashtiani, A Nik-Khah, Impulsive waves caused by subaerial landslides. *Environmental Fluid Mechanics*, **8** (2008), 263–280.
36. NJ Balmforth, J von Hardenberg, A Provenzale, R Zammett, Dam breaking by wave-induced erosional incision. *Journal of Geophysical Research*, **113** (2008), F01020.
37. NJ Balmforth, J von Hardenberg, RJ Zammett, Dam-breaking seiches. *Journal of Fluid Mechanics*, **628** (2009), 1–21.
38. MW Morris, M Hassan, A Kortenhaus, P Geisenhainer, PJ Visser, Y Zhu, Modelling breach initiation and growth. In P Samuels, S Huntington, W Allsop, J Harrop (eds) *Flood Risk Management: Research and Practice*. London: Routledge, (2009), pp. 581–591.
39. DL Fread, BREACH: An erosion model for earthen dam failures. Hydrologic Research Laboratory, Office of Hydrology, NWS, NOAA (1988).
40. DL Fread, NWS FLDWAV model: the replacement of DAMBRK for dam-break flood prediction. Hydrologic Research Laboratory, Office of Hydrology, NWS, NOAA (1993).
41. G Meon, W Schwarz, Estimation of glacier lake outburst flood and its impact on a hydro project in Nepal. In GJ Young (ed.), *Snow and Glacier Hydrology: Proceedings*

of a Symposium Held at Kathmandu, Nepal, in November 1992. IAHS (1993), pp. 331–339.

42. AB Shrestha, M Eriksson, P Mool, P Ghimire, B Mishra, NR Khanal, Glacial lake outburst flood risk assessment of Sun Koshi basin, Nepal. *Geomatics, Natural Hazards and Risk*, **1** (2010), 157–169.

43. R Osti, S Egashira, Hydrodynamic characteristics of the Tam Pokhari Glacial Lake outburst flood in the Mt. Everest region, Nepal. *Hydrological Processes*, **23** (2009), 2943–2955.

44. MAA Mohamed, PG Samuels, MW Morris, GS Ghataora, Improving the accuracy of prediction of breach formation through embankment dams and flood embankments. In D Bousmar, Y Zech (eds) *River Flow 2002: 1st International Conference on Fluvial Hydraulics*, Louvain-la-Neuve, Belgium, September 3–6, 2002.

45. JL Carrivick, Dam break: outburst flood propagation and transient hydraulics – a geosciences perspective. *Journal of Hydrology*, **380** (2010), 338–355.

46. JL Carrivick, AGD Turner, AJ Russell, T Ingeman-Nielsen, JC Yde, Outburst flood evolution at Russell Glacier, western Greenland: effects of a bedrock channel cascade with intermediary lakes. *Quaternary Science Reviews*, **67** (2013), 39–58.

47. DA Cenderelli, EE Wohl, Peak discharge estimates of glacial-lake outburst floods and 'normal' climatic floods in the Mount Everest region, Nepal. *Geomorphology*, **40** (2001), 57–90.

48. PA Carling, I Villanueva, J Herget, N Wright, P Borodavko, H Morvan, Unsteady 1D and 2D hydraulic models with ice dam break for Quaternary megaflood, Altai Mountains, southern Siberia. *Global and Planetary Change*, **70** (2010), 24–34.

49. JL Carrivick, 2D modelling of high-magnitude outburst floods: an example from Kverkfjöll, Iceland. *Journal of Hydrology*, **321** (2006) 187–199.

50. JL Carrivick, Hydrodynamics and geomorphic work of jökulhlaups (glacial outburst floods) from Kverkfjöll volcano, Iceland. *Hydrological Processes*, **21** (2007), 725–740.

51. JL Carrivick, R Jones, G Keevil, Experimental insights towards geomorphic processes within dam break outburst floods. *Journal of Hydrology*, **408** (2011), 153–163.

52. JL Carrivick, V Manville, A Graettinger, S Cronin, Coupled fluid dynamics–sediment transport modelling of a crater lake break-out lahar: Mt. Ruapehu, New Zealand. *Journal of Hydrology*, **388** (2010), 399–413.

53. R Worni, C Huggel, M Stoffel, Glacial lakes in the Indian Himalayas: from an area-wide glacial lake inventory to on-site and modeling based risk assessment of critical glacial lakes. *Science of the Total Environment*, **468–469** (2013), S71–S84.

54. T Bolch, M Buchroitner, J Peters, M Baessier, S Bajracharya, Identification of glacier motion and potentially dangerous glacial lakes in the Mt. Everest region/Nepal using spaceborne imagery. *Natural Hazards and Earth System Science*, **8** (2008), 1329–1340.

55. J Komori, Recent expansions of glacial lakes in the Bhutan Himalayas. *Quaternary International*, **184** (2008), 177–186.

56. ICIMOD, *Glacial Lakes and Glacial Lake Outburst Floods in Nepal*, Kathmandu: ICIMOD (2011).

57. JM Reynolds, On the formation of supraglacial lakes on debris-covered glaciers. In M Nakawo, CF Raymond, A Fountain (eds), *Debris-Covered Glaciers, Proceedings of a Workshop held at Seattle, Washington, U.S.A., September 2000*, Oxford: IAHS, (2000), pp. 153–161.

58. DJ Quincey, SD Richardson, A Luckman, *et al.*, Early recognition of glacial lake hazards in the Himalaya using remote sensing datasets. *Global and Planetary Change*, **56** (2007), 137–152.

59. H Frey, W Haeberli, A Linsbauer, C Huggel, F Paul, A multi-level strategy for anticipating future glacier lake formation and associated hazard potentials. *Natural Hazards and Earth System Science*, **10** (2010), 339–352.

60. M Carey, Living and dying with glaciers: people's historical vulnerability to avalanches and outburst floods in Peru. *Global and Planetary Change*, **47** (2005), 122–134.

61. Reynolds GeoSciences Ltd, Development of glacial hazard and risk minimisation protocols in rural environments, DFID project R7816. Available at www.bgs.ac.uk/research/international/dfid-kar/add046_col.pdf (last accessed April 16, 2014).

62. RJ McKillop, JJ Clague Statistical, Remote sensing-based approach for estimating the probability of catastrophic drainage from moraine-dammed lakes in southwestern British Columbia. *Global and Planetary Change*, **56** (2007), 153–171.

63. MJ Westoby, NF Glasser, J Brasington, MJ Hambrey, DJ Quincey, JM Reynolds, Modelling outburst floods from moraine-dammed glacial lakes. *Earth-Science Reviews*, **134** (2014), 137–159.

64. J Gardelle, Y Arnaud, E Berthier, Contrasted evolution of glacial lakes along the Hindu Kush Himalaya mountain range between 1990 and 2009. *Global and Planetary Change*, **75** (2011), 47–55.

65. JM Reynolds, A Dolecki, C Portocarrero, The construction of a drainage tunnel as part of glacial lake hazard mitigation at Hualcán, Cordillera Blanca, Peru. In J Maund, M Eddleston (eds) *Geohazards in Engineering Geology*. London: Geological Society (1998), pp. 41–48.

66. A Emmer, V Vilímek, J Klimeš, A Cochachin, Glacier retreat, lakes development and associated natural hazards in cordilera blanca, Peru, In W Shan, Y Guo, F Wang, H Marui, A Strom (eds) *Landslides in Cold Regions in the Context of Climate Change*, Berlin: Springer (2014), pp. 231–252.

67. W Haeberli, A Kääb, A Mühll, D Vonder, P Teysseire, Prevention of outburst floods from periglacial lakes at Grubengletscher, Valais, Swiss Alps. *Journal of Glaciology*, **47** (2001), 111–122.

68. B Hubbard, A Heald, JM Reynolds, *et al.*, Impact of a rock avalanche on a moraine-dammed proglacial lake: Laguna Safuna Alta, Cordillera Blanca, Peru. *Earth Surface Processes and Landforms*, **30** (2005), 1251–1264.

69. B Rana, AB Shrestha, JM Reynolds, R Aryal, AP Pokhrel, KP Budhathoki, Hazard assessment of the Tsho Rolpa Glacier Lake and ongoing remediation measures. *Journal of Nepal Geological Society*, **22** (2000), 563–570.

70. A Kääb, C Huggel, S Barbero, *et al.*, Glacier hazards at Belvedere Glacier and the Monte Rosa east face, Italian Alps: processes and mitigation. *Proceedings, Interpraevent*, **1** (2004), 67–78.

71. C Vincent, S Auclair, E Le Meur, Outburst flood hazard for glacier-dammed Lac de Rochemelon, France. *Journal of Glaciology*, **56** (2010), 91–100.

72. L Canelli, AM Ferrero, M Migliazza, A Segalini, Debris flow risk mitigation by the means of rigid and flexible barriers: experimental tests and impact analysis. *Natural Hazards and Earth System Science*, **12** (2012), 1693–1699.

73. D Schneider, C Huggel, A Cochachin, S Guillén, J Garcia, Mapping hazards from glacier lake outburst floods based on modelling of process cascades at Lake 513, Carhuaz, Peru. *Advances in GeoSciences*, **35** (2014), 145–155.

74. M Eriksson, J Fang, J Dekens, How does climate change affect human health in the Hindu Kush-Himalaya region? *Regional Health Forum*, **12** (2008), 11–15.

75. W Xin, L Shiyin, G Wanqin, X Junli, Assessment and simulation of glacier lake outburst floods for Longbasaba and Pida Lakes, China. *Mountain Research and Development*, **28** (2008), 310–317.

76. GKC Clarke, WH Mathews, Estimates of the magnitude of glacier outburst floods from Lake Donjek, Yukon Territory, Canada. *Canadian Journal of Earth Sciences*, **18** (1981), 1452–1463.
77. FS Tweed, AJ Russell, Controls on the formation and sudden drainage of glacier-impounded lakes: implications for jökulhlaup characteristics. *Progress in Physical Geography*, **23** (1999), 79–110.
78. MJ Westoby, The development of a unified framework for low-cost Glacial Lake Outburst Flood hazard. PhD thesis. Aberystwyth University, (2013).

13

Ecosystem change in high tropical mountains

KENNETH R. YOUNG

13.1 Introduction

High tropical mountains support biologically diverse ecosystems, have long human histories, neighbor on or include large rural and urban populations, and are currently being altered by glacier retreat. The cryosphere of tropical and subtropical latitudes is located in high mountains, often on isolated peaks that rise above diverse vegetation types in East Africa, New Guinea, and Mesoamerica, but it is also associated with large mountain chains and plateaus, particularly in the Andes Mountains where topography allows for the existence of hundreds of mountain glaciers, especially in Peru and Bolivia.

As noted elsewhere in this volume, the high elevations and high latitudes of the world are undergoing dramatic environmental alterations forced by global climate change; see also [1–3]. High tropical mountains change due to warming temperatures, altered precipitation regimes, and negative glacial mass balances [4–6]. However, these are also montane landscapes shaped in part by their human and biodiversity legacies, which include (1) alterations in land cover made over centuries by the activities of pastoralists and farmers [7,8]; and (2) changes in today's water resource availability that involve not only discharge in glacier-fed streams and affect local land use, but that are further shaped by interactions with social actors demanding water for growing cities, electricity production, expanding mines, and for irrigated agriculture [9]. Ecologically, these high-mountain areas have a 365 days per year growing season for plants, with cold extremes at night, rather than during a long winter season, with many implications for species adaptations, for biodiversity patterns, and for ecosystem processes [10–12]. High relief produces steep environmental gradients, dramatic changes in vegetation structure in response to topographic features, and disjunctions in biota at major biogeographical barriers [13–15].

The High-Mountain Cryosphere, ed. Christian Huggel, Mark Carey, John J. Clague and Andreas Kääb. Published by Cambridge University Press. © Cambridge University Press 2015.

The goal of this chapter is to evaluate how these physical, ecological, and social processes interact to drive ecosystem change in landscapes of the tropical cryosphere. It updates and expands on Young [16], who used landscape ecology as a disciplinary framework to evaluate the biodiversity implications of change in high tropical mountains. This is done by summarizing some of the effects of glacier retreat on ecological processes, directly through altering substrates and moisture conditions, and indirectly by changing human land use, sometimes in places remote from glaciers. Ecological change occurs as primary succession on newly exposed substrates, and as shifts in locations of wetlands, of altitudinal treelines, and in the abundances of woody plants. This chapter has a primary focus on the high elevations nearest the cryosphere, but also mentions concomitant ecological changes occurring at lower elevations. It concludes by comparing the respective contributions of biophysical and human-caused drivers of change. Most of the examples and discussions of landscape dynamism are based on fieldwork and experiences in the Andes of Peru, although mention is made of other tropical sites when feasible.

13.2 Glacier retreat

Pastoruri Glacier was the most accessible mountain glacier for visitors to Peru's Huascaran National Park [17], and supported a nascent recreation industry based on trekking, skiing, and sledding (Figure 13.1). In 2013, the tour agencies were instead offering day trips oriented around a climate-change theme. Ecologically, the once nival landscape had become mostly a series of mounds of rocky substrates

Figure 13.1 Approach to Pastoruri Glacier in the Cordillera Blanca of north-central Peru.
Source: photograph by Molly Polk.

of different sizes, dissected by dozens of small streams. Landscape ecologists refer to the dominant land cover type as the "matrix," with other isolated cover types in the landscape recognized as habitat patches, unless they are long and narrow and provide connectivity, in which case they are viewed as habitat corridors [16,18]. Thus, the Pastoruri landscape consists of a landscape matrix that was transformed from snow and ice to one dominated by rocks and cobbles, with moraines and tills left by the receding ice, and dotted with small patches of remnant ice or of colonizing plants, and corridors formed by streams and emerging peatlands.

Glaciers constantly change, with snowfall increasing their mass and the loss of ice to melting or sublimation reducing that mass. Gravity shifts mountain glaciers downhill, creating tension crevasses on the surface and bedrock erosion below [19,20]. That motion also generates a conveyer-belt type movement of particles picked up by the glacier, which get carried downward to the snout of the glacier. The rates of these various erosive and depositional processes will be modulated by whether the glacier is growing, retreating or stable, which in turn is a function of, respectively, a positive mass balance, a negative one, or an equilibrium in the input–output of water to the glacier, connecting to ecosystem responses (e.g. [21]). For the past several decades, the glaciers of the tropical cryosphere have had a negative mass balance [22–24].

The glaciers and associated ecosystems of tropical mountains show an inherent east–west asymmetry [25,26], with moist air masses coming in an easterly direction, bringing moisture from the humid lowlands. Seasonal shifts in precipitation track movements of the Intertropical Convergence Zone, in some places augmented by monsoonal circulations [27,28]. Often the topographic aspects that are sheltered from these influences or that are located on the leeward side of mountains have smaller extents of ice, as snow input is less. These aspect-related differences have consequences for glacier loss. For example, Mark and Seltzer [29] found that for glaciers on the Queshque massif in the Cordillera Blanca of Peru the most extensive melting was on southwest aspects and the least on east-facing ones. Over millennia the valley bottoms fill in with glacial tills further buried under rocky debris, creating low-gradient slopes and relatively flat-bottomed valleys: the classic U-shaped post-glacial valleys (Figure 13.2). The underlying edaphic heterogeneity causes spatial variation in vegetation.

Many additional ecosystem consequences below the snowline are mediated through indirect land use/land cover changes or are associated with increased water flows from the initial melting process, some of which may move through the valley system catastrophically [30]. Total annual stream discharge will tend to drop after reaching a peak, perhaps several decades under negative mass balance [31], as has been recorded for most of the watersheds of the Cordillera Blanca [9] (see also Chapter 11 in this book). Because dry season flow is sustained to a great

Figure 13.2 U-shaped valley in the Cordillera Blanca of Peru, inside Huascaran National Park. Shown also are valley bottom wetlands and high Andean forests on the slopes.
Source: photograph by Kenneth Young.

extent by glacial meltwater in seasonal tropical climates, streams not influenced by the glaciers will have low discharges or even be intermittent in flow during dry seasons. Thus, glacier retreat removes not only the millennial-scale environmental subsidy provided by having precipitation from previous centuries entering the downhill system, but it also eliminates at least part of the seasonal subsidy present when dry season flow is maintained by hydrological connectivity to the glacier. Predictions of future landscape dynamism can be made by comparing glaciated landscapes to others. Postigo and others [32] used Landsat satellite imagery to quantify change of high elevation Huancavelica in central Peru from 1990 to 2000, when virtually all of the small mountain glaciers of the study region had disappeared. The respective landscapes went from matrices of glaciers and tropical alpine vegetation types, to expanded lakes, wetlands, and grasslands.

13.3 Ecological change

Species move within changing landscapes in relationship to their habitat needs and dispersal abilities. Plants with wind-dispersed seeds and animals with wings, such as birds and some insects, can move easily over many landscape features; their successful colonization of those sites, however, will require the presence of the necessary substrate, soil moisture, temperature regime, and solar radiation for the plants, and the needed habitat structure and food available for the animal. Other species will have more limited dispersal; also the occurrence of a species may depend on the presence of other species, and hence it would be limited to places where those other species already exist to serve as shelter, nutrients, or other basic needs for the new colonist. Because there are always disturbances, some of that species sorting is also related to post-disturbance ecological succession. Global climate change adds additional directional shifts in temperature and humidity parameters [3]. Regional modeling by Tovar and colleagues [33] suggests that many sites in the tropical Andes will no longer support glaciers, periglacial zones, or tropical alpine vegetation in the future. Much of the tropical cryosphere will be in ecological flux.

Those current and likely future environmental shifts are acting upon assemblages of species organized over ecological time but that evolved over millennia to millions of years, that is, in evolutionary time. The vascular plants to be found today are often small statured, and frequently have leaves protected by thick cuticles and/or abundant hairs [10,11]. There are also cushion plants and small shrubs that keep photosynthetic tissue near to the ground. Another common plant life-form is that of the graminoids, that is the grasses and sedges, which grow from their bases and use narrow leaves for photosynthesis. These are characteristics that allow survival despite the nightly freezing above 4000 m elevation and often persistent winds. Other plants of the high mountains are large statured, like the rosette-forming plants *Puya raimondii* of Peru and Bolivia or the *Lobelia* species of highland Kenya, and there are some trees that can survive in exposed high elevations, such as *Polylepis* and *Buddleja* species; it is assumed that these taller plant forms are surviving under severe constraints on photosynthesis [11,34]. Although the native species are relatively diverse, they often do not have adaptations for persistent snow. Exceptions include, for instance, a finch *Diuca speculifera* [35], which builds its nests on the lower end of the Quelccaya Ice Cap in southern Peru. Instead, most tropical species are limited in their distributions at uppermost elevations by the permanent snowline.

The predominant vegetation type below the snowline and above the timberline, typically forming the landscape matrices, are grasslands of the puna from Peru to the south, and páramo in the northern Andes [15], extending also in isolated areas in Mesoamerica [36]. Wetter substrates are frequently covered by sedges. Similar vegetation formations occur in these same topographic positions in East Africa and New Guinea [11,37]. Close inspection of each landscape reveals ecosystem heterogeneity, along with substrates that vary from rocky slopes, to gravels, loams, and peatlands. These form vegetation mosaics (e.g., [38]) that are dynamic, with patches and corridors shifting as moisture changes, disturbances occur, and long-term edaphic and geomorphic change takes place.

13.3.1 Primary succession

Ecosystem recovery without an initial substrate is termed "primary succession" as the species that colonize the recently available site need to either prosper without soil or they include species that can initiate soil formation. This would be the kind of succession found on landslides and river cobbles of lower elevations, but also includes the ecological systems occurring in sites opened by glacier retreat [39]. The rate of ecological succession is set in part by the rate of soil formation. Typically this is a slow process, represented in a standard textbook example by lichens growing on rocks over centuries.

Observations in southern Peru suggest a somewhat more complicated and rapid scenario is likely. Figure 13.3 shows primary succession near the receding Quelccaya Ice Cap, which begins not on pure glacial till, but on a substrate that also includes the much smaller grain size of glacial loess originally deposited as dry season dust on the snow over centuries; located further from the ice margins are large wetlands with organic soils. In addition, this ice cap is renowned for the exposure of well-preserved plant materials at least 5200 years old by the ice's retreat [22], in many cases with the plants still rooted into their original substrate (Figure 13.4): Apparently the glacier advanced rapidly five millennia ago in such a way that it left the plants preserved intact under the ice, until recent glacial loss has exposed them. One result is that a small amount of organic matter is also available for the species colonizing the new substrate from this subfossil input. The species of vascular plants found during primary succession on Quelccaya include grasses (Poaceae: *Anthochloa*, *Calamagrostis*, *Dielsiochloa*), composites (Asteraceae: *Baccharis*, *Perezia*, *Senecio*), mustards (Brassicaceae: *Draba*), pinks (Caryophyllaceae: *Colobanthus*, *Pycnophyllum*), and valerians (Valerianaceae: *Valeriana*).

Figure 13.3 Quelccaya Ice Cap landscape in southern Peru, with retreating glacier limits, newly exposed substrate, and wetlands in topographic depressions.
Source: photograph by Kenneth Young.

Figure 13.4 Subfossil plant material belonging to *Distichia muscoides*, exposed by recent glacier retreat, is here inspected by Blanca León and Lonnie Thompson on the edge of the Quelccaya Ice Cap in southern Peru.
Source: photograph by Kenneth Young.

Researchers elsewhere on Peru's Vilcanota Cordillera, however, have shown that the more important first stage of post-glacial succession belongs to fungal, algal, and microbial organisms, which begin nitrogen cycling, photosynthesis, and biomass formation such that a living crust develops, probably within a decade or two. Vascular plants only become conspicuous on substrates that have several decades of exposure. Seimon and colleagues [40] describe these landscapes, mentioning newly formed ponds in deglaciated areas and plant species colonizing up to around 5500 m elevation. Schmidt and colleagues [41] examined periglacial soils there at 5400 m with temperatures at 5 cm depth that ranged from 27 °C in the day to –12 °C at night. They demonstrated nitrogen cycling mediated by fixation by *Nitrospira* nitrifiers and oxidation by Betaproteobaceria. Schmidt and colleagues [42] also showed that phosphorus is limiting microbial growth in these same substrates. Freeman and colleagues [43] found that the fungal diversity in water-saturated periglacial soils in this site were dominated by chytrids (Chytridiocycota), which also probably dominated the decomposition processes. The fungal communities in this site share some features with other periglacial sites worldwide, and all evidence points to the importance of fungi and microbes for initial ecological succession.

13.3.2 Treeline shifts

Treelines represent spatial limits to arboreal growth, such that they can refer to the upper (and sometimes lower) limit at which a tree-forming individual of a particular plant species may be found. Alternatively, there are also ecologically important upper (and lower) limits to where forests can be found in mountains, referred to as timberlines (Figure 13.2). Both kinds of tree limits are potentially important markers for assessing the effects of climate change [34], but they may be controlled in their spatial distributions by different sets of landscape controls [44]. The species treeline can reveal physiological limits to the survival of that species at high elevations, including both water and carbon constraints. Sometimes the highest individuals found are seedlings, whose presence suggests either future upward shifts for the species, or perhaps the incapacity of the species to reach maturity at those extreme sites despite successful dispersal; if the highest individuals are instead adult plants but without associated seedlings, then this might be a result of past establishment under milder conditions in the past. The timberline is different because forests alter their own microclimate to some extent, with lower wind speeds, higher relative humidity, increased litter inputs to surface soils, and lower solar radiation among the trees. Thus, forest altitudinal limits include these sets of moderating feedbacks, along with the influence of external disturbances such as frost events or fire, which may lower the timberline, as seen in the northern Peruvian Andes [44] and Mt. Kilimanjaro in Kenya [37].

Young and León [44] summarized some of the landscape implications of treeline/timberline dynamics for the Andes Mountains. They noted the hybrid nature of high tropical landscape mosaics that contain shifting species treelines and upward expansions of timberline with climate warming, but with the potential for individualistic treeline responses among the many different woody species, and the lowering of timberlines in places with an increased fire regime in the grassland matrix caused by burning by people. Theory would seem to suggest that forests should shift upslope with changing biophysical conditions [44–46], and indeed that can happen on the Eastern Cordillera of the Peruvian Andes [47]. Some forest expansion there is due to the timberline shifting upslope, but other cases are caused by shrub colonization into former grasslands and by woody plant expansion from protected microsites where relictual trees had survived. Bader and Ruijten [48] showed in Ecuador that this kind of timberline heterogeneity was related to interactions of elevation, topographic position, and aspect. But there also will be expectations of even more complex ecosystem shifts [33], for example on dry mountain slopes where lower tree and forest limits are set by soil moisture, such that intermontane valleys have shrubland matrices rather than extensive forests [7,15–16].

In fact, there are many reasons to expect treelines to shift with climate warming, but not necessarily timberlines, except with additional human impacts. For example, Byers [49] reported stable margins for high-elevation forests in the Cordillera Blanca, and the forest patches studied by Jameson and Ramsay [50] in southern Peru maintained their sizes for 50 years, although they lost canopy density. Most land cover loss of *Polylepis* forest in Ecuador, Peru, and Bolivia is caused by people, often as mediated through burning [51], with resulting loss of habitat for specialist high Andean forest species. Gosling and colleagues [52] use paleoecological information to point to biophysical conditions that constrain *Polylepis* species, suggesting that predictions for a warmer and drier Andes appear likely to further restrict places where these species could prosper in the future.

13.3.3 Shrubland expansion

Many tropical mountains have shrubland matrices where native forests are restricted to isolated patches, exotic tree plantations are found, and agricultural fields form patches of crops and fallow (Figure 13.5). Worldwide there is evidence that increasing amounts of carbon dioxide in the atmosphere may tip the competitive balance between woody plants with C3 photosynthesis and graminoids, often with C4 photosynthesis, towards the former, explaining trends of increased woody plant invasions into grasslands [53]. Other explanations of pervasive shrubland expansion include land-use drivers of change, including release from overgrazing and altered fire regimes [54]. Woody plants would presumably be more efficient under higher CO_2 levels in regards to their moisture use; once established, they tend to be taller than graminoids and so would shade them out. Naito and Cairns [55] show that shrub expansion in arctic environments is also driven in part by climate warming. Some encroachment elsewhere is due to the introduction of invasive woody species [56]. Eldridge and colleagues [54] report that shrub

Figure 13.5 Andean landscape in the Cordillera Blanca area of Peru. Most forest patches shown are plantings of non-native eucalypt trees.
Source: photograph by Kenneth Young.

encroachment tends to change ecosystem functions, with increases in soil carbon and nitrogen, while grass cover and soil pH decrease.

The expectation of an expansion of woody plants in high tropical mountains is supported by remote sensing and modeling studies, and appears to involve all of these possible mechanisms. Tovar and colleagues [33] used modeling to suggest dramatic shifts in ecological zones in the future, with expansions in environments supporting montane shrublands and seasonally dry forest; recent mapping using MODIS data found many dry and steep areas with reforestation or the expansion of woody plants over the last ten years [57]. There are increases in woody plants in former grasslands in the Peruvian Andes [47,58] as detected in Landsat imagery. Given these lines of evidence, it would be important for researchers to also examine the human dimensions of change in tropical shrublands.

13.3.4 Wetlands and aquatic systems

With the initial pulse of water moving down off of retreating glaciers, and an early increase in water availability based on the hypothetical glacier-retreat hydrograph [59], lakes and wetlands adjacent to the glacier should increase in size. This can be seen for the Cordillera Blanca, where Lipton [58] noted an increase in both proglacial lakes and wetlands in Landsat imagery from 1989 to 2001. Again, following the expectation of the non-linear hydrograph, after some time there should be a reduction in water availability, and indeed Bury and others [9] found that for one valley studied in detail the wetlands showed a decrease of 17% in area from 2000 to 2011, giving independent evidence that peak water from glacier retreat had already passed, at least in this particular location.

These kinds of landscape changes in extent and location of wetland and aquatic ecosystems will have implications for the plant and animal species that require those kinds of ecological spaces. For example, in the Vilcanota Cordillera, Seimon and colleagues [40] report that current climate change is causing the retreating ice to expose an ice-free corridor that allows terrestrial species to move across the divide, while expanding aquatic environments. They also showed the related upward shifts of distributions of three frog species, and similar shifts in the chytrid fungus that kills those same amphibians [40].

Over hundreds of years, proglacial lakes fill in with organic sediments, replacing the inorganic sediments that characterized the Late Pleistocene to Early Holocene [60]. Buytaert and Beven [61] used hydrological models to elucidate the temporal and spatial characteristics of high tropical watersheds, showing stores of water that augment dry season flows. There are also many implications for ecosystem services [9,62], as the old carbon stored in wetland peats represents photosynthetic products from decades to centuries ago. High water levels keep the organic matter in an anoxic state, slowing decomposition rates and storing the carbon within the substrate and the ecosystem. Drying of wetlands and lakes would increase carbon dioxide inputs to the atmosphere.

The development of streams at high elevations has been studied in the tropics by researchers who found a surprisingly high diversity of aquatic invertebrates, whose species composition is affected by distance from glacier influences. Kuhn and others [63] examined the downstream changes in stream channel stability, temperature, and substrate type as a function of distance from an actively retreating glacial margin in highland Ecuador, thereby using space as an analytical substitute for successional time. They link physical and chemical changes in the water to increases in the density and diversity of aquatic invertebrates. Boyero and colleagues [64] singled out the tropical stream detrivores as being particularly at risk with ongoing climate changes, given high local uniqueness in species composition, while Jacobsen and others [65] further demonstrated a potential loss of both alpha and beta diversity with glacial retreat, affecting particularly specialist aquatic species in tropical uplands. High-elevation drivers of change will cause changes far downstream in terms of river ecology through hydrological connectivity.

13.4 Land use responses

Land use alters ecosystem dynamics and landscape mosaics by creating new cover types, by increasing some cover types while decreasing others, and by changing the shapes, sizes, and locations of patches and corridors. Large-scale drivers of change may even alter the type of background matrix. All of these kinds of land-use related changes can be found in the high tropical mountains of the world, although the tropical Andes in particular have a long history of human-caused alterations in land cover, as related to land use [16], with ancient deforestation and with iconic and charismatic large fauna surviving nowadays only in nature reserves. Significant direct feedback from glacier recession to land cover/ecosystem change appears to be limited to the highest elevations; instead, most impacts will be from glacier to land cover as mediated through hydrological and human influences (Figure 13.6). Relatively little ecosystem ecology has been done in the inhabited and utilized landscapes of tropical mountains, so this is a topic needing more empirical studies. Note that some of the influences on land use in tropical mountains originate as distant or global socio-economic processes. For example, as discussed in Chapter 5, the values of ores and minerals drive many new mining operations in the Andes and elsewhere [66], which in turn affects water resource use and land cover; many new mining claims are being made in mountains affected by deglaciation.

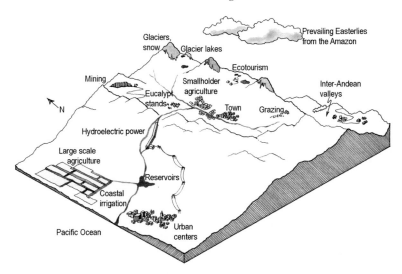

Figure 13.6 Representation of land cover changes in response to biophysical changes affecting glaciers, plus socio-economic changes altering land use and resource extraction. The figure is based on the situation in the River Santa drainage basin in Peru [9].
Source: designed by Molly Polk and drawn by Molly O'Halloran.

Changes in the cryosphere have many implications for people living downslope. As their livelihoods are changed, there are then feedbacks from human actions to altered land cover [67,68], some of which have repercussions for landscapes near and within the cryosphere. Probably the biggest influence comes from the combined actions of thousands of smallholders who manage multiple fields and pastures across a range of elevations [69]. They, and other social actors, affect land cover, with implications for ecosystems. For example, the retreat of glaciers opens up ecological space that can potentially serve as pasture and for establishing new agricultural fields. If so, then the switch from a nival to a tropical alpine landscape would include land use effects due to the altered grasslands and wetlands utilized by grazing livestock up to the snowline, and the insertion of patches corresponding to fields at lower sites. High Andean wetlands can be transformed into potato fields by farmers with hand tools [70]; wetlands expanding with the initial flow of water from glacier recession could be targeted for drainage and conversion to agriculture. Postigo [71] documents an interesting case near the Quelccaya Ice Cap wherein pastoralists have recently begun planting crops on lands that traditionally had only been used to produce products from their alpacas, llamas, and sheep, thus adapting their land use to take advantage of new possibilities.

The designation of lands for nature conservation is an important land use that is often done at a national level, although increasingly lands may also be set aside by regional and local governments. Historically, the highest elevations were often judged to have scenic value, but little usefulness for farming, making it politically feasible to establish national parks or their equivalents in high mountains. Climate change brings existential challenges to protected areas, as land use that threatens biodiversity may shift up into altitudinal zones originally judged to have negligible agronomic value. In addition, often protected areas receive political and economic credit for the ecosystem services they provide, which may be lessened under climate change conditions [62].

The world has been experiencing a general rural-to-urban shift (e.g. [72]) and high tropical mountains have also been affected (Figure 13.6). Large cities have grown in both lowland and highland sites. The ecological footprints of those urban areas reach far out into the countryside, diverting water resources from upslope, including hydroelectric energy produced by high-mountain water flows [9], and providing markets for commodities. Indirectly, cities alter land use goals of rural people, and hence more subtly change vast hinterland areas. For example, land use may intensify with increased markets, leading to the establishment of tree planta-tions and crop monocultures [57]. In other situations, land use may become less intense, with rural abandonment.

All these economic and demographic trends would increase extraction of natural resources: water, minerals from the subsurface [66], and timber and other forest products from the surface (e.g. [73]). Forest plantations are usually visible as isolated patches in landscapes (Figure 13.5); because many tree plantings are done with non-native species, there is a decrease in native biological diversity in these landscapes. Trees transpire more than grasses or shrubs, so tree plantings use more water, which may lead to less discharge in nearby streams [12]. Harden and colleagues [74], for example, showed afforestation in highland Ecuador altered soil moisture regimes, leaving less water available. As regards mining, open-pit mines leave behind a rocky matrix that may persist despite efforts to revegetate post-mine closure. Because the state typically controls access to subsurface resources [66], often land use of the surface will be negatively affected, for example if soil and vegetation are removed. Older mines have left a legacy of tailings in the Andes and other mineral-rich mountains, with continuing negative effects on water quality and human health [75], and causing the loss of sensitive aquatic invertebrates [76], and hence altered ecosystem functioning.

13.5 Conclusions

Change in the landscapes of the tropical cryosphere is only in part driven by climate shifts affecting the extent of snow, ice, and glacial features. Other import-ant land cover changes affect nearby vegetation, and also influence natural hazards and ecosystem properties and services, in some cases far downslope (e.g., Figure 13.6). The expectation for the biogeography of the future is not simple upslope movements of vegetation, fauna, and land use systems into areas once covered by snow and ice. Species that are controlled in their distributions by humidity rather than temperature may shift along changing precipitation gradients, rather than with elevation. Furthermore, those species must disperse and colonize across inhabited and utilized landscapes.

Ecosystems and landscapes change due to the interacting influences of all the drivers of change, namely atmospheric, climatic, ecological, and socio-economic processes. Assessments that include feedbacks are more likely to be able to predict complex trajectories of change, for example with thresholds or hysteresis. For example, demands for products may alter smallholder decisions, leading to land use changes that counteract or even accelerate changes due solely to climate change. As a result, socio-ecological feedbacks and interactions explain landscape dynamism, but may need to be examined at multiple scales, both temporal and spatial, to clarify the processes involved. Often social conflicts originate at the interface of contradictory demands for products or services from ecosystems and landscapes. In turn, some drivers are not only slow-acting, but may not easily change states, or once changed may be put on an irreversible trajectory.

Most high tropical mountains have direct human actions affecting the landscapes, whether through burning, grazing of domesticated livestock, mining, planting trees, or trekking (Figures 13.1, 13.5). A panorama of additional social and economic influences affect land use goals, for example whether landscapes are utilized for dairy or energy production or nature conservation or ores or wool (Figures 13.2, 13.6). If biophysical changes shift ecosystems upslope, then while land use will tend to track along with those changes, in fact, land tenure and access to that land may not be able to adaptively move along those same environmental gradients, or at the same velocity. Hobbs and others [77], for example, show that fragmentation of rangelands may limit adaptive strategies of pastoralists. Some kinds of institutional, legal, and political change can be much slower to materialize than ecological alterations.

Are the biophysical changes currently affecting high tropical mountains outside of the historical range of variability of these socio-ecological systems? If so, what does this mean for plants, animals, ecosystems, and people? Granted, the native species evolved over millions of years, and doubtless the mountain ecosystems assembled from their species have similarly changed over very long time spans in terms of the evolution of new traits, taxa, and ecosystem properties. However, these crucial questions must be addressed in reference to mountains nowadays that are altered in their land covers by people, with rates of temperature increases and losses of glacier mass significantly faster than many to be found in the geological record [3]. Mora and others [78] conclude that tropical countries will be the soonest to be exposed to the full consequences of climate change, based on shifting climatic variables. Research from the Alps suggests great risk for species extinctions [79]; the findings of Williams and colleagues [80] caution that novel climates and species assemblages are to be expected. Watershed approaches can link glacial and ecological change to resulting discharge and water resource availability [9,12,61]. Less tangible changes are best understood through the socio-ecological interactions that serve to connect distant economic and political decisions with the changing land cover of high mountains [9,67].

A long-term perspective on the ice-capped archipelagos of tropical mountains could be that biophysical change creates many opportunities for the formation of new species. In fact, speciation rates recorded for Andean plants such as the lupines (*Lupinus*) [81] and birds such as tanagers [82] are relatively high over evolutionary time. The reciprocal effects of lineage evolution and habitat accessibility over millions of years can explain the presence of biodiversity hot spots in particular places, such as in the tropical Andes, the highlands of Mesoamerica, and the mountains of Africa and New Guinea [83]. However, that innate capacity for innovation must be matched to the rates and kinds of rather different socio-ecological processes happening in today's and tomorrow's tropical landscapes. Some species extinctions are inevitable, affected by limited dispersal, habitat fragmentation, and habitat alterations; constraints are due to past land use, which altered landscape mosaics, and to present land use, which is tied to human decision making and economic goals [16].

The changes affecting species and landscapes also have ecosystem implications, including for environmental services that are directly or indirectly useful to people. Those services, in some cases, are recognized and even rewarded with subsidies and payments (e.g. [12,62]). The changes occurring in the high tropics thus have economic and political consequences. Historically, high mountains have been a hinterland, often with little representation in national and international policy making. The changing attributes of the tropical cryosphere establish important constraints, while environmental governance at local, regional, and global scales will need to adapt at the velocity of ecological and social change.

Acknowledgments

Funding for research in Peru has come from the National Science Foundation (CNH 1010381; DEB 1146446; GSS 1333141). Help with fieldwork and analyses were received from Asunción Cano, Blanca León, and Molly Polk, and helpful comments on the manuscript came from Mark Carey and several anonymous reviewers.

References

1. GM MacDonald, Global warming and the Arctic: a new world beyond the reach of the Grinnellian niche? *The Journal of Experimental Biology*, **213** (2010), 855–861.
2. AH MacDougall, CA Avis, AJ Weaver, Significant contribution to climate warming from the permafrost carbon feedback. *Nature Geoscience*, **5** (2012), 719–721.
3. NS Diffenbaugh, CB Field, Changes in ecologically critical terrestrial climate conditions. *Science*, **341** (2013), 486–492.
4. RS Bradley, FT Keimig, HF Diaz, DR Hardy, Recent changes in freezing level heights in the Tropics with implications for the deglacierization of high mountain regions. *Geophysical Research Letters*, **36** (2009), L17701. doi:10.1029/2009GL037712
5. R Urrutia, M Vuille, Climate change projections for the tropical Andes using a regional climate model: temperature and precipitation simulations for the end of the 21st century. *Journal of Geophysical Research*, **114** (2009), D02108. doi:10.1029/2008JD011021
6. I Mahlstein, R Knutti, S Solomon, RW Portmann, Early onset of significant local warming in low latitude countries. *Environmental Research Letters*, **6** (2011), 034009. doi:10.1088/1748-9326/6/3/034009
7. KR Young, Deforestation in landscapes with humid forests in the central Andes: patterns and processes. In *Nature's Geography: New Lessons for Conservation in Developing Countries*, KS Zimmerer, KR Young, eds. (Madison, WI: University of Wisconsin Press, 1998), pp. 75–99.
8. DW Gade, *Nature and Culture in the Andes* (Madison, WI: University of Wisconsin Press, 1999).
9. J Bury, BG Mark, M Carey, *et al.*, New geographies of water and climate change in Peru: coupled natural and social transformations in the Santa River watershed. *Annals of the Association of American Geographers*, **103** (2013), 363–374.
10. PW Rundel, AP Smith, FC Meinzer, *Tropical Alpine Environments: Plant Form and Function* (Cambridge: Cambridge University Press, 1994).
11. C Körner, *Alpine Plant Life: Functional Plant Ecology of High Mountain Ecosystems* (Berlin: Springer, 1999).
12. AG Ponette-González, E Marín-Spiotta, KA Brauman, KA Farley, KC Weathers, KR Young, Hydrologic connectivity in the high-elevation tropics: heterogeneous responses to land change. *Bioscience*, **64** (2014), 92–104.
13. F Vuilleumier, M Monasterio, eds., *High Altitude Tropical Biogeography* (Oxford: Oxford University Press, 1986).
14. KR Young, C Ulloa Ulloa, JL Luteyn, S Knapp, Plant evolution and endemism in Andean South America: an introduction. *Botanical Review*, **68** (2002), 4–21.
15. KR Young, B León, PM Jørgensen, C Ulloa Ulloa, Tropical and subtropical landscapes of the Andes Mountains. In *The Physical Geography of South America*, TT Veblen, KR Young, AR Orme, eds. (Oxford: Oxford University Press, 2007), pp. 200–216.
16. KR Young, Andean land use and biodiversity: humanized landscapes in a time of change. *Annals of the Missouri Botanical Garden* **96**, (2009), 492–507.
17. J Sevink, *The Cordillera Blanca Guide* (Peru: The Mountain Institute, 2009).
18. MG Turner, Landscape ecology: what is the state of the science? *Annual Review of Ecology and Systematics*, **36** (2005), 319–344.
19. DI Benn, DJA Evans, *Glaciers and Glaciation*, 2nd edn (London: Hodder Arnold, 2010).
20. G Kaser, H Osmaston, *Tropical Glaciers* (Cambridge: Cambridge University Press, 2002).

21. AG Fountain, JL Campbell, EAG Schuur, SE Stammerjohn, MW Williams, HW Ducklow, The disappearing cryosphere: impacts and ecosystem responses to rapid cryosphere loss. *BioScience*, **62** (2012), 405–415.
22. LG Thompson, E Mosley-Thompson, H Brecher, *et al.*, Abrupt tropical climate change: past and present. *Proceedings of the National Academy of Science*, **103** (2006), 10536–10543.
23. V Radić, R Hock, Regionally differentiated contribution of mountain glaciers and ice caps to future sea-level rise. *Nature Geoscience*, **4** (2011), 91–94.
24. A Rabatel, B Francou, A Soruco, *et al.*, Current state of glaciers in the tropical Andes: a multi-century perspective on glacier evolution and climate change. *The Cryosphere*, **7** (2013), 81–102.
25. KR Young, The tropical Andes as a morphoclimatic zone. *Progress in Physical Geography*, **13** (1989), 13–22.
26. IS Evans, NJ Cox, Global variations of local asymmetry in glacier altitude: separation of north–south and east–west components. *Journal of Glaciology*, **51** (2005), 469–482.
27. S Hastenrath, *Climate Dynamics of the Tropics* (Netherlands: Springer-Kluwer, 1991).
28. SE Metcalfe, DJ Nash, eds., *Quaternary Environmental Change in the Tropics* (Chichester: John Wiley, 2012).
29. BG Mark, GO Seltzer, Evaluation of recent glacier recession in the Cordillera Blanca, Peru (AD 1962–1999): spatial distribution of mass loss and climatic forcing. *Quaternary Science Reviews*, **24** (2005), 2265–2280.
30. M Carey, *In the Shadow of Melting Glaciers: Climate Change and Andean Society* (Oxford: Oxford University Press, 2010).
31. M Baraer, BG Mark, JM McKenzie, *et al.*, Glacier recession and water resources in Peru's Cordillera Blanca. *Journal of Glaciology*, **58** (2012), 134–150.
32. JC Postigo, KR Young, KA Crews, Change and continuity in a pastoralist community in the high Peruvian Andes. *Human Ecology*, **36** (2008), 535–551.
33. C Tovar, CA Arnillas, F Cuesta, W Buytaert, Diverging responses of tropical Andean biomes under future climate conditions. *PLoS ONE*, **8** (5) (2013), e63634. doi:10.1371/journal.pone.0063634
34. C Körner, *Alpine Treelines: Functional Ecology of the Global High Elevation Tree Limits* (Basel: Springer, 2012).
35. DR Hardy, SP Hardy, White-winged diuca finch (*Diuca speculifera*) nesting on Quelccaya Ice Cap, Peru. *The Wilson Journal of Ornithology*, **120** (208), 613–617.
36. JL Luteyn, *Páramos: A Checklist of Plant Diversity, Geographical Distribution, and Botanical Literature* (Bronx: New York Botanical Garden, 1999).
37. A Hemp, Climate change and its impact on the forests of Kilimanjaro. *African Journal of Ecology*, **47** (1) (2009), 3–10.
38. C Tovar, JF Duivenvoorden, I Sánchez-Vega, AC Seijmonsbergen, Recent changes in patch characteristics and plant communities in the jalca grasslands of the Peruvian Andes. *Biotropica*, **44** (2012), 321–330.
39. LR Walker, R del Moral, *Primary Succession and Ecosystem Rehabilitation* (Cambridge: Cambridge University Press, 2003).
40. TA Seimon, A Seimon, P Daszak, *et al.*, Upward range extension of Andean anurans and chytridiomycosis to extreme elevations in response to tropical deglaciation. *Global Change Biology*, **13** (2007), 288–299.
41. SK Schmidt, DR Nemergut, AE Miller, KR Freeman, AJ King, A Seimon, Microbial activity and diversity during extreme freeze–thaw cycles in periglacial soils, 5400 m elevation, Cordillera Vilcanota, Peru. *Extremophiles*, **13** (2009), 807–816.

42. SK Schmidt, CC Cleveland, DR Nemergut, SC Reed, AJ King, P Sowell, Estimating phosphorus availability for microbial growth in an emerging landscape. *Geoderma*, **163** (2011), 135–140.

43. KR Freeman, AP Martin, D Karki, *et al.*, Evidence that chytrids dominate fungal communities in high-elevation soils. *Proceedings of the National Academy of Science*, **106** (2009), 18315–18320.

44. KR Young, B León, Tree-line changes along the Andes: implications of spatial patterns and dynamics. *Philosophical Transactions of the Royal Society B: Biological Sciences*, **362** (2007), 263–272.

45. F-K Holtmeier, *Mountain Timberlines: Ecology, Patchiness, and Dynamics* (Dordrecht: Springer, 2009).

46. GP Malanson, LM Resler, MY Bader, *et al.*, Mountain treelines: a roadmap for research orientation. *Arctic, Antarctic, and Alpine Research*, **43** (2011), 167–177.

47. DB Kintz, KR Young, KA Crews-Meyer, Implications of land use/land cover change in the buffer zone of a national park in the tropical Andes. *Environmental Management*, **38** (2006), 238–252.

48. MY Bader, JJA Ruijten, A topography-based model of forest cover at the alpine tree line in the tropical Andes. *Journal of Biogeography*, **35** (2008), 711–723.

49. AC Byers, Contemporary landscape change in the Huascarán National Park and buffer zone, Cordillera Blanca, Peru. *Mountain Research and Development*, **20** (2000), 52–63.

50. JS Jameson, PM Ramsay, Changes in high-altitude *Polylepis* forest cover and quality in the Cordillera de Vilcanota, Peru, 1956–2005. *Biological Conservation*, **138** (2007), 38–46.

51. A Cierjacks, S Salgado, K Wesche, I Hensen, Post-fire population dynamics of two tree species in high-altitude *Polylepis* forests of central Ecuador. *Biotropica*, **40** (2008), 176–182.

52. WD Gosling, JA Hanselman, C Knox, BG Valencia, MB Bush, Long-term drivers of change in *Polylepis* woodland distribution in the central Andes. *Journal of Vegetation Science*, **20** (2009), 1041–1052.

53. SI Higgins, S Scheiter, Atmospheric CO_2 forces abrupt vegetation shifts locally, but not globally. *Nature*, **488** (2012), 209–213.

54. DJ Eldridge, MA Bowker, FT Maestre, E Roger, JF Reynolds, WG Whitford, Impacts of shrub encroachment on ecosystem structure and functioning: towards a global synthesis. *Ecology Letters*, **14** (2011), 709–722.

55. AT Naito, DM Cairns, Patterns and processes of global shrub expansion. *Progress in Physical Geography*, **35** (2011), 423–442.

56. PW Rundel, IA Dickie, DM Richardson, Tree invasions into treeless areas: mechanisms and ecosystem processes. *Biological Invasions*, (2014). doi: 10.1007/s10530-013-0614-9

57. TM Aide, ML Clark, HR Grau, *et al.*, Deforestation and reforestation of Latin America and the Caribbean (2001–2010). *Biotropica*, **45** (2013), 262–271.

58. JK Lipton, Human dimensions of conservation, land use, and climate change in Huascaran National Park, Peru. PhD dissertation, University of Texas, 2008).

59. M Baerer, BG Mark, JM McKenzie, *et al.*, Glacier recession and water resources in Peru's Cordillera Blanca. *Journal of Glaciology*, **58** (2012), 134–150.

60. ND Stansell, DT Rodbell, MB Abbott, BG Mark, Proglacial lake sediment records of Holocene climate change in the western Cordillera of Peru. *Quaternary Science Reviews*, **70** (2013), 1–14.

61. W Buytaert, K Beven, Models as multiple working hypotheses: hydrological simulation of tropical alpine wetlands. *Hydrological Processes*, **25** (2011), 1784–1799.

62. W Buytaert, F Cuesta-Camacho, C Tobón, Potential impacts of climate change on the environmental services of humid tropical alpine regions. *Global Ecology and Biogeography*, **20** (2011), 19–33.

63. J Kuhn, P Andino, R Calvez, *et al.*, Spatial variability in macroinvertebrate assemblages along and among neighbouring equatorial glacier-fed streams. *Freshwater Biology*, **56** (2011), 2226–2244.

64. L Boyero, RG Pearson, D Dudgeon, *et al.*, Global patterns of stream detritivore distribution: implications for biodiversity loss in changing climates. *Global Ecology and Biogeography*, **21** (2012), 134–141.

65. D Jacobsen, AM Milner, LE Brown, O Dangles, Biodiversity under threat in glacier-fed river systems. *Nature Climate Change*, **2** (2012), 361–364.

66. AJ Bebbington, JT Bury, eds., *Subterranean Struggles: New Geographies of Extractive Industries in Latin America* (Austin, TX: University of Texas Press, 2013).

67. KR Young, Stasis and flux in long-inhabited locales: change in rural Andean landscapes. In *Land-Change Science in the Tropics: Changing Agricultural Landscapes*, A Millington, W Jepson, eds. (New York: Springer, 2008), pp. 11–32.

68. KR Young, Change in tropical landscapes: implications for health and livelihoods. In *Ecologies and Politics of Health*, BH King, KA Crews, eds. (New York: Routledge, 2013), pp. 55–72.

69. KR Young, JK Lipton, Adaptive governance and climate change in the tropical highlands of western South America. *Climatic Change*, **78** (2006), 63–102.

70. KS Zimmerer, Wetland production and smallholder persistence: agricultural change in a highland Peruvian region. *Annals of the Association of American Geographers*, **81**, (1991), 443–463.

71. JC Postigo, Responses of plants, pastoralists, and governments to social environmental changes in the Peruvian Southern Andes. PhD dissertation, University of Texas, 2012).

72. NL Álvarez-Berríos, IK Parés-Ramos, TM Aide, Contrasting patterns of urban expansion in Colombia, Ecuador, Peru, and Bolivia between 1992 and 2009. *Ambio*, **42** (2013), 29–40.

73. KR Young, Wildlife conservation in the cultural landscapes of the central Andes. *Landscape and Urban Planning*, **38** (1997), 137–147.

74. CP Harden, J Hartsig, KA Farley, J Lee, LL Bremer, Effects of land-use change on water in Andean páramo grassland soils. *Annals of the Association of American Geographers*, **103** (2013), 375–384.

75. JC Postigo, M Montoya, KR Young, Natural resources in the subsoil and social conflicts on the surface: perspectives on Peru's subsurface political ecology. In *Subterranean Struggles: New Geographies of Extractive Industries in Latin America*, AJ Bebbington, JT Bury, eds. (Austin, TX: University of Texas Press, 2013), pp. 223–240.

76. RA Loayza-Muro, R Elías-Letts, JK Marticorena-Ruíz, *et al.*, Metal-induced shifts in benthic macroinvertebrate community composition in Andean high altitude streams. *Environmental Toxicology and Chemistry*, **29** (2010), 2761–2768.

77. NT Hobbs, KA Galvin, CJ Stokes, *et al.*, Fragmentation of rangelands: implications for humans, animals, and landscapes. *Global Environmental Change*, **18** (2008), 776–785.

78. C Mora, AG Frazier, RJ Longman, *et al.*, The projected timing of climate departure from recent variability. *Nature*, **502** (2013), 183–187.

79. S Dullinger, A Gattringer, W Thuiller, *et al.*, Extinction debt of high-mountain plants under twenty-first-century climate change. *Nature Climate Change*, **2** (2012), 619–622.

80. JW Williams, ST Jackson, JE Kutzbach, Projected distributions of novel and disappearing climates by 2100 AD. *Proceedings of the National Academy of Sciences*, **104** (2007), 5738–5742.
81. C Hughes, R Eastwood, Island radiation on a continental scale: exceptional rates of plant diversification after uplift of the Andes. *Proceedings of the National Academy of Sciences*, **103** (2006), 10334–10339.
82. RE Sedan, KJ Burns, Are the northern Andes a species pump for Neotropical birds? Phylogenetics and biogeography of a clade of Neotropical tanagers (Aves: Thraupini). *Journal of Biogeography*, **37** (2010), 325–343.
83. N Myers, RA Mittermeier, CG Mittermeier, GAB da Fonseca, J Kent, Biodiversity hotspots for conservation priorities. *Nature*, **403** (2000), 853–858.

Part III

Consequences and responses

14

The honour of the snow-mountains is the snow

Tibetan livelihoods in a changing climate

HILDEGARD DIEMBERGER, ASTRID HOVDEN, AND EMILY T. YEH

14.1 Introduction

The ice and snows of the Himalayas and the Tibetan Plateau have been widely perceived as indicators of the global phenomenon of anthropogenic climate change or global warming. Receding snowlines and retreating glaciers, powerfully illustrated by contrasting the images taken by early explorers and mountaineers with recent ones, often accompany articles on the subject. But what effect do these changes have on the livelihoods of those who live at the foot of these snow-capped mountains? How do they respond to ancient and new climate change induced hazards? And what do they think about what they see unfolding in their landscapes? This chapter addresses these questions through three case studies from Tibetan communities on both sides of the Himalayas.

Debates about vanishing ice and climate change have highlighted the importance of the cryosphere for human communities at the local, regional, and global levels. Whereas the notion of the cryosphere is commonly defined as the areas of the earth in which the water is in solid form, it does not only comprise ice and snow in multifarious forms, but also feedback loops that shape climate and ecosystems – and all of this is tightly entwined with human perceptions that drive interaction with the environment. Tibetan snow-mountains are not only topographical features, but are also regarded as lords of the territory and masters of the weather. Although based on notions of causality and attribution that may not coincide with models developed within the natural sciences, this has deep moral and political implications. A study of the effect of a changing climate on livelihoods in the cryosphere therefore necessitates a broad approach, taking social life and cultural understandings of how natural phenomena shape livelihoods and human responses to environmental hazards into account.

The High-Mountain Cryosphere, ed. Christian Huggel, Mark Carey, John J. Clague and Andreas Kääb. Published by Cambridge University Press. © Cambridge University Press 2015.

Livelihoods are often defined, following Chambers and Conway, as comprising 'the capabilities, assets (including both material and social resources), and activities required for a means of living' [1]. Livelihood approaches recognize that people draw upon a complex diversity of activities to make their livings, and examine how different strategies have affected livelihood trajectories [2]. Also central to livelihood analysis is a focus on households' access to assets and activities and on how this access is influenced by institutions and by social relations such as gender, class, kin, and belief systems [3,4]. Chambers and Conway define sustainable livelihoods as those in which households and communities can cope with and recover from various stresses and shocks, through their access to capacities, knowledge, and assets [1].

These assets are not merely material. As Tony Bebbington argues in discussing livelihoods in rural South America, 'peoples' assets are not merely means through which they make a living: they also give meaning to the person's world' [5]. Compared to technical assessments of coping and short-term adaptation, cultural and political dimensions of risks and hazards have been relatively neglected in the livelihoods literature [2]. Yet material loss is clearly not the only consequence of exposure to hazards. Perception and responses to environmental hazards that may be linked to climate change are mediated by culturally specific understandings, by the interface of different forms of environmental knowledge, and by the social practices that are seen as appropriate to address them. For Tibetans who live in these landscapes, mountains are not just assemblages of ice and rocks with the relevant vegetation and wildlife; they are the 'owners of the land' (Tib. *sadag, shibdag*)[1] that control what enables human livelihood in the widest sense. Snow and ice are the honour (Tib. *uphang*) of the snow-mountains while hazards such as exceptional sandstorms reflect inauspiciousness (Tib. *tendrel ngenpa*), which is predicated on the Buddhist understanding that all phenomena are interconnected and morally significant.

Exhortations from climate scientists to 'think globally, assess regionally and act locally'[6] point to the need to understand the different scales of the environmental phenomena entailed in climate change in order to be able to respond effectively. But there are often tensions between the scales, and many questions emerge when looking at specific vulnerable communities. How is causality and attribution of environmental hazards conceptualized? How do international ideas of risk map onto local ideas of risk and risk management? Which kind of knowledge informs local decision-making processes? Which coping strategies are deployed and what influences their effectiveness?

---

[1] Tibetan terms have been transliterated phonetically.

Addressing these questions, this chapter brings together highlights of separate studies conducted by the three authors – an anthropologist, a historian of religion, and a human geographer – each in conjunction with interdisciplinary teams. The experiential dimensions of risks and vulnerability to environmental hazards can only be partially captured by quantification and computer models, but are essential to understanding human responses to climate change. Thus, we demonstrate the importance of qualitative research methods for understanding differing cultural perspectives and social dynamics of climate change in the cryosphere. Furthermore, given the relative lack in the livelihoods literature of historical analyses of long-term changes in livelihoods [2], we pay particular attention to historical experiences with these phenomena. In the communities discussed here, the frequency and intensity of extreme weather events such as snowstorms, droughts, and glacial lake outburst floods (GLOFs) are predicted to increase (see also Quincey and Carrivick's chapter in this volume), but the events themselves are not qualitatively new. Households and communities thus draw at least partially on their historical memory and established socio-cultural practices to cope with these events, while also navigating new socio-political contexts.

The first case, based on work by Yeh and colleagues [7], shows the always present link between livelihoods, politics, and environmental change through a study of how the effectiveness of snowstorm coping strategies of pastoralists in the Nagchu area in the north of the present-day Tibet Autonomous Region (TAR) has been altered by broader political change. The second case, from Diemberger [8] on pastoralists and agriculturalists from the Himalayan slopes in the southwest of the TAR, focuses more on cultural understandings and historical references to cryosphere environmental hazards, particularly flooding from glacial rivers. The last case, drawing on Hovden's fieldwork among agropastoralists in a high-altitude Himalayan valley in western Nepal, examines attribution, risk assessment, and past and present strategies to deal with GLOFs that threaten a community and its monastery. The two cases in the TAR look at communities in which environmental strategies have been shaped by state intervention and radical Chinese land management reforms. The third case, on the other hand, focuses on a community that has been more left to its own devices by the administrative framework of the Nepalese state. In their responses to hazards, community resources are combined with what can be obtained from outside through administrative bodies and/or wider networks. This variously entails an increasing degree of dependency on national state structures and external agencies as well as deliberate attempts to counter this trend through self-reliance and resistance to interference (Figure 14.1).

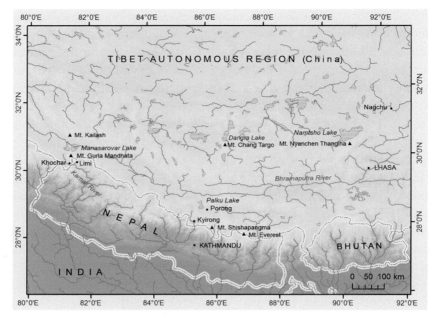

Figure 14.1 Map of the Himalayan region, showing the three field sites discussed in this chapter.
Source: J. Kropáček and A. Hovden.

14.2 Hazards and risks as cultural concepts

Efforts to promote a global governance of the responses to climate change have given rise to a distinctive vocabulary that currently shapes the dominant discourse on environmental hazards and human responses. Among these are the ideas of hazard and risk. The former is defined as the interaction of human exposure and vulnerability to natural extremes that cause adverse impacts, where vulnerability is recognized to be a result of 'historical, social, economic, political, cultural, institutional, natural resource, and environmental conditions and processes' [9]. Risks are defined in the international scientific epistemic community as the probability that a hazard will have a particular negative consequence. This probabilistic definition is not, of course, how people understand risks in their everyday lives, as a large body of scholarship on 'natural hazards' has shown. Furthermore, as pointed out by John Ash, divergent risk perception is not only a question of knowledge discrepancy, but is also largely dependent on trust, which in the Tibetan case is tightly interlinked with the moral framework within which the environment is perceived [10]. Some differences between risk perception and technical risk assessment are relatively constant across societies and cultures, while others are shaped by factors ranging from life history and age to cultural meanings and historical memories.

The English term 'risk' is translated into Tibetan with the same word as for 'danger' or 'hazard': *nyen kha*. The distinction between risk and hazard is thus not clear. At the same time, however, Tibetan language is extremely sophisticated in the use of verbal constructions to indicate different levels of probability of something happening and the word '*nyen*' itself can be used as a verb indicating the probability (risk) of something happening, e.g. *me shor nyen* meaning 'there is the risk that it may catch fire'. This grammatical construction leads to a range of expressions indicating risk-averse behaviour that implies a subject that calculates cause and effect determined on the basis of empirical knowledge. For example, '*khe chewa le nyen chungwa ga*' means 'it is better to take less risk than to go for greater profit' (literally 'than great profit, low danger/risk is better') [11]. On the other hand, ideas of risk and danger are linked to a wider understanding of karmic connections interlinking all phenomena as well as the relationship between the human community and the spirits inhabiting the natural environment. Rituals and morality in observing environmental management rules and social interconnectedness linked to a religious sense of community are therefore central to Tibetan understandings of risk and its management.

By highlighting the experiential dimension of the human response to hazards we move away from the propensity to describe social life as if ethics and beliefs were a superficial or illusory dimension of it. Instead, they are often at the very root of 'why we disagree about climate change' [12]. The case studies presented demonstrate how different understandings of causality, risk, and accountability linked to different forms of environmental knowledge coexist and interface in Tibetan communities living in the cryosphere managing their livelihoods and the hazards to which they are exposed.

14.2.1 The central Tibetan Plateau: pastoralists coping with snowstorms

Two great snow-covered mountains have shaped Tibetan perceptions of what is currently Nagchu Prefecture in the north of the TAR: Nyanchen Thanglha and Chang Targo, both brothers of the more famous Mt. Kailash, and married to adjacent holy lakes [13]. The two mountains are considered territorial deities in control of the weather. In fact, when Nyanchen Thanglha was first tamed and converted to Buddhism by the great tantric master Padmasambhava, he reportedly put up a fierce fight with snowstorms and thunder [14]. Climate science also suggests that these mountains and lakes together shape the local moisture circulation system and the interfaces with regional climatic patterns, especially the monsoon regime [8,15]. Throughout the history of human habitation on this harsh terrain livelihood has been strongly shaped by extreme events such as snowstorms and drought, the frequency and intensity of which are projected to increase with climate change, which is also expected to negatively affect rangeland health. Herders in Nagchu report that snowstorms are particularly challenging after drought, as the weakened livestock are more vulnerable to harsh winter conditions in which snow coverage and ice prevent access to forage.

Some of the most devastating snow disasters of the past have entered local oral tradition, as an official in Amdo County, Nagchu Prefecture explained:

Here we are particularly exposed because if clouds full of moisture reach the Tangula range [north of Mt. Nyanchen Thanglha], this is where most precipitation occurs. In the 1920s there was the most disastrous snowstorm we can remember: the snow covered everything, the animals were completely buried and when people went out to try to find their herds they were caught and frozen on the spot. As the snow melted the corpses remained standing on top of a pedestal of ice protected by their shadow [like *penitentes* in the Alps]. They were like horsemen on white horses.

This powerful warning image notwithstanding, herders in Nagchu have long had a variety of coping strategies to deal with snowstorms, the effectiveness of which are changing in new political-economic and policy contexts.

14.2.1.1 Coping strategies before 1959

Available historical sources provide little information about government responses to the snowstorms in Nagchu before the 1950s, though they do report seven severe snowstorms with high mortality rates between 1827 and 1956. A record from 1827 showed that the Tibetan government in Lhasa provided roasted barley flour (Tib. *tsampa*) to herders, whereas the Chinese Qing Empire provided some monetary support. Moreover, we found two accounts of the postponement and reduction of taxes in 1901 and 1927. Overall, though, the historical archive and oral history suggest that except for in the most extreme cases, there were few external interventions by the Chinese or Tibetan governments. Unfortunately, the lack of available records does not currently allow for a more nuanced picture of how local or tribe-based coping strategies may have changed over time, or what implications they had for power and inequality [7].

Oral histories suggest that the most important local strategies used by Nagchu herders in snowstorms involved mobility, storage of feed, and covers and shelter for livestock. As in all pastoral societies, mobility was central to communities' overall snowstorm coping strategies because it gave the herders the ability to move livestock to less affected areas during extreme snow events, while also improving livestock health and thus resilience throughout the year. Storage entailed the capacity to store and feed the animals grain, hay, and leftover food during times of scarcity. However, the ability to store grain and hay was limited and most households had little leftover food to spare. Finally, pastoralists covered weak animals with old sheepskin robes, and in areas that focused mainly on sheep herding, constructed pens to provide shelter from wind and low temperatures.

14.2.1.2 Changes in coping strategies from the 1950s until present

Rangeland use rights underwent dramatic transformations after the failed uprising in Lhasa in 1959 and the forcible implementation of 'democratic reforms'. Both livestock and grasslands were collectivized and then merged into communes; livestock became commune property and herders became commune workers. In the 1980s livestock were again privatized while grasslands remained in common use. Then beginning in the 1990s and into the 2000s, the Rangeland Household Responsibility System (RHRS), which contracts pasture use rights, generally to the scale of the household, was implemented. This has led to the fragmentation of land and an overall decrease in mobility and flexibility. In various parts of the Tibetan Plateau, there is also evidence that where household contracting has been implemented, it has contributed to increased strain on labour resources, increased conflicts, and weakening of the traditional reciprocal relationships between communities. All of these effects have disrupted the efficacy of traditional snowstorm coping strategies.

At the same time as new rangeland policies have decreased mobility, other political-economic and policy changes have strengthened other coping mechanisms. Storage capacity has been improved significantly since the abolition of rural taxation, dissolution of communes, and restoration of private livestock ownership from the 1980s onwards. Reserve fencing programmes and increased market access have led to increased availability of grain and fodder. The use of covers and shelters to protect livestock from cold temperatures has also become widespread, especially since the 1990s. However, all of these strategies are ultimately less significant than mobility in preventing livestock loss.

Pastoralism is labour intensive and the ability to employ different coping strategies successfully depends on labour availability. Demographic and socio-economic transformations including subsidized housing, part-time migration of family members to urban areas in search of income opportunities, and the enforcement of nine-year compulsory education have resulted in decreased labour availability within the household. The recent policy push toward centralized schools [16] has often forced one of the adult family members to move to the townships together with the children, further reducing household labour availability. Together with the fragmentation of rangeland use rights and past communal cooperation mechanisms, this has increased the vulnerability of the livestock to snowstorms (Figure 14.2).

Figure 14.2 Yaks grazing on snow-covered pasture after a snowstorm hit eastern Nagchu in early October 2009.
Source: photo courtesy of Yonten Nyima.

14.2.1.3 Implications for local livelihoods

Mobility and labour are critical in determining a successful response to an extreme event. Consider the case of a young herding household in eastern Nagchu in the snowstorm of 1989–1990. Whereas many other families in their village were able to split up and take their herds to different pastures, this household had three very small children, all too young to work. As a result, the entire family had to move together with their livestock in search of forage, rather than splitting up herds of different livestock according to their different needs. The sheep were not able to go very far, and the yaks had trouble finding forage. They therefore tried to cover them with blankets and feed them stored supplies. However, the household still lost 80% of their sheep and goats, and 89% of their yaks. As the young head of the household explained, 'You can imagine that even if we'd had all the food and covers in the world, we simply did not have enough capable people to give all the livestock such treatment' [7]. Herders understand their loss of livestock partially through idioms of fortune and karma, but simultaneously through the importance of labour and mobility relative to other strategies. This and other families also predict difficulties during future snowstorms because of the division of rangelands. Another herder stated:

Because rangeland has now been allocated to individual families, it will become more difficult for us to graze in other places. . . . Without the government arranging it, it will be hard for us to migrate.

[7]

Contemporary political-economic changes have led to declines in both labour power and mobility, which in turn increase herders' dependency on external support, particularly from the state, to cope with snowstorms. From a managerial point of view, this case study suggests that a finer scale, ethnographic understanding of Tibetan herders' historical and continuously adapting strategies to cope with snowstorms would improve well-being and make sustainable livelihoods in the Tibetan cryosphere more likely. At the same time, the observation that the various transformations and interventions to Tibetan livelihoods have led to greater dependence on the state – in the context of the extremely fraught relationships between Tibetans and the Chinese state over cultural and political rights – suggests the always cultural and political nature of adaptation and livelihoods vis-à-vis climate change.

14.2.2 The northern Himalayan slopes: historical perspectives and current challenges

Mt. Shishapangma, an 8000 m peak lying northwest of Mt. Everest, is a giant of rocks, ice, and snow. According to the son of a former official of the Tibetan government, it is known locally as Pholha Gomchen (lit. ancestral deity Great Meditator), but the more recent name 'Shishapangma' is interesting in its direct connection to the hazards that this powerful mountain god can bring to the human communities inhabiting its foothills. Literally translated it means 'dead meat' (Tib. *shisha*) and 'soaked [fields]' (Tib. *bangma*), a reference to the corpses of the animals killed by the large snowfalls and the flooded fields inundated by torrential rains and flash floods that characterize the cryosphere. At the same time, this holy mountain and the adjacent range are the sources of the most important rivers that enable human livelihood in a wide region. For centuries the nomads of Porong and the high-altitude farmers of Dingri and the Kyirong valley have inhabited the Himalayan slopes next to Mt. Shishapangma. The plain that stretches northwards from the main range is touched by the turquoise waters of Palku Lake and is crossed by nomads' pathways, ancient trading routes and pilgrims' walks, as well as the much more recent tarmac road connecting Kyirong County to the Qomolangma Highway.

Pastoralists and farmers in the region perceive changes in the crysopheric environment around them as an inauspicious omen. Porong Dawa observed:

Tsangla is almost completely black now because of glacier retreat and so are many snow mountains of Porong. The Khala pass that leads straight from western Porong to Kyirong, which was used by many pilgrims travelling to India, used to be surrounded by snow-mountains but now they have almost all turned to slate (*yari*).

Dawa Dargye, now living in Kathmandu but originally from the Penche area (traditionally under the Porong rulers), commented that the brother of the Tsanglha peak, the wrathful Takyong, always had a permanent white patch on the top. According to the elders, if this disappeared disaster would hit the community and its leadership. The snow on the snow-mountains is therefore not just a source of water and a factor in local weather patterns; it is also a powerful indicator of the overall well-being of the land and of political stability (see also [8]). Weather and water have been linked to Tibetan politics throughout the history of the 'Land of Snow' and the interpretation of environmental hazards at times contributed to regime change [13]. It is therefore not surprising that snow is celebrated as the honour of the snow-mountains in songs and poems, and addressed in rituals. For communities that have developed coping strategies over centuries, the ominous darkening of the peaks seems to announce an unpredictable future that may demand adaptation to unprecedented conditions with a wide range of consequences for their livelihoods, politics, and social organization.

14.2.2.1 Porong pastoralists coping with extraordinary snowstorms, sandstorms, and droughts

Like Nagchu, the Porong region has a history of snow disasters claiming the lives of many animals, but drought and sandstorms seem to have been an even more acute problem here, especially when drought preceded extraordinary snowfall. The narrative of 'climate change' (Tib. *namshi gyurba*) is sometimes deployed when talking about recent environmental hazards. However, the sense of 'change' is far from clear here. Often it is more a dismayed observation that weather is no longer quite 'right', and different forms of causality are entangled in the relevant narratives.

Dawa, a nomad from Porong, commented:

In the history of Porong, there were some very bad sandstorms. I think big sandstorms are symbols of inauspicious circumstances (*tendrel ngenpa*). Just before the death of Bodong Panchen [in 1451], an extraordinary windstorm set out from the Palkhu Lake, engulfed Porong, and affected the whole of Central Tibet [see 17].

Dawa was referring to a historical account that highlights what Huber and Pedersen have called the 'moral climate', according to which meteorological events are tightly linked to human morality and fortune [18]. Dawa continued:

In the 1980s, there was a really bad drought in Porong, the grassland was very much reduced, the water sources were depleted and many livestock were killed. More generally, compared to earlier times, rain became scarce, the animals became weak and the conditions of the grassland deteriorated significantly. . . . This could have been partially caused by the excess number of animals on the grassland.

These observations echoed comments made by many other Porong residents when carrying out rituals around desiccated springs or commenting on grassland conditions, that see recent socio-political transformations as connected simultaneously to material improvement and moral and environmental decline. Dawa's final comment, however, resonates with scientific research about the ways in which overgrazing can affect the local moisture circulation system and weather patterns on the Tibetan plateau [19]. Global and local processes of environmental change are often linked in feedback loops, and are difficult to disentangle [20]. Different ways of understanding causality and attributing responsibility may not only shape the actual responses, i.e. reduction of livestock numbers or asking for government assistance, but can also lead, as a large body of scholarship in political ecology has shown, to blame games between rural communities and distant urban centres, as well as interethnic tensions or even scapegoating within communities.

A member of the Porong community living in Switzerland, the late Tsering Damchoe, mentioned the same drought and the unprecedented sandstorm that then covered the pastures with a layer of sand; only after a couple of years following a snow-rich winter would the grass recover, and then only partially. His accounts of past snowstorm livestock losses and external support are similar to those told by herders in Nagchu. Drought preceding large snowfalls increased livestock mortality, and movement to least affected pastures was a key strategy for mitigating losses, particularly because the most severe snowfalls tend to be highly localized. As a result, moving livestock is the first and most important response to sudden extreme snowfalls. Further, he recalled that in the pre-1950s 'old society', not only was taxation sometimes suspended during environmental emergencies, but monasteries such as Porong Palmo Choeding (Figure 14.3) also acted as a source of food and assistance.

The traditional management of pastureland, called *khyusa*, entailed a high level of flexibility, a form of periodic land re-distribution that took into account the conditions of the pastures over time and allowed for porous boundaries [8,21,22]. Some of this flexibility persisted through administrative changes, but has been progressively reduced by land reforms and fencing [7,23]. As in Nagchu, mobility, pooling of resources, and skilful deployment of adequate labour are critical factors for mitigating livestock loss and thus maintaining livelihoods when snowstorms hit in Porong. These long-established strategies to cope with hazards have been rendered more difficult by recent transformations of the Tibetan countryside. Local

260   *H. Diemberger* et al.

Figure 14.3 Porong Palmo Choeding Monastery with Mt. Shishapangma in the background.
Source: photo by H. Diemberger.

governments and non-governmental organizations (NGOs) can now more easily distribute emergency food and clothing because of better communication, transportation, and availability of resources, but this entails greater external dependency (Figure 14.4).

Community members' direct experience also plays an important role in designing successful coping strategies, as responses to cryosphere hazards are a matter of trial and error and learning from past mistakes. Dawa commented that in a storm in 2013, residents learned from their experiences in 1995–1996, the biggest snow disaster in living memory, and were thus in a better position to make use of all of the resources available to them. The use of mobile shelters for livestock, the greater availability of trucks, socially shared knowledge and a thorough understanding of the landscape, and the timing of the snowfall vis-à-vis the location of livestock, all made a difference between the 2013 and 1995 events.

Rituals, which can engage both the invisible interconnectedness of phenomena and community social networks, are also seen as an essential part of risk management. They work at different levels and are not incompatible with local empirical knowledge and with scientific understandings of the environment coming from outside through government cadres and experts. Dawa explained that performing a *gangkhor*, a ritual that goes back to Domarpa, a great spiritual master born in this region in 1695, is one of the most efficacious means of managing risks by addressing the mountains. By performing offerings and cleansing the relationship between human beings and the spirits of their landscape, as well as, on a more properly Buddhist plane, promoting auspicious circumstances, rituals have both

Figure 14.4 Food deliveries after snow disaster in Porong.
Source: photo by H. Diemberger.

religious and social efficacy. Open to a variety of interpretations, they certainly have been part of the management of the uncertainty that is intrinsically part of pastoral life in the high-altitude environment and unsurprisingly have survived a wide range of religious and political reforms.

14.2.2.2 Coping with floods

Floods have historically been a well-known hazard in these areas, especially in the deep glacial river gorges. To the west of Porong, the valley that leads from Dzonkha towards Kyirong and further down into Nepal characteristically has villages located high above the river as flooding occurs regularly, especially during the monsoon. Glacial outbursts can bring about sudden flooding of unexpected proportion and at unexpected times. Historical sources such as biographies of spiritual masters give descriptions of flash floods and other weather events. For example, the biography of Chokyi Dronma, the fifteenth-century princess who was recognized as the first of the Samding Dorje Phagmo reincarnations, mentions a flash flood that occurred in the 1430s in the upper Kyirong Valley:

When the army arrived at Minkyu Dibma Dzong all frozen waters and snow melted suddenly and a big flash flood occurred in the gorge. Many stones were falling left and right from the steep slopes and everybody tried to escape. As the flood was about to reach the troops they seemed doomed. However, the young Vajra Queen, by pointing her finger, magically made the water flow upwards so that the army managed to proceed on its path without any damage.

[24]

How the army of the Gungthang king escaped the flash flood is not clear since the text is mainly preoccupied with celebrating the magic deed of the holy princess Chokyi Dronma. What is relevant here is that the localisation and the geographic detail in the passage seem to point to a real and sudden event beyond the standard narrative tropes of hagiography. In the areas above the locality described in the text are several side-valleys descending from snow-mountains with glaciers and permanent snowfields – presumably the source of the flash flood which was obviously unseasonal as the army would not have otherwise marched into the valley. Tibetan biographies and histories are interspersed with accounts like this, which are distant in time but represent a vivid historical memory of the possible hazards affecting these steep landscapes – events that may occur again, without a holy princess to save the day.

In this valley the location of villages and monasteries high enough to be out of reach of floods is probably part of the more general risk-management strategy. The construction of the new road at the bottom of the valley certainly presents a wide range of maintenance challenges, currently being addressed with modern engineering technology. Can information contained in local historical narratives contribute to the awareness of rare but not impossible events with potentially disastrous consequences? That is, can they help frame the perception of risk so as to take into account the possibility of 'outliers' that challenge risk assessment computation?

As we shall see again below, Buddhist responses to environmental risk, and more generally livelihood management in and around the cryosphere, have included not only culturally specific forms of ritual power but also 'skill' in dealing with the specific environment as well as engineering enterprises, including creation of water channels, bridges, and embankments.

14.2.3 The Nepal Himalaya: coping with flood risks to monasteries and livelihoods

At 7694 m asl, Gurla Mandhata towers above the frontier between western Tibet and Nepal. To the north, the massif faces the sacred Mt. Kailash; the western flank is drained by the Karnali River; the southern and eastern slopes are bordered by the Limi and Chakpalung valleys in Nepal's Humla district. Local people regard Gurla Mandhata as the abode of Drabli Gyalmo, 'Queen of the warrior deities', also referred to as Gangkar Lhamo, 'Goddess of the white snow'. In Tibetan the

Figure 14.5 GLOF in Halji, June 30, 2011.
Source: photo by A. Hovden.

mountain is called Memo Nagnyil; the Memo are female wrathful spirits constitut-
ing the retinue of the goddess, whereas Nagnyil means 'black fringes' [25],
referring to the many dark, eroded valleys carved out by meltwater streams from
the glacier.

In June 2004 a sudden outburst from a glacial lake in the Gurla Mandhata
Massif transformed a small glacial stream to the west of Halji village into a torrent.
The flood, which has repeated itself in almost every summer since, has destroyed
houses, watermills, paths, and bridges, as well as a large number of fields and
pasture. In the most recent flooding event of June 30, 2011, approximately
100 fields were washed away and another 100 rendered useless by debris and silt
(Figure 14.5). Halji, which is the largest village in Limi, is clustered around the
eleventh-century Rinchenling Monastery, one of the oldest Tibetan Buddhist
monasteries in Nepal [26]. The monastery is still standing, but the floods have
eroded so much of the riverbank that it is currently only 20 m from the monastery
walls. These floods have been too small to fit into the large grids of the national
surveys and GLOF vulnerability in Humla is still ranked as 'very low' [27], but the
recurring floods constitute a serious threat to local livelihoods.

14.2.3.1 Flood management in historical accounts

Because Halji village is located at an altitude of 3740 m in a precarious natural
environment, its residents have a long history of adapting to changing weather
conditions and coping with natural hazards such as landslides, untimely precipitation, snowstorms, and glacial floods. Detailed information about management of

natural resources is recorded in administrative documents, and accounts of hazards and local coping strategies are found in oral histories, biographies, monastic catalogues, chronicles, and other historical works from the region.

The sixteenth-century *Catalogue of the statue of the three silver brothers* from neighbouring Khochar provides a particularly rich example of flood management [28]. The tenth-century Khochar Monastery, which is famous for housing a triple Jowo statue [29], is located at the bank of the Karnali River in the eastern slopes of the Gurla Mandhata Massif and has been threatened by several floods throughout history. Although the main purpose of the inclusion of flood accounts in the monastic catalogue must have been to praise the sacred statue, the descriptions offer quite a bit of detail of how the work was organized, embankment construction techniques, and the combination of ritual and practical measures to reduce the risk of further destruction, showing some interesting parallels with the present. One of the worst floods occurred in 1493:

The silty river flooded with such force that all the previous embankments were completely destroyed and carried away without leaving a trace behind. The monks convened to recite religious texts and performed offering and propitiation rituals to Buddhas and Dharma protectors, whereas all the laypeople worked diligently day and night on the embankments. The chief of Mustang provided the community with written authorisation to collect wooden material for the embankments from Humla to the south. In addition, a large number of workers from surrounding villages were summoned for repair work. Under the lamas' and the nobles' leadership the workers built an embankment measuring 85 arm spans to divert the water. The work was undertaken at great cost of fields and labour.

[28]

The account unfortunately does not provide information about the causes of the floods, but their strength suggests that they were flash floods. They may have been GLOFs from a branch of the same glacier in the Gurla Mandhata Massif as the Halji floods, but this will have to remain speculation, as a period of intense precipitation could have a similar effect. While no information is given about individual household losses, the text is quite explicit that repeated construction work imposed a great strain on local livelihoods. Embankments had to be rebuilt every year, a Sisyphean task, and after a particularly large flood in 1503, villagers reached the point where they felt they had no choice but to move the temple. However, the decision lay with an external chief residing far away from the community at risk, who decided that embankments would be sufficient [28]. The villagers had to comply; work groups were subsequently organized and people from Limi were ordered to assist. A large embankment was then built, but it did not take more than three years before the next large flood struck.

Similar floods have continued to haunt both Khochar and Limi until the present day. Accounts of these floods, the causality attributed, and the successes and failures of strategies used to deal with them constitute an important part of the cultural memory people carry in the face of the current hazards. As we will see below, a number of parallels can be found in the way people cope with GLOFs today, but the different socio-political circumstances of contemporary Nepal, along with the availability of scientific explanations, technological inventions, and globalized networks provide a new set of challenges and opportunities.

14.2.3.2 Explanations of the contemporary floods in Halji

In 2009, just after a GLOF had struck the village, a team of local villagers walked up to the glacier in the mountains above Halji to locate its source. After a half-hour climb up the glacier the villagers discovered a lake and their preliminary theory was that ice sheets must have broken off, causing the lake to discharge. With some technical modifications these observations have recently been confirmed by scientists from the universities of Tübingen and Dresden in Germany, who are currently investigating the cause of the floods and the capacity of the lake in order to assess the risk for future outbursts [30].

The lake has been identified as a supra-glacial lake and is located at an altitude of 5350 m (Figure 14.6). Kropáček *et al.*'s analysis of data from the High Asia Reanalysis (HAR) data set shows an increase in temperature over the last decade, and their analysis of data from Ice, Cloud, and Land Elevation Satellite (ICESat) for 2003–2009 indicates that there has been some glacial retreat throughout this

Figure 14.6 Part of the glacial lake basin, November 2013.
Source: photo courtesy of J. Kropáček.

period. The glacier drains mainly through a number of subglacial channels that freeze in the winter season, but crack open during summer due to melting snow and the movements of the glacier. These movements combined with increased melt may explain the outbursts, while the amount and timing of precipitation, solar radiation, wind, and other factors may also contribute to the size and timing of the floods [30].

Most villagers in Limi have not been exposed to the global climate change narrative, and when asked specifically about causes for the floods, they tend to give explanations related to the disruption of the 'moral climate' [8,18,31]. Divergent attributions on this level occasionally lead to scapegoating and some villagers speculated that the floods could have been caused by intruders from outside disturbing the territorial deities, whereas others claimed that the territorial deities must have been offended because of local conflict. The recent involvement of researchers and development organizations has led to an increased focus on scientific empirical explanations, but these are not seen as contradictory to religious explanations, which are placed beyond the empirical explanations in a chain of causality.

14.2.3.3 Risk assessment and coping strategies

The villagers in Halji employ a broad range of knowledge and resources to evaluate risk and make decisions about strategies to reduce their vulnerability and exposure to the floods. Monks have performed extensive rituals directed at local territorial deities, dharma protectors, Buddhas and Bodhisattvas. These form part of a larger repertoire that can be employed more generally to avert misfortunes and obstacles and more specifically to manipulate the weather and cope with glacial floods. Whereas most of the rituals have been performed locally, external religious authorities have often decided the type of ritual response. The villagers have also sent requests to Rinpoches in Nepal and India for religious predictions (Tib. *mo*, *tagpa*) to assess the probability of the occurrence and impact of future floods. These predictions are used to inform the choice of coping strategies ranging from rituals to movement of houses and property and seeking of external help. In addition to its moral and ritual function, the monastery plays an important role in the local economy and has significant mobilizing capacity both locally and externally through connections with larger monasteries abroad. The monastery is thus a key institution in the local responses to the flood.

As we have seen in the cases above, mobility is a common strategy for coping with natural hazards [9,32,33]. As a concept, mobility is very broad, covering a range of practices at different scales, from short-term evacuation to permanent resettlement, and from voluntary relocation to forced displacement [33,34]. Its outcome is therefore largely dependent on context. In Limi, the families who lost their houses to the last GLOF have been forced to rebuild them on safer ground, and because of the recurring floods relocation is an option that is discussed by all. Divergent risk evaluation has led some to choose to relocate pre-emptively, whereas most have so far opted for other strategies.

There is currently no insurance system, leaving individual households to cope with losses to private property and other livelihood assets. As Goldstein observed in the 1970s, and as the livelihood literature has suggested is usually the case, the local livelihood strategies in Limi involve spreading economic risk between agriculture, pastoralism, trade, crafts, and manual labour [35]. Because the floods have caused losses to large portions of farmland, a logical response would be to increase the focus on pastoralism, which traditionally constituted an important part of the local economy. However, after the demarcation of the Sino-Nepali border in 1961, herders from Limi gradually lost access to their traditional winter pastures on the Tibetan plateau and had to reduce their livestock numbers, an illustration of how geopolitics shapes local livelihoods. Most families make up for these losses by relying more heavily on trade, crafts, and manual labour. These alternative sources of income, however, are unstable and the ability to spread economic risk varies between the social strata (see also [35]). Another option is to borrow grain from the monastery or village assembly, but loans have to be repaid with interest and function only as a short-term strategy.

Longer-term strategies to adapt to the recurring floods include both institutional and structural measures. The local organisational skill is of vital importance in flood management, and this seems to be one of Limi's strengths. The construction of embankments in remote communities without proper infrastructure and access to heavy vehicles demands a significant amount of manual labour. Like in the historical example from Khochar, each household in Limi has the duty to provide one man for communal labour, and this system has been used to mobilize workers for the construction of embankments after the recent floods (Figure 14.7). However, purchase and transport of gabion wires used to strengthen and stabilize the walls is costly, and though embankments are among the most common structural measures used for flood protection, their efficiency is disputed [36,37]. Evidence from both Khochar and Halji suggest that although the embankments seem to have delayed erosion, the floods have repeatedly washed the walls away.

Because the villagers are coping with a recurring hazard, the notion of global warming has not been a necessary driver for action. Like those in the cases at the Tibetan Plateau discussed above, the villagers attribute the floods to local causalities, but in their negotiations with Nepali authorities and development organizations, the concept of global climate change and the ability to present the floods according to the contemporary disaster narrative has been vital. Although flood management has been initiated at the local level, it has also involved

Figure 14.7 Building new embankments in Halji, December 2011.
Source: photo by A. Hovden.

lobbying at multiple scales in order to get funding for longer-term protection measures. GLOFs have received considerable attention in Nepal, but the community's remote location and political marginalization have made the process challenging. The special committee, which was formed to negotiate with Nepali authorities and NGOs, has eventually managed to secure some funding to protect the village and monastery from future floods, but one of the side effects has been closer integration into the Nepali administrative system and more dependency on external assistance.

14.3 Conclusion

The three ethnographic examples of historical and contemporary high-altitude Tibetan communities described here provide insight into how livelihoods in the Earth's cryosphere are being reshaped by changing hazard-coping strategies and the articulation of ecological, political, social, and cultural processes at multiple scales. Ecological factors that affect rural livelihoods are perceived in culturally mediated ways that shape social responses, including risk management. At the same time, geopolitical transformations and socio-economic processes are rendering local communities increasingly dependent on national and global structures and processes. However, the cases show that there have been tensions between internal and external responses to environmental hazards, which have played out in different ways in the three localities. As climate change is only one of the potential factors contributing to environmental hazards in the cryosphere, and attribution is often disputed, dealing with the relevant risks means taking into account this complexity and finding ways for scientific data and projections to interface with the experiential dimension of the phenomena. As Julie Cruikshank has pointed out:

Climate science presents a more comprehensive picture than weather. But similarly, oral traditions convey understandings that are more comprehensive than data. The two cannot always be conflated, but both reveal a great deal about the human experience of environmental change. One primary value of local traditions about weather is to deploy authoritative local traditions in problem solving during unexpected weather events.

[38]

The inclusion of culturally and socially specific understandings and forms of adaptation to hazards constitutes an important aspect in what has been defined as an emerging polycentric governance of the environment. Challenges posed by a changing climate cannot be solved only by international and national institutions, but must also involve actors at multiple scales; local decision making is of key importance to effective management of the environment [39]. Our case studies demonstrate that communities living in the cryosphere have significant histories of experimentation with ways of adapting to environmental hazards. These include ritual repertoires, different types of mobility, spreading the economic risk, as well as structural engineering. The case studies also suggest, however, that the current dominant direction of change in Tibet and the Himalayas is a scaling up from locally to nationally directed and enabled strategies for coping and adaptation. These not only have uncertain effects for the sustainability of livelihoods, but also have important implications for the sometimes fraught relationship between states and their citizens.

Acknowledgements

Hildegard Diemberger's research was supported by a string of grants funded by the AHRC, the Austrian Science Fund, the Ev-K2-CNR Project, and the Mongolia and Inner Asia Studies Unit of the University of Cambridge. Special thanks go to the colleagues of the Tibetan Academy of Social Sciences and the people of Porong. Astrid Hovden's study was supported by the Faculty of Humanities, University of Oslo and the Institute for Comparative Research in Human Culture. We are grateful to the villagers in Halji for their hospitality and for sharing their knowledge. The work of Emily T. Yeh was supported by NSF SBE-0624315. We thank the many Tibetan herders in Nagchu who took the time to share their experiences and insights with us. Finally, we thank Yonten Nyima, Julia Klein, and Kelly Hopping, the co-authors of the article on which part of this chapter is based.

References

1. R Chambers, G Conway, *Sustainable Rural Livelihoods: Practical Concepts for the 21st Century* (Brighton: IDS, 1992).
2. I Scoones, Livelihoods perspectives and rural development. *Journal of Peasant Studies*, **36**:1 (2009), 171–196.
3. F Ellis, The determinants of rural livelihood diversification in developing countries. *Journal of Agricultural Economics*, **51**:2 (2000), 289–302.
4. F Ellis, *Rural Livelihoods and Diversity in Developing Countries* (Oxford: Oxford University Press, 2000).
5. A Bebbington, Capitals and capabilities: a framework for analysing peasant viability, rural livelihoods and poverty. *World Development*, **27**:12 (1999), 2012–2044.
6. CF Kennel, Think globally, assess regionally, act locally. *Issues in Science in Technology*, **25**:2 (2009), 46–52.
7. ET Yeh, Y Nyima, KA Hopping, JA Klein, Tibetan pastoralists' vulnerability to climate change: a political ecology analysis of snowstorm coping capacity. *Human Ecology*, **42**:1 (2014), 61–74.
8. HGM Diemberger, Anticipating the future in the land of the snows. In *The Social Life of Climate Change Models: Anticipating Nature*. eds K Hastrup, M Skrydstrup (New York: Routledge, 2013).
9. IPCC SREX. *Managing the Risks of Extreme Events and Disasters to Advance Climate Change Adaptation: A Special Report of Working Groups I and II of the Intergovernmental Panel on Climate Change* (Cambridge: Cambridge University Press, 2012).
10. J Ash, New nuclear energy, risk, and justice: regulatory strategies for an era of limited trust. *Politics & Policy*, **38**:2 (2010), 255–284.
11. MC Goldstein, TN Shelling, JT Surkhang, *The New Tibetan–English Dictionary of Modern Tibetan*. (Berkeley, CA: Berkeley University of California Press, 2001).
12. M Hulme, *Why We Disagree about Climate Change: Understanding Controversy, Inaction and Opportunity* (Cambridge: Cambridge University Press, 2009).
13. JV Bellezza, *Divine Dyads: Ancient Civilization in Tibet* (Dharamsala: Library of Tibetan Works and Archives, 1997).
14. P Wangdu, H Diemberger, *dBa' bzhed: The Royal Narrative Concerning the Bringing of the Buddha's Doctrine to Tibet*. (Wien: Verlag der Österreichischen Akademie der Wissenschaften, 2000).
15. T Gerken, T Biermann, W Babel, *et al.*, A modelling investigation into lake-breeze development and convection triggering in the Nam Co Lake basin, Tibetan Plateau. *Theoretical and Applied Climatology*, **117** (2014), 149.
16. G Postiglione, B Jiao, L Xiaoliang, Education change and development in nomadic communities of the Tibetan Autonomous Region (TAR). *International Journal of Chinese Education*, **1**:1 (2012), 89–105.
17. J Bang, *Biography of Bodong Panchen*. (Lhasa: Ancient Tibetan Books Publishing House of the TAR, 1991).
18. T Huber, P Pedersen, Meterological knowledge and environmental ideas in traditional and modern societies: the case of Tibet. *Journal of Royal Anthropological Institute*, **3**:3 (1997), 577–598.
19. X Cui, HF Graf, B Langmann, W Chen, R Huang, Climate impacts of anthropogenic land use changes on the Tibetan Plateau. *Global and Planetary Change*, **54**:1–2 (2006), 33–56.
20. A Marin, Riders under storms: contributions of nomadic herders' observations to analysing climate change in Mongolia. *Global Environmental Change*, **20**:1 (2010), 162–176.

21. K Bauer, Common property and power: insights from a spatial analysis of historical and contemporary pasture boundaries among pastoralists in central Tibet. *Journal of Political Ecology*, **13** (2006), 24–47.
22. MC Goldstein, *Nomads of Western Tibet* (Berkeley, CA: University of California Press, 1990).
23. K Bauer, Y Nyima, Laws and regulations impacting the enclosure movement on the Tibetan Plateau of China. *Himalaya*, **30**:1–2 (2010), 23–38.
24. Dpal 'Chi med grub pa, *Ye shes mkha' 'gro bsod nams 'dren gyi sku skyes gsum pa rje btsun ma chos kyi sgron ma'i rnam thar* (folio 10b) [Tibetan manuscript].
25. T Lama, *The Kailash Mandala: A Pilgrim's Trekking Guide.* (Kathmandu: Humla Conservation and Development Association, 2nd edition, 2012).
26. A Hovden, Who were the sponsors? Reflections on recruitment and ritual economy in three Himalayan village monasteries. In *Tibetans Who Escaped the Historian's Net: Studies in the Social History of Tibetan-speaking Societies.* eds C Ramble, P Schwieger, A Travers (Kathmandu: Vajra Publications, 2013).
27. Ministry of Environment, *National Adaptation Programme of Action (NAPA) to Climate Change.* (Kathmandu: Ministry of Environment, Government of Nepal, 2010).
28. Nga dbang phrin las (Wa gindra karma, 16th century), *The catalogue of the three silver brothers. Jo bo rin po che ngul sku mched gsum rten dang brten pa bcas pa'i dkar chag rab dga'i glu byangs.* Edited by RI Vitali (Dharamsala: Tho ling gtsug lag khang lo gcig stong 'khor ba'i rje dran mdzad sgo'i go sgrigs tshogs chung, 1996).
29. T Gyalpo, C Jahoda, C Kalantari, P Sutherland, *Khorchag.* (Lhasa: Bod ljongs bod yi dpe rnying dpe skrun khang, 2012).
30. J Kropáček, N Neckel, B Tyrna, *et al.*, *Exploration of a periodic GLOF in Halji, West Nepal using modeling and remote sensing.* Himalayan Karakorum Tibet Workshop and International Symposium on Tibetan Plateau; Tübingen 2013.
31. A Byg, J Salick, Local perspectives on a global phenomenon: climate change in Eastern Tibetan villages. *Global Environmental Change*, **19**:2 (2009), 155–166.
32. A Agrawal, Local institutions and adaptation to climate change. In *Social Dimensions of Climate Change: Equity and Vulnerabilty in a Warming World.* eds R Mearns, A Norton (Washington, DC: World Bank, 2009) pp. 173–198.
33. K Hastrup, KF Olwig, eds, *Climate Change and Human Mobility: Challenges to the Social Sciences* (Cambridge: Cambridge University Press, 2012).
34. J Sward, S Codjoe, *Human Mobility and Climate Change Adaptation Policy: A Review of Migration in National Adaptation Programmes of Action (NAPAs)* (Brighton: Migrating out of Poverty RPC, University of Sussex, 2012).
35. MC Goldstein, Tibetan speaking agro-pastoralists of Limi: a cultural ecological overview of high altitude adaptation in the northwest Himalaya. *Objets et Mondes*, **14**:4 (1974), 259–268.
36. A Dixit, Kosi embankment breach in Nepal: need for a paradigm shift in responding to floods. *Economic and Political Weekly*, **44**:6 (2009), 70–78.
37. AB Shrestha, SH Shah, R Karim, *Resource Manual on Flash Flood Risk Management. Module 1: Community-based Management* (Kathmandu: ICIMOD, 2008).
38. J Cruikshank, Melting glaciers and emerging histories in the Saint Elias Mountains. In *Indigeneous Experience Today.* eds M de la Cadena, O Starn (Oxford: Berg, 2007) pp. 335–378.
39. E Ostrom, Polycentric systems for coping with collective action and global environmental change. *Global Environmental Change*, **20** (2010), 550–557.

15

Ice-clad volcanoes

RICHARD B. WAITT, BENJAMIN R. EDWARDS, AND ANDREW G. FOUNTAIN

An icy volcano, even if called extinct or dormant, may be active at depth. Magma creeps up, crystallizes, and releases gas. After decades or millennia the pressure from magmatic gas exceeds the resistance of overlying rock and the volcano erupts. Repeated eruptions build a cone that pokes one or two kilometers or more above its surroundings – a point of cool climate supporting glaciers. Ice-clad volcanic peaks ring the northern Pacific and reach south to Chile, New Zealand, and Antarctica. Others punctuate Iceland and Africa (Figure 15.1). To climb is irresistible – if only "because it's there" in George Mallory's words. Among the intrepid ascents of icy volcanoes we count Alexander von Humboldt's attempt on the 6270 m Chimborazo in 1802 and Edward Whymper's success there 78 years later. By then, Cotopaxi steamed to the north.

15.1 Explosive eruptions

Glacier-clad Cotopaxi (5897 m) erupted one June morning in 1877, foaming across the crater rim like a rice pan boiling over. Water floods poured down the flanks. From the church tower at Mulalo, many kilometers south, the preacher watched many travelers trek the old road along the river plain. A closer group of 20 men, women, children, and their servants spurred their animals toward higher ground. The flood arrived and devoured them [1].

Cotopaxi had erupted at least 17 times since the Spanish conquest, five since 1853 [2,3]. The 1877 flood was smaller than many earlier but still took 1000 lives. Climbing 2.5 months after eruption, geologist Theodor Wolf found that hot pumice flowing thickly down steep glaciers had gullied them deeply. From here came the floodwater, he wrote, and in one hour wiped out the work of generations.

The High-Mountain Cryosphere, ed. Christian Huggel, Mark Carey, John J. Clague and Andreas Kääb. Published by Cambridge University Press. © Cambridge University Press 2015.

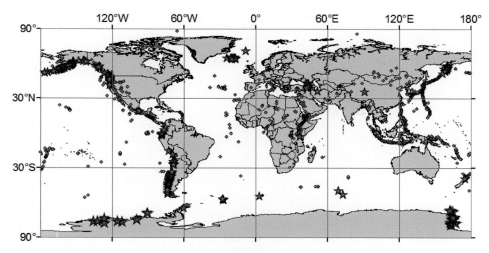

Figure 15.1 Distribution of active volcanoes, ice-clad ones starred.
Source: Volcano locations from Global Volcanism Program, glacier database from GLIMS, and National Snow and Ice Data Center, 2005 updated 2012. GLIMS Glacier Database. Boulder, Colorado USA: National Snow and Ice Data Center: http://dx.doi.org/10.7265/N5V98602.

Hekla, including its fissure swarm, in south Iceland (Figure 15.2) has erupted 23 times in the ten centuries since Iceland's settlement [4]. In a March 1947 burst, floods ran from the northwest brow down into Rangá River, peak flow far deeper and swifter than the most violent spring snowmelt runoffs. Near the summit, weeks later, Guðmundur Kjartansson saw the creeping lava there couldn't melt snow and glacier ice fast enough to flood. Yet oceans away several violent eruptions – Mt. Pelée in 1902 and 1929, Katmai in 1912 – had pulverized molten lava into ash whose incandescent clouds hugged the ground downslope [5]. An eruption at Taal in 1965, and from it the concept of "base surge" showed how exploding lava can shatter and speed over ground as turbulent hot clouds [6]. Reflecting later on Hekla's eruption, Sigurður Thórarinsson opined its flood came from a hot surge swiftly melting firn and glacier ice [7].

Hot erupted material on any ice-clad volcano (Figure 15.3) can cause rapid melting. By the 1970s geologists could helicopter to icy vents just after eruption. A great swelling of the north brow of Mount St. Helens in spring 1980 slid off suddenly one May morning. A hot ashy surge churned across summit snow, firn, and glaciers, and floods poured off east and west down valleys. The great landslides had carried off not just rock and glaciers but much of the volcano's groundwater. Muddy water welled from the new deposit, entrained much sand, and raged downvalley. A March 1982 blast of pumice and hot dome rock melted much crater snow, and water then flooded down a valley (Figure 15.4a). Other

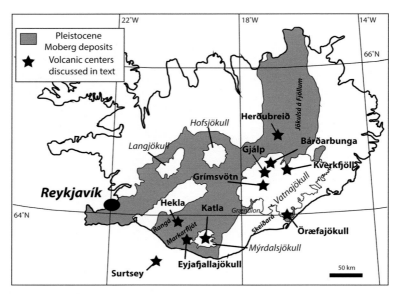

Figure 15.2 Map of Iceland showing places mentioned in the text. Holocene lava and deposits covering Pleistocene subglacial moberg omitted.

Figure 15.3 Ice-clad Pavlof volcano in Alaska during a small burst June 2, 2014.
Source: photograph by William Yi.

explosions of hot rock into snow released a common sequence of flows: dry dirty-snow avalanche, big dirty slushflow, then watery flood (Figure 15.4b) [8]. Each flow seemed a single plug of three parts, not a succession of three flows. The belt of drier outer snow is analogous to debris at the front and sides of a debris flow moving as a plug [9]. The slush and slushy water coming behind pushes ahead and aside resistant drier snow at the front.

Figure 15.4 Mount St. Helens eruptions activating snow in the early 1980s. (a) View southward of Mount St. Helens crater in March 1982. Two separate arms of flood and avalanche deposits merge on the floor of the breach. Crater at dome about 800 m wide. (b) View up-flow of deposits just after a May 1984 eruption. Outer microhummocky snowflow, followed by intermedial sheared slushflow, and inner watery flow. Entire flow about 200 m wide.
Source: photographs by R.B. Waitt.

Scorched tree snags and deposits lacking mud showed Mount St. Helens'
1980 ash surge had been hot and dry. The 1982–1986 dome bursts proved that
quick-melted snow could have formed the 1980 floods, and flowing slush could
have caused their speed down the cone [10].

Nevado del Ruiz (5390 m) in Colombia erupted in November 1985. Though
fairly small, its turbulent hot flows melted snow and glacier ice into slushy and
watery flows. They flowed down steep canyons and bulked with debris and
streamwater and trees to four times their initial volumes (Figure 15.5) [11]. Raging
through towns at night, they took 23 100 lives.

Figure 15.5 Aerial view down Rio Azufrado about 35 km downvalley from the
summit ice cap of Nevado del Ruiz. Bulked up with debris and trees scoured from
the valley, the lahar flowed 30 m deep here.
Source: photograph December 9, 1985 (26 days after lahar) by T.C. Pierson.

Ashflows have swiftly melted snow and ice at other volcanoes. Hot, blocky flows from Redoubt in Alaska in late 1989 and early 1990, and again in spring 2009, mixed with meters of loose snow to form great avalanches and slushflows; trailing watery flows deeply gullied Drift Glacier. Ice-grain flows freighted large rocks and ice blocks out 14 km, and slushy floods swept 34 km down to the sea [12,13]. Eruptions in 1995 and 2007 at Ruapehu in New Zealand swiftly melted snow and firn, and icy floods ran down into the plains [14,15]. Except at vents deep beneath ice (as in Iceland), turbulent hot ashflows across snow and glaciers have caused most large primary floods during historic eruptions of snowy stratovolcanoes [16].

Deposits around most ice-clad stratovolcanoes reveal prehistoric floods much larger than recent ones. Huge ashy floods from Popocatépetl in 800–900 CE wounded if not ended Mexico's "classic" civilization anchored by the temple cities of Teotihuacán and Cholula [17]. Mount Rainier has shed at least 40 lahars far down its large valleys in the past 10 000 years. Mount St. Helens has sent at least 14 large lahars far down its north valley in the past 5000 years. Ice-clad Mount Baker, Glacier Peak, and Mount Hood have all released large lahars during Holocene eruptions. Cotopaxi in the Andes has shed at least 12 great lahars in 800 years [3,18–20].

When water gains access to molten magma, powerful steam explosions can pulverize the lava at a vent. Ash convecting into the atmosphere may reach out thousands of kilometers. The progress of an observed maritime eruption illustrates a sequence inferred at many a once ice-covered terrestrial volcano. In November 1963 Surtsey emerged from the sea off Iceland's coast in explosive bursts. With water in the vent, powerful explosions pulverized basaltic magma. Once a wide tephra cone surrounded the vent and kept out most seawater, drier eruptions jetted up ash continuously. By April 1964 the ash cone had grown enough above sea level to exclude water altogether, and fluid lava poured out. Cooling to hard rock, the lava guaranteed the permanence of the island against the relentless sea [21]. Many a once ice-clad vent in Iceland shows a comparable progress in eruptive style.

Iceland holds many active volcanoes, and glaciers crown the higher ones. When the icy cone Öræfajökull erupted in 1362, ferocious floods burst from its glaciers on the southwest and southeast and buried many farms (Figure 15.2) [22].

Through most of the twentieth century, most of Iceland's huge outburst floods – jökulhlaups – originated at the frozen-over caldera lake Grímsvötn within Vatnajökull ice cap and emerged from beneath the ice margin at Skeiðará, 50 km farther south. Grímsvötn has erupted often since medieval times, ten times from 1902 to 1948 [23–25]. Siguður Thórarrinsson noted that the slow rise and abrupt cessation of jökulhlaups from Grímsvötn resembled outbursts from Grænalón and other ice-dammed lakes [26]. Storage and release of water at a glacier bed must stem from the physical properties of ice and water. As geothermal heat melts ice, Grímsvötn's level gradually rises 80–100 m. When pressure from the heightened water column nears the ice pressure at the base of surrounding ice, the lake begins to drain along the glacier bed. The friction of flowing water in a small tunnel heats the water slightly and melts the tunnel walls. The wider tunnel funnels more water, and more heat further enlarges the tunnel. And so on – a runaway flood [27]. Grímsvötn eruptions in 1922, 1934, and 2004 *followed* a jökulhlaup as if the lake's draining – lower pressure on the vent – triggered an eruption. Ten floods between 1938 and 1982 entailed no eruption [27,28].

Yet most great jökulhlaups begin with eruptions. Erupting lava when it fragments can melt five to eight times its volume of glacier ice. In October 1996 a subglacial fissure named Gjálp erupted 0.4 km^3 of basaltic andesite beneath ice as thick as 600–700 m. Cauldrons sinking into the ice are a measure of the volume of ice melted from below. A heat budget calculated from the missing ice shows most Gjálp lava erupting into subglacial water had shattered into fragments – as at Surtsey's submarine vent. Gjálp's meltwater drained south. Grímsvötn rose to well above its usual limit, 1430–1440 m, until in November some 3.5 km^3 of water escaped to Skeiðará, a jökulhlaup discharging up to 45 000 m^3 s^{-1} (Figure 15.6) [29].

Between 1477 and 1903 ten Vatnajökull floods drained north down Jökulsá á Fjöllum and buried farms on its coastal plain. Yet these were tiny compared to

Figure 15.6 View toward the southwest, over the edge of Grímsvötn caldera, showing the ice cauldron at 600 m wide, and tephra cone formed during the eruption of Grímsvötn.
Source: photograph, November 7, 2004 by B.R. Edwards.

Figure 15.7 View southeast across Vesturdalur area. Post-glacial intracanyon lava flow carved into a deep cataract alcove (center) by the Late Holocene flood from eruption beneath Vatnajökull. The whole scabland is about 1 km wide.
Source: photograph 1986 by R.B. Waitt.

prehistoric floods. Huge dry cataracts, channels cut into rock, water-fluted basalt, and gravel bars capped by giant current dunes reveal an ancient discharge 700 000 $m^3 s^{-1}$ had filled and overflowed Jökulsá valley (Figure 15.7) [30]. The presence or absence of dated ashes [31] atop flood surfaces show that the great flood came 2500 to 2000 years ago. The ashes also date many earlier gigantic Jökulsá floods between 8000 and 4000 years ago. Melting at high Kverkfjöll caldera sent some floodwater to Jökulsá, but the huge Bárðarbunga caldera has erupted frequently for millennia [32]. Subglacial water gradients [27] show that a fissure from Bárðarbunga erupting northeastward would drain meltwater to Jökulsá á Fjöllum. A northward eruption from Bárðarbunga began mid-August 2014 and as of mid-December 2014 has erupted more than 1 km^3 of basaltic lava beyond the Dyngjujökull ice margin. Such an eruption closer to Bárðarbunga beneath the glacier would melt and pond water that could escape down Jökulsá as a flood.

Eruption from Katla caldera beneath Mýrdalsjökull ice cap in south Iceland in October 1918 released an enormous jökulhlaup. A discharge above 300 000 $m^3 s^{-1}$, and volume of 8 km^3 swept 20 km wide southeast across the coastal plain. Witnesses watched turbulent deep water float thousands of gigantic blocks of glacier ice [33,34]. But the flow's base must have been fairly dense; much of its deposit is poorly sorted and weakly bedded sandy gravel that rafted huge slabs of soil and other fragile clasts [35,36]. The flood deposited 8 m of sandy debris across the plain and built the coast out hundreds of meters. Icy floods had also swept the plain during eruptions in 1660, 1721, and 1755; a medieval eruption aimed its flood south [35,37]. Of nearly 400 km² swept by these Katla floods, half remains uninhabited and unfarmed.

In April 2010 huge explosions fragmented the lava and melted up through 200 m of ice in the glacier-covered Eyjafjallajökull caldera, west of Katla. Ice cauldrons marked three vents (Figure 15.8a). The eruptions' melting released

Figure 15.8 Melting of summit ice cap of Eyjafjallajokull, July 9, 2010. (a) The melt crater about 100 m wide. The ice cap is covered by tephra about 20 m thick, pocked by small craters from bombs (the image does not show the main craters). (b) View northward down ash-covered Gigjökull glacier, north side of Eyjafjallajö-kull ice cap. The trough is about 200 m wide and holds lava flow melted into the ice. Source: photographs taken July 9, 2010 by B.R. Edwards.

occasional floods down the river Markarfljót that bulked up with sediment to reach peak discharges of 5000–15 000 m^3 s^{-1} [38], but diminished quickly across the wide sandur. Far downstream the debris partly filled a ferry harbor and severed the national highway. Eastward ash clouds stranded millions of European air travelers.

While most volcano–ice fireworks have been Iceland's, Deception Island in Antarctica – a caldera mostly below sea level – burst through its glaciers in early 1969. Several fissures along the east caldera wall erupted, fragmenting lava that melted tunnels in the glaciers. Meltwater flooded beneath the glaciers and piped through fractures to flow over the glacier tops. Even in this remote setting the floods smashed buildings – two research stations and a former whaling station. The next year eruptions perforated glaciers on the caldera's north wall [39].

15.2 Effusive lava eruptions

During Hekla's 1947 eruption geologists saw lava flowing over ice and snow with oddly little melting. Shouldn't 1100 °C lava melt snow rapidly? Yet in later eruptions on four continents, hot lava oozed over ice and snow, also with little drama. The 2010 Icelandic eruption had begun quietly as a flank eruption in the pass between high Katla and Eyjafjallajökull. Basaltic lava soon melted through 3 m of snowpack. But melting was slow enough to make only minor floods [40]. During the main eruption from the summit caldera, lava flows burrowed into and through Gigajökull and melted dramatic canyons in the ice – again slowly enough to make no flood (Figure 15.8b). During a 2013 eruption at Tolbachik in Russia, red-hot lava flows also melted snow slowly (Figure 15.9). In the twentieth century only 15 known lava flows on glaciated volcanoes caused significant flooding [16]. Mount Etna on Sicily supports no glaciers, but its 3350 m summit and upper flanks are snowpacked in winter. In 1755 lava melted enough snow that a watery flood descended Valle de Bove to the coast.

During several Chilean eruptions in the 1980s and 1990s, lava flowed partly beneath ice. At Villarica in 1984–1985 lava flows down the mountain flanks burrowed beneath ice and melted channels in ice more than 1 km long and 50 m wide, but released only small lahars [41]. A larger lava flow in 1971 had caused floods that killed 15 people. Mount Hudson in southern Chile sprang to life in 1991 with explosions and lava flows from its ice-filled summit caldera. Steam rising from cracks in ice tracked lava flowing beneath ice down the volcano's west flank. In 1994 Llaima volcano in central Chile erupted as fire fountains that melted snow and ice fast enough that its flood destroyed five bridges. Lava melted a canyon tens of meters deep and 2 km long in ice then flowed 2 km subglacially down the west flank in paths sporadically marked by steam and small explosions [42].

Figure 15.9 Small lobe of pahoehoe basaltic lava flowing over snow during the eruption at Tolbachik volcano in Kamchatka, Russia on April 4, 2013. The main lobe shown is about 3 m long. There is no steam or obvious surface water.
Source: photograph by B.R. Edwards.

The steep snow-capped Alaskan volcanoes of Westdahl in 1991 and Pavlof in 2013, and the Russian volcano Klyuchevskoi in 1974, 1983, and 1994 erupted lava that melted much snow and ice, but in those remote settings the water drained off harmlessly [43]. In the past 30 years Veniaminof in Alaska has thrice erupted short basaltic lava flows that descended on and into ice in its caldera but caused no floods. A 1983–1984 lava flow melted an open ice cauldron 100 m deep and 1000 m across, but the ponded lake drained slowly into porous debris and fractured ice [44].

The fronts of lava flowing down steep water-saturated slopes of Klyuchevskoy volcano in 1994 and Etna in 2006–2007 broke apart to form lava–snow avalanches 2–5 km downslope of erupting vents. Mixing with snow and water, they exploded into clouds of ash. Violent secondary explosions churned into snow, and it melted into small lahars [45].

Laboratory experiments help show what must happen when molten lava flows sink through ice. Once lava reaches the bottom of laboratory snow or ice, melting slows and the lava flows beneath the ice with little effect [46].

15.3 Pleistocene ice-clad volcanoes

Most lava flows have sloping rough lobate sides and vertical cooling joints. But some flows in formerly glaciated terrains instead have steep and smooth sides, glassy margins, and irregular to horizontal cooling joints. Some flows are brecciated or pillowed, some deposited among till or gravel. Such features in British Columbia, on Mauna Kea in Hawaii, on Mount Rainier, South Sister, and Mount Mazama volcanoes in northwestern United States, and in many areas of Iceland reveal materials erupted against or dammed by Pleistocene glaciers [47–49].

Normal ropey lava flows from Clinker Peak in British Columbia thicken down-slope to over 300 m and now tower over valley floors. They grew thick by ponding against something since disappeared – a former valley glacier. Such lava cliffs stand 200 m above the flanks of Hoodoo Mountain (Figure 15.10a). Nearby glaciers are the shriveled survivors of thick Pleistocene ice that had dammed the lava [50].

In the 1940s Bill Mathews in British Columbia, and separately Guðmundur Kjartansson in Iceland, inferred that oddly steep-sided volcanoes had formed when ice sheets covered the landscape. Tuya Butte and The Table in BC stand above the tundra-like volcanic dinner tables (Figure 15.10b) [47]. Flat-topped Herðubreið ("broad-shouldered one") illustrates in north Iceland the sizable cones that Mathews calls tuyas (Figure 15.10c). Basalt erupting under a thick glacier may form pillow lavas at deep levels where pressure of overlying ice and water inhibits explosions. A sustained eruption melts a hole in the glacier – a lake held in by ice. The vent builds into shallower water and lower pressure. Exploding magmatic gases, steam bursts, and contraction from cold shatter the lava, as at Surtsey. The fine debris chills – some of it breccia, some of it glass, deposits sometimes called hydroclastite or hyaloclastite [51–54]. If the eruption builds above lake level and isolates the vent from water, lava pours out. Crystallizing to hard rock, it gives a tuya its mesa-like cap. In British Columbia tuyas formed during many glaciations back to more than 2.6 million years ago. They dot other once-glaciated areas in Iceland, Russia, Alaska, and the South Cascades of the United States [55,56].

Fluctuations in ocean oxygen isotopes within dated deep-sea strata show that great Northern Hemisphere ice sheets have formed and melted away dozens of times in the past three million years [57]. Volcanoes erupting through ice leave distinctive forms and record episodes of thicker glaciers in past climates. Antarctica encompasses several score volcanoes, most of them beneath ice. Their moberg-like deposits show that volcanoes as far apart as 4500 km in West Antarctica have from time to time erupted into water held in by glaciers as far back as 12 million years [58–60].

Geologic maps of Iceland's active volcanic belts record 11 000 km^2 of Pleistocene basaltic pillow lava and altered glassy breccia – in Iceland called moberg (Figure 15.2). Clearly erupted in water held in by huge glaciers, moberg reaches 15 km and more beyond the margins of Hofsjökull and Langjökull ice caps and tens of kilometers north and southwest of Vatnajökull ice cap [61–63]. Extensive deposits of fragmented lava – some in steep-sided ridges called

284   *R. B. Waitt* et al.

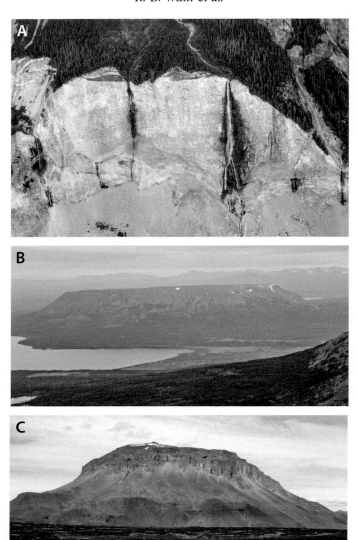

Figure 15.10 Tuyas or table mountains. (a) Aerial view down the cliff of
ice-dammed Pleistocene trachyte lava nearly 200 m high on the southwest
side of Hoodoo Mountain. (b) View southwestward showing the northeast
side of Tuya Butte, northern British Columbia. The upper surface of the tuya
is ~500 m above the level of Tuya Lake to the left. (c) Herðubreið tuya,
north Iceland, view south. The top above the cliffs is about 2.3 km wide.
Sources: (a) and (b) photograph by B.R. Edwards; (c) photograph July 1986 by R.B.
Waitt.

tindars – reveal eruptions from ice-clad fissures or central vents during several Pleistocene glaciations.

Decades of work in Iceland pioneered by Guðmundur Kjartansson and J.G. Jones document sequences of lavas and volcanic breccia revealed that eruptions beneath Pleistocene ice. More recent studies reveal details in Antarctica and Iceland. Some hydroclastite overlies older rocks that glaciers have gouged and smeared with till. Some of these volcanoes are marked by ice that closed over the vents after the eruption. Some interbedded lava and breccia reveal eruptions beneath fairly thin glaciers, others beneath immense ice sheets more than 1 km thick [64,65].

Icelandic volcanoes erupted far more often while last-glacial ice sheets thinned and retreated than they did before or since [66,67]. Decreasing weight of thinning glaciers exerts less pressure on vents and the mantle, and so magma chambers exolve gas and tend to erupt. Eruptions add carbon dioxide to Earth's atmosphere, and that causes climate to warm faster.

15.4 Collapse

Peculiar hummocky topography at and beyond the bases of some volcanoes – Galunggung in Indonesia, Ruapehu in New Zealand, Bandai-san in Japan, Mount Rainier – had been thought some sort of mudflow. When the top of Mount St. Helens slid off as great landslides in 1980, it filled upper Toutle valley with rocky debris in the form of high-relief hummocks. Soon volcanoes around the world – those named just above, Socompa in Chile, Shasta in California, Yatsugatake in Japan, scores of others – were identified as having collapsed as landslides [68–71]. The odd hummocky topography is now explained by faults pulling apart within a mass expanding sideways while sliding forward [72].

Ice-clad volcanoes seem to fail as big avalanches and muddy floods more than bare ones. An ice cap supplies a persistent downward seep of water that circulates in the cone. Mixing with sulfur gases rising from below, the water forms acidic fluids that attack hard minerals and glass in rocks. They turn gradually into soft clay minerals that adsorb water. The upper core of a stratovolcano grows weak and water-saturated. On the steep slopes some of it eventually slides off as a great clayish avalanche and lahar [73–75]. Some now-ice-free volcanoes collapsed when cloaked in Pleistocene ice – Citlaltépetl and Cofre de Perote in Mexico [76].

15.5 Volcano glaciers and climate

Glaciers on volcanoes sustain runoff in hot, dry periods and reduce it during cool periods [77]. Local topography affects solar radiation, air temperature, and wind-drift of snow. Glacial cirques cut into a volcano with weak strata, and sometimes fail as rock avalanches. Thick debris on a glacier's lower reaches insulates the ice and suppresses melting: the glacier may advance slower than a clean one [78,79]. Yet Chimborazo's glaciers melted faster after thin ashfall from 1999–2000 eruptions of Tunguarahua volcano, the darkened snow adsorbing more solar radiation [80]. Geothermal heat on some volcanoes reduces any glaciers. Increased melting atop Mount Baker in Washington, United States, revealed a tenfold jump in heat flow in 1975 [81]. Geothermal heat melts the bases of several large Icelandic glaciers – enough at Grímsvötn, as we've seen, to release episodic jökulhlaups.

Earth's alpine glaciers expanded from 1275 to 1850 CE, a cool period called the Little Ice Age. Long-term cycles of glaciation and deglaciation respond to cyclical variations in Earth's orbit – variations in the roundness in Earth's orbit around the Sun on a 100 000-year cycle and in the tilt of Earth's axis on a 41 000-year cycle [82,83]. Yet several big volcanic eruptions such as Rinjani 1257, Quilotoa 1280, Oræfajökull 1362, Kuwae 1452, Sakurajima 1471, Bárðarbunga-Veiðivötn 1477, Mount St. Helens 1480 and 1482, Huaynaputina 1600, Lakagigar 1783, and Tambora 1815 may have pumped enough fine ash and sulfur into the atmosphere to sustain Little Ice Age glaciers [4,32,84–87]. By the early twentieth century most of the world's mountain glaciers were retreating. They stabilized in mid-century, then resumed retreat [88,89].

Glaciers recede when climate warms. One cause of warming is increasing carbon dioxide (CO_2) in Earth's atmosphere. Cores through ice caps in Greenland and Antarctica show that atmospheric CO_2 ranged between 180 and 280 parts per million for 20 000 years, but since the late eighteenth century the burning of carbonaceous fuels has added more and more carbon dioxide to the atmosphere – monitored since 1958 at Mauna Loa Observatory on Hawaii. CO_2 reached 295 ppm by about 1900. By 2000 it was about 370 ppm and is now about 400 ppm. This trend could bring carbon dioxide to 500–600 ppm by the year 2100 – levels not seen in tens of millions of years [83,90]. Atmospheric CO_2 is rising when, by orbital variations, it should be falling.

Glaciers on most high volcanoes respond somewhat less to climatic warming than do glaciers on lower mountains. At Hoodoo Mountain volcano in British Columbia, low-valley glaciers have retreated more than 2 km in 90 years, but changes to the ice cap on the 2000 m summit are barely measurable [91]. In the Pacific Northwest of the United States, glaciers in the non-volcanic Cascade Range lost an average 46% in area over the past century, but those on the high cones of Mounts Baker, Rainier, and Hood volcanoes shrank 24–30% [79]. Yet Mount Rainier's glaciers shrank 6.7% in area (but 14% in volume) just between 1970 and

2007 [92]. In the Three Sisters volcanoes of the South Cascades, glaciers had reached their limits against sharp moraines in the 1850s–1860s. They were visibly in retreat by 1906 and in rapid retreat since (Figure 15.11) [93].

The shrinking of volcano glaciers is global. Glaciers on Popocatepetl in Mexico and on Villarica in Chile have shrunk [94,95]. A glacier on Heard Island volcano in the Indian Ocean lost 29% of its ice area from 1947 to 2004 [96]. Glaciers on

Figure 15.11 Matched views southward of Collier Glacier toward Middle Sister, North Sister volcano to the left (from [93]), images arranged by him and used here with permission. Top: August 1910. Bottom: August 2010. Much of the moraine in the foreground has slid away during the century between the two photographs. People in lower right of both images show scale.
Source: top: photograph by Clarence Winter of Kiser Photography; scan of original print courtesy of Mazamas. Bottom: photograph by Jim E. O'Connor.

Cotopaxi and Antizana volcano in Ecuador lost 30% of their area from 1956 to 1997; large glaciers on much higher Chimborazo also shrank [97,98].

Glaciers on tropical Kilimanjaro in Africa are in spectacular retreat – the summit lost 85% of its ice cover between 1912 and 2007. This high summit (~5800 m) is rarely above freezing, and ice wastes by sublimation. Having since the early 1900s receded back into deeply embayed, crenulated vertical walls, the glacier fronts expose vast surface areas and retreat especially rapidly. The air is apparently drier than it had been before the mid twentieth century – there is more solar radiation now and less snowfall. Shrinking swiftly in recent decades, glaciers on some tropical volcanoes including Kilimanjaro seem unsustainable relics of the Little Ice Age [99,100].

A retreating glacier yields more water than one of the same size in equilibrium with climate. Once restabilized the smaller glacier will then yield less runoff [77,101]. If current retreat rates on Mount Hood in Oregon continue for some decades, the shrunken glaciers will yield much less late-summer river flow than they do now.

While the world's glaciers shrank, three new ones formed – but they only replaced glaciers lost when volcanoes collapsed. The sudden escape of 12 km^3 of magma in the 1912 Katmai eruption in Alaska caused Mount Katmai's summit to fall in. The new caldera accumulated snow and eventually harbored two small glaciers [102]. Landslides removed Mount St. Helens' summit in 1980, taking most of its glaciers. Snow accumulated in the new lower crater, the snowfield thickened rapidly and, by 1992, it had compressed partly to ice. The glacier's two arms swelled and lengthened around the 1980s dome. A new dome eruption in 2004–2008 partly melted and shoved Crater Glacier aside, but since then both arms have continued to thicken and lengthen [103,104].

15.5.1 Recapitulation

An overall quantification of ideas explored in this chapter isn't very meaningful, for too many disparate processes affect ice-clad volcanoes. A big vertical eruption column may do little to summit ice. But a small, hot surge flowing turbulently laterally can melt much snow and ice swiftly and can make big floods – as at Mount St. Helens in 1980 and Nevado del Ruiz in 1985.

Processes that extract heat from volcanic eruptions and melt the frozen tops of ice-clad volcanoes are complex. Even for a given volume of volcanic material erupted, it is difficult to predict how much snow and ice will be melted and how much flooding will result. A simple comparison of volume of volcanic material erupted to the estimated maximum flood discharges for several historic eruptions shows no pattern at all (Figure 15.12).

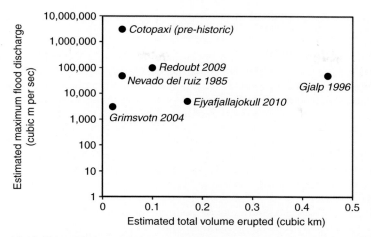

Figure 15.12 Plot of estimated volume erupted vs. estimated peak-flood discharge of a few eruptions that have made big floods.

References

1. T Wolf, Geognostische Mitteilungen aus Ecuador; Der Cotopaxi und seine letzte Eruption am 26 Juni 1877. *Neues Jahrbuch für Mineralogie, Geologie, und Palaeontologie,* **1878** (1878), 113–167. [Authors used unpublished translation to English by Jane Crandell, 1976].
2. M Hall, P Mothes, The rhyolitic–andesitic eruptive history of Cotopaxi volcano, Ecuador. *Bulletin of Volcanology,* **70** (2008), 675–702.
3. M Pistolesi, R Cioni, M Rosi, KV Cashman, A Rossotti, E Aguilera, Evidence for lahar-triggering mechanisms in complex stratigraphic sequences: the post-twelfth century eruptive activity of Cotopaxi volcano, Ecuador. *Bulletin of Volcanology,* **75** (2013), 698. doi: 10.1007/s00445-013-0698-1.
4. T Thordarson, G Larsen, Volcanism in Iceland in historical time: volcano types, eruption styles, and eruptive history. *Journal of Geodynamics,* **43** (2007), 118–152.
5. FA Perret, *The Eruption of Mt. Pelée 1929–1932* (Washington, DC: Carnegie Institution of Washington, 1935).
6. JG Moore, Base surge in recent volcanic eruptions. *Bullein Volcanologique,* **30** (1967), 337–363.
7. S Thórarinsson, Course of events. In *The Eruption of Hekla 1947–1948,* ed. T Einarsson, G Kjartansson, S Thórarinsson (Reykjavík: Visindafélag Islendinga [Societas Scientiarum Islandica], 1976).
8. RB Waitt, TC Pierson, NS MacLeod, RJ Janda, B Voight, RT Holcomb, Eruption-triggered avalanche, flood, and lahar at Mount St. Helens: effects of winter snowpack. *Science,* **221** (1983), 1394–1397.
9. RM Iverson, Debris flows: behavior and hazard assessment. *Geology Today,* **30**, 1 (2014), 15–20.
10. RB Waitt, Swift snowmelt and floods (lahars) caused by great pyroclastic surge at Mount St. Helens, Washington, 18 May 1980. *Bulletin of Volcanology,* **52** (1989), 138–157.
11. TC Pierson, RJ Janda, J-C Thouret, CA Borrero, Perturbation and melting of snow and ice by the 13 November 1985 eruption of Nevado del Ruiz, Colombia, and

consequent mobilization, flow and deposition of lahars. *Journal of Volcanology and Geothermal Research*, **41** (1990), 17–66.

12. RB Waitt, CA Gardner, TC Pierson, JJ Major, CA Neal, Unusual ice diamicts emplaced during 15 December 1989 eruption of Redoubt Volcano, Alaska. *Journal of Volcanology and Geothermal Research*, **62** (1994), 409–428.

13. CF Waythomas, TC Pierson, JJ Major, WE Scott, Voluminous ice-rich and water-rich lahars generated during the 2009 eruption of Redoubt Volcano, Alaska. *Journal of Volcanology and Geothermal Research*, **239** (2013), 389–413.

14. SJ Cronin, VE Neall, JA Lecointre, AS Palmer, Unusual "snow slurry" lahars from Ruapehu volcano, New Zealand, September 1995. *Geology*, **24** (1996), 1107–1110.

15. G Kilgour, V Manville, F Della Pasqua, A Graettinger, KA Hodgson, GE Jolly, The 25 September eruption of Mount Ruapehu, New Zealand: directed ballistics, surtsean jets, and ice-slurry lahars. *Journal of Volcanology and Geothermal Research*, **191** (2010), 1–14.

16. JJ Major, CG Newhall, Snow and ice perturbations during historic volcanic eruptions and the formation of lahars and floods: a global review. *Bulletin of Volcanololgy*, **52** (1989), 1–27.

17. C Siebe, M Abrams, J Luis Macias, J Obenholzner, Repeated volcanic disasters in prehispanic time at Popocatépetl, central Mexico: past key to the future?. *Geology*, **24** (1996), 399–402.

18. DR Crandell, *Postglacial Lahars from Mount Rainier Volcano*, (Washington, DC: US Geological Survey, 1971).

19. KM Scott, *Magnitude and Frequency of Lahars and Lahar-Runout Flows in the Toutle-Cowlitz River System*. (Washington, DC: US Geological Survey, 1989).

20. TW Sisson, JW Vallance, Frequent eruptions at Mount Rainier over the last 2,600 years. *Bulletin of Volcanology*, **71** (2009), 595–618.

21. S Thórarinsson, Th Einarsson, G Sigvaldasson, G Elisson, The submarine eruption off the Vestmann Islands 1963–64: a preliminary report. *Bulletin Volcanologique*, **27** (1964) 435–445.

22. S Thórarinsson, *The Öræfajökull Eruption of 1362. Acta Naturalia Islandica*, **2** (1958).

23. K Grönvold, H Jóhannesson, Eruption of Grímsvötn 1983. *Jökull*, **34** (1984), 1–11.

24. H Björnsson, *Hydrology of Ice Caps in Volcanic Regions* (Reykjavík: Vísingdafélag Íslendinga, 1988).

25. L Siebert, T Simkin, P Kimberly, *Volcanoes of the World* (3rd edition) (Berkeley, CA: University of California Press, 2010).

26. S Thórarinsson, Some new aspects of the Grímsvötn problem. *Journal of Glaciology*, **14** (1953), 267–274.

27. H Björnsson, Subglacial lakes and jökulhlaups in Iceland. *Global and Planetary Change*, **35** (2002), 255–271.

28. MT Guðmundsson, H Björnsson, Eruptions in Grímsvötn,Vatnajökull, Iceland 1934–1991. *Jökull*, **41** (1991), 21–45.

29. MT Guðmundsson, F Sigmundsson, H Björnsson, T Högnadóttir, The 1996 eruption at Gjálp, Vatnajökull ice cap, Iceland: efficiency of heat transfer, ice deformation and subglacial water pressure. *Bulletin of Volcanology*, **66** (2004), 46–65.

30. RB Waitt, Great Holocene floods along Jökulsá á Fjöllum, north Iceland. In *Flood and Megaflood Processes, Recent and Ancient Examples*, ed. IP Martini, VR Baker, G Garzón. (Oxford, MA: Blackwell Science, 2002), pp. 37–52.

31. G Larsen, MT Guðmundsson, H Björnsson, Eight centuries of periodic volcanism at the center of the Iceland hotspot revealed by glacier tephrostratigraphy. *Geology*, **26** (1998), 943–946.

32. BA Óladóttir, G Larsen, O Sigmarsson, Holocene volcanic activity at Grímsvötn, Bárdarbunga and Kverkfjöll subglacial centers beneath Vatnajökull, Iceland. *Bulletin of Volcanology*, **73** (2011), 1187–1208.

33. S Thórarinsson, The jökulhlaup from the Katla area in 1955 compared with other jökulhlaups in Iceland. *Jökull*, **7** (1957), 21–25.

34. H Tómasson, The jökulhlaup from Katla in 1918. *Annals of Glaciology*, **22** (1996), 249–254.

35. J Jónsson, Notes on the Katla volcanoglacial debris flows. *Jökull*, **32** (1982), 61–68.

36. J Maizels, Lithofacies variations within sandur deposits: the role of runoff regime, flow dynamics, and sediment supply characteristics. *Sedimentary Geology*, **85** (1993), 299–325.

37. H Björnsson, F Pálsson, MT Guðmundsson, Surface and bedrock topography of the Mýrdalsjökull ice cap, Iceland: the Katla caldera sites and routes of jökulhlaups. *Jökull*, **49** (2001), 29–46.

38. E Magnússon, MT Guðmundsson, MJ Roberts, G Sigurðsson, F Hökuldsson, FB Oddson, Ice–volcano interactions during the 2010 Eyjafjallajökull eruption, as revealed by airborne imaging radar. *Journal of Geophysical Research*, **117** (2012), B07405. doi: 10.1029/2012/JB009250, 2012.

39. JL Smellie, The 1969 subglacial eruption on Deception Island (Antarctica): events and processes during an eruption beneath a thin glacier and implications for volcanic hazards. In *Volcano-Ice Interaction on Earth and Mars*, ed. JL Smellie, MG Chapman (London: Geological Society of London, 2002), pp. 59–79.

40. BR Edwards, MT Gudmundsson, T Thordarson, *et al.*, Interactions between lava and snow/ice during the 2010 Fimmvorduhals eruption, south-central Iceland. *Journal of Geophysical Research*, **117** (2012), B04302. doi: 10.1029/2011JB008985.

41. JA Naranjo, HR Moreno, Laharic debris-flows from Villarica Volcano. *Boletín – Servicio Nacional de Geología y Minería*, **61** (2004), 28–38.

42. HR Moreno, GC Fuentealba, The May 17–19 1994 Llaima volcano eruption, southern Andes (38°42'S–71°44'W). *Andean Geology (Revista geológia de Chile)*, **21** (1994), 167–171.

43. AY Ozerov, GA Karpov, VA Droznin, *et al.*, The September 7 – October 2, 1994 eruption of Klyuchevskoi volcano. *Kamchatka, Volcanology and Seismology*, **18** (1997) 501–516.

44. ME Yount, TP Miller, RP Emanuel, FH Wilson, Eruption in the ice-filled caldera of Mount Veniaminof. *US Geological Survey Circular*, **945** (1985), 58–60.

45. A Belousov, B Behncke, M Belousova, Generation of pyroclastic flows by explosive interaction of lava flows with ice/water-saturated substrate. *Journal of Volcanology and Geothermal Research*, **202** (2011), 60–72.

46. BR Edwards, J Karson, B Wysocki, E Lev, I Bindeman, U Kueppers, Insights on lava–ice/snow interactions from large-scale basaltic melt experiments. *Geology*, **41** (2013), 851–854. doi:10.1130/G34305.1.

47. WH Mathews, The Table, a flat-topped volcano in southern British Columbia. *American Journal of Science*, **249** (1951), 830–841.

48. SC Porter, Pleistocene subglacial eruptions on Mauna Kea. In *Volcanism in Hawaii*, ed. RW Decker, TL Wright, PH Stauffer (Washington, DC: US Geological Survey, 1987), pp. 587–598.

49. DT Lescinsky, JH Fink, Lava and ice interaction at stratovolcanoes: use of characteristic features to determine past glacial extents and future volcanic hazards. *Journal of Geophysical Research*, **105** (2000), 23711–23726. doi:10.1029/2000JB900214.

50. WH Mathews, Ice-dammed lavas from Clinker Mountain, southwestern British Columbia. *American Journal of Science*, **250** (1952), 553–565.

51. JG Jones, Intraglacial volcanoes of the Laugarvatn region, south-west Iceland, I. *Quarterly Journal of the Geological Society of London*, **124** (1968), 197–211.

52. P Kokelaar, Magma–water interactions in subaqueous and emergent basaltic volcanism. *Bulletin of Volcanology*, **48** (1986), 275–289.

53. R Werner, H-U Schmincke, G Sigvaldasson, A new model for the evolution of table mountains: volcanological and petrological evidence from Herdubreid and Herdubreidatögl volcanoes (Iceland). *Geologische Rundschau*, **85** (1996), 390–397.

54. BR Edwards, JK Russell, K Simpson, Volcanology and petrology of Mathews Tuya, northern British Columbia, Canada: glaciovolcanic constraints on interpretation of the 0.730 Ma Cordilleran paleoclimate. *Bulletin of Volcanology*, **73** (2011), 479–496.

55. CA Wood, J Kienle, *Volcanoes of North America, United States and Canada* (Cambridge: Cambridge University Press, 1990).

56. G Komatsu, SG Arzhannikov, AV Arzhannikova, K Ershov, Geomorphology of subglacial volcanoes in the Azas Plateau, the Tuva Republic, Russia. *Geomorphology*, **88** (2007), 312–328.

57. LE Lisiecki, ME Raymo, A Pliocene–Pleistocene stack of 57 globally distributed benthic δ^{18}O records. *Paleoceanography*, **20** (2005). doi: 10.1029/2004PA001071

58. W Hamilton, *The Hallett Volcanic Province, Antarctica* (Washington, DC: US Geological Survey, 1972).

59. JL Smellie, MJ Hole, PAR Nell, Late Miocene valley-confined subglacial volcanism in northern Alexander Island, Antarctic Peninsula. *Bulletin of Volcanology*, **55** (1993), 273–288.

60. JL Smellie, JS Johnson, WC McIntosh, *et al.*, Six million years of glacial history recorded in volcanic lithofacies of the James Ross Island Volcanic Group, Antarctic Peninsula. *Paleogeography, Paleoclimatology, Paleoecology*, **260** (2008), 122–148.

61. G Kjartansson, *Geological Map of Iceland, Sheet 5, Central Iceland (scale 1:250,000)*. (Reykjavík: Museum of Natural History Iceland, 1983).

62. H Jóhannesson, S Jakobsson, K Sæmundsson, *Geological Map of Iceland, Sheet 6, South Iceland (scale 1:250,000)*. (Reykjavík: Museum of Natural History, 1977).

63. SP Jakobsson, GI Johnson, Intraglacial volcanism in the Western Volcanic Zone, Iceland. *Bulletin of Volcanology*, **74** (2012), 1141–1160.

64. SC Loughlin, Facies analysis of proximal subglacial and proximal volcaniclastic successions at the Eyjafjallajökull central volcano, southern Iceland. In *Volcano–Ice Interaction on Earth and Mars*, ed. JL Smellie, MG Chapman (London: Geological Society of London, 2002), pp. 149–178.

65. JL Smellie, Basaltic subglacial sheet-like sequences: evidence for two types with different implications for the inferred thickness of associated ice. *Earth-Science Reviews*, **88** (2008), 60–88.

66. P Huybers, C Langmuir, Feedback between deglaciation, volcanism, and atmospheric CO_2. *Earth and Planetary Science Letters*, **286** (2009), 479–491.

67. P Schmidt, B Lund, C Hieronymus, J Maclennan, T Arnadottir, C Pagli, Effects of present-day deglaciation in Iceland on mantle melt production rates. *Journal of Geophysical Research – Solid Earth*, **118** (2013), 3366–3379.

68. T Ui, H Yamamoto, K Suzuki-Kamata, Characterization of debris avalanche deposits in Japan. *Journal of Volcanology and Geothermal Research*, **29** (1986), 231–243.

69. L Siebert, H Glicken, T Ui, Volcanic hazards from Bezymianny- and Bandai-type eruptions. *Bulletin of Volcanology*, **49** (1987), 435–459.

70. P Francis, S Self, Collapsing volcanoes. *Scientific American*, **256** (1986), 90–97.

71. DR Crandell, *Gigantic Debris Avalanche of Pleistocene Age from Ancestral Mount Shasta Volcano, California, and Debris-Avalanche Hazards Zonation* (Washington, DC: US Geological Survey, 1989).
72. RMR Paguican, B van Wyk de Vries, A Lagmay, Hummocks: how they form and how they evolve in rockslide-debris avalanches. *Landslides*, **11** (2014), 67–80.
73. JW Vallance, KM Scott, The Osceola mudflow from Mount Rainier: sedimentology and hazard implications of a huge clay-rich debris flow. *Geological Society of America Bulletin*, **109** (1997), 143–163.
74. JW Vallance, Volcanic debris flows. In *Debris-Flow Hazards and Related Phenomena*, ed. M Jakob, O. Hungr (Berlin: Springer-Praxis, 2005), pp. 247–274.
75. ME Reid, TW Sisson, DI Brien, Volcano collapse promoted by hydrothermal alteration and edifice shape, Mount Rainier, Washington. *Geology*, **29** (2001), 779–782.
76. G Carrasco-Núñez, L Siebert, R Díaz-Castellón, L Vázquez-Selem, L Capra, Evolution and hazards of a long quiescent compound shield-like volcano: Cofre de Perote, eastern trans-Mexican volcanic belt. *Journal of Volcanology and Geothermal Research*, **197** (2010), 209–224.
77. MF Meier, Glaciers and water supply. *Journal of the American Water Works Association*, **61** (1969), 8–12.
78. LE Mattson, The influence of a debris cover on the mid-summer discharge of Dome Glacier, Canadian Rocky Mountains. *International Association of Hydrological Sciences*, **264** (2000), 25–34.
79. KM Jackson, AG Fountain, Spatial and morphological change on Eliot Glacier, Mount Hood, Oregon, USA. *Annals of Glaciology* **46** (2007), 222–226.
80. P Ginot, U Schotterer, W Stichler, MM Godoi, B Francou, M Schwikowski, Influence of the Tungurahua eruption on the ice cores records of Chimborazo, Ecuador. *The Cryosphere*, **4** (2010), 561–568.
81. D Frank, MF Meier, DA Swanson, *Assessment of Increased Thermal Activity at Mount Baker, Washington, March 1975 – March 1976* (Washington, DC: US Geological Survey, 1977).
82. J Imbrie, A Berger, EA Boyle, *et al.*, On the structure and origin of major cycles, 2: the 100,000-year cycle. *Paleoceanography*, **8** (1993), 699–735.
83. WF Ruddiman, *Plows, Plagues, and Petroleum: How Humans took Control of Nature* (Princeton, NJ: Princeton University Press, 2005).
84. SL de Silva, GA Zielinski, Global influence of the AD 1600 eruption of Huaynaputina, Peru. *Nature*, **393** (1998), 455–458.
85. A Robock, Volcanic eruptions and climate. *Reviews of Geophysics*, **38** (2000), 191–219.
86. JB Witter, S Self, The Kuwae (Vanuatu) eruption of AD 1452: potential magnitude and volatile release. *Bulletin of Volcanology*, **69** (2007), 301–318.
87. GH Miller, A Giersdóttir, Y Zhong, *et al.*, Abrupt onset of the Little Ice Age triggered by volcanism and sustained by sea-ice/ocean feedbacks. *Geophysical Research Letters*, **39**, 2 (2012), L02708. doi:10.1029/2011GL050168.
88. MB Dyurgerov, MF Meier, Twentieth century climate change: evidence from small glaciers. *Proceedings of the National Academy of Sciences USA*, **97**, (2000), 1406–1411.
89. S Yamaguchi, R Naruse, T Shiraiwa, Climate reconstruction since the Little Ice Age by modelling Koryto glacier, Kamchatka Peninsula, Russia. *Journal of Glaciology*, **54** (2008), 125–130.
90. PD Ward, *Under a Green Sky: Global Warming, the Mass Extinctions of the Past and What They Can Tell Us about Our Future* (New York: Harper Collins, 2007).

91. J Kargel, G Leonard, R Wheate, B Edwards, ASTER and SEM change assessment of changing glaciers near Hoodoo Mountain, British Columbia, Canada. In *Global Land Ice Measurements from Space*, eds. J Kargel, GJ Leonard, MP Bishop, A Kààp, BH Raup (Berlin: Springer-Praxis, 2014) pp. 354–373.

92. TW Sisson, JE Robinson, DD Swinney, Whole-edifice ice volume change A.D. 1970 to 2007/2008 at Mount Rainier, Washington, based on LiDAR surveying. *Geology*, **39** (2011), 639–642.

93. JE O'Connor, Our vanishing glaciers: one hundred years of glacier retreat in the Three Sisters area, Oregon Cascade Range. *Oregon Historical Quarterly*, **114** (2013), 402–427.

94. HD Granados, The glaciers of Popocatépetl volcano (Mexico): changes and causes. *Quaternary International*, **43** (1997), 53–60.

95. A Rivera, F Bown, R Mella, *et al.*, Ice volumetric changes on active volcanoes in southern Chile. *Annals of Glaciology* **43** (2006), 111–122.

96. DE Thost, M Truffer, Glacier recession on Heard Island, southern Indian Ocean. *Arctic, Antarctic, and Alpine Research*, **40** (2008), 199–214.

97. E Jordan, L Ungerechts, B Caceres, A Penafiel, B Francou, Estimation by photogrammetry of the glacier recession on the Cotopaxi volcano (Ecuador) between 1956 and 1997. *Hydrological Science Journal*, **50** (2005), 949–961.

98. M Vuille, B Francou, P Wagnon, *et al.*, Climate change and tropical Andean glaciers: past, present and future. *Earth-Science Reviews*, **89** (2008), 79–96, doi:10.1016/j.earscirev.2008.04.002

99. PW Mote, G Kaser, The shrinking glaciers of Kilimanjaro: can global warming be blamed? *American Scientist*, **95** (2007), 318–325.

100. LG Thompson, HH Brecher, E Mosley-Thompson, DR Hardy, BG Mark, Glacier loss on Kilimanjaro continues unabated. *Proceedings of the National Academy of Sciences* **106** (2009), 19770–19775.

101. RD Moore, SW Fleming, B Menounos, *et al.*, Glacier change in western North America: influences on hydrology, geomorphic hazards and water quality. *Hydrological Processes*, **23** (2009), 42–61.

102. W Hildreth, J Fierstein, Eruptive history of Mount Katmai, Alaska. *Geosphere*, **8** (2012), 1527–1567.

103. SP Schilling, PE Carrara, RA Thompson, EY Iwatsubo, Posteruption glacier development within the crater of Mount St. Helens, Washington, USA. *Quaternary Research*, **61** (2004), 325–329.

104. JS Walder, SP Schilling, JW Vallance, RG LaHusen, Effects of lava-dome growth on the Crater Glacier of Mount St. Helens, Washington. In *A Volcano Rekindled: The Renewed Eruption of Mount St. Helens, 2004–2006*, eds. DR Sherrod, WE Scott, PH Stauffer (Washington, DC: US Geological Survey, 2008), pp. 257–276.

16

Debris-flow activity from high-elevation, periglacial environments

MARKUS STOFFEL AND CHRISTOPH GRAF

16.1 Introduction

Extreme rainstorms in headwater catchments will typically trigger flash floods, debris floods or debris flows [1]. The nature of the actual process will depend on the interplay of hydrological, geomorphometrical and geotechnical slope parameters, composition of source materials, availability of sediments, and on frequency–magnitude characteristics of the precipitation itself [2]. These different process types develop at temporal and spatial scales that conventional observation systems for rainfall, torrent and sediment discharge are unable to monitor systematically [1]. This results in limited knowledge of their controls and leads to considerable uncertainty in early warning and risk management. Given the small size of high-elevation headwater catchments (<100 km^2) and proximal sediment sources, their response to rainfall can be very sudden. Flash flood systems are known to be hotspots in terms of the number of persons affected, and the proportion of fatalities for individual events [3]. A comparable detail in data cannot readily be retrieved for debris flows, as indications on fatalities and damage are often reported in combination with information on landslide or flood damage [4].

Debris flows in high mountains are typically initiated by the sudden input of considerable quantities of water into channels which in turn mobilize stored sediment. In addition to triggering by extreme rainstorms, debris flows have also been reported to be released by rapid snowmelt, rain-on-snow storms, the sudden emptying of glacier water bodies (water pockets, proglacial/supraglacial lakes), or through the rupture of landslide dams [5,6]. More frequently, however, high-elevation debris flows occur as a result of high-intensity, convective rainstorms of short duration, or low-intensity advective precipitation events over several days [7,8].

The High-Mountain Cryosphere, ed. Christian Huggel, Mark Carey, John J. Clague and Andreas Kääb. Published by Cambridge University Press. © Cambridge University Press 2015.

With ongoing climate change, air temperatures have increased considerably in the last few decades, such that they were warmer than in any comparable period in Europe during at least the last 2000 years [9]. Based on the results of IPCC AR5 [10] or its special report on extremes (SREX [11]), mean and extreme temperatures will likely continue to rise over the next decades. Under the A1B emission scenario from IPCC AR4 [11], Gobiet *et al.* [12] project a decadal warming of 0.25 °C for the European Alps until the mid twenty-first century. This warming trend is expected to accelerate during the latter half of the century, for which an average decadal increase of 0.36 °C is assessed. In addition, Gobiet *et al.* [12] also project shifts in precipitation seasonality and intensity. Changes in precipitation, in particular heavy precipitation, have been demonstrated to affect the occurrence of debris flows, and a significant increase in high-intensity precipitation has been documented over Europe for the recent past [13]. Projections for future precipitation are consistent with past changes, but future changes tend to be less uniform and more hidden in natural variability and model errors [14]. Nevertheless, several studies indicate that future heavy precipitation might increase in fall and winter, whereas models portray a less consistent picture for future summers [15].

Such changes in temperature and precipitation are likely to alter the nature of debris-flow-triggering precipitation events in many ways. A warmer climate is likely to result in higher 0 °C isotherms, allowing for more precipitation to fall in its liquid form and thus potentially increasing the area contributing effectively to direct runoff [16]. In addition, a warmer atmosphere will translate – at least in theory – into higher air moisture content, which in turn might increase the potential for increased frequency and intensity of heavy precipitation events over the European Alps [17].

Climate change is also expected to have a range of secondary effects, with indirect linkages to debris-flow activity. Changes in subsurface temperatures and distribution of permafrost are expected to promote in certain zones its downwasting in the medium to long term [18], which in turn is likely to liberate additional sources of unconsolidated material for mass transfers [19,20]. As a result of the temperature increase of the twentieth century, the lower permafrost limit has already risen by 150–250 m in the Swiss Alps, thus increasing the probability for slopes and rockwalls to become more unstable and to provide sediment into torrential systems [21]. Conversely, increasing temperatures are also thought to allow vegetation to grow at higher altitudes and to stabilize loose material [22], provided the ground is sufficiently stable for plant colonization.

While many of the above considerations point to a potentially enhanced debris-flow activity in a future climate, changes in activity are so far difficult to detect in observational records. Uncertainties also remain considerable due to error margins inherent to scenario-driven global projections, and due to the coarse spatial resolution of available downscaled model data in mountain environments. A series of studies has addressed possible consequences of climate changes on permafrost bodies and the associated occurrence of debris flows, with a research focus on the European Alps. Keiler *et al.* [23] assumed that glacier melt and permafrost thaw could lead to widespread surface destabilization and an increase in the frequency and magnitude of geomorphic hazards. Sattler *et al.* [24] compared zones of marginal permafrost with initiation zones of debris flows, and concluded that no links exist between atmospheric warming, permafrost degradation and debris-flow occurrence in South Tyrol. In the Dolomites, Floris *et al.* [25] observed an increase in high-intensity, short-duration rainfalls since 1990 and suggest that such changes could lead to more rainfall events triggering debris flows. The diverging conclusions on future debris-flow activity primarily reflect (1) the error margins inherent in scenario-driven global climate projections, and (2) lacking spatial resolution of downscaled climate projections and limited information on past process behaviour in high-elevation catchments [26].

16.2 Periglacial debris flows in the Zermatt valley: an overview

In the Swiss Alps, the majority of unstable slopes are located at altitudes of ~2700 m asl, where sediment transfers typically occur outside the reach of humans or their infrastructure. The situation is somewhat different in the Zermatt Valley (Figure 16.1), a high-elevation, north–south oriented glacial valley in the Swiss Alps. There, the detachment of thawing permafrost material causes frequent rockfalls on steep slopes and debris flows in high-gradient gullies [20,27] through which material is transferred to the valley floor at elevations between 1100 (N) and 1600 m asl (S). As a result of the excellent database on past disasters, recent developments in the local rock glacier bodies and the availability of highly resolved, long-term projections of climatic changes, we hereafter illustrate data from eight (Table 16.1) debris-flow torrents to document impacts of past, ongoing and possible future changes in the cryosphere and atmosphere on periglacial debris flows.

Extensive till, scree slopes and rock glaciers represent the principal and extensive sediment sources for debris flows which are commonly triggered at elevations between 2000 and 3000 m asl. Here, high annual and daily thermal ranges favour frost weathering and regolith production delivered to scree slopes. Slopes in the initiation zones range from angles of 27° to 41° (mean 33°) and are dominated by permafrost in most catchments. The frozen debris is thought to form poorly permeable layers which promote drainage along preferential paths in the source areas of debris flows. Debris flows are triggered either through the wetting of material continuously delivered by the permafrost body to the channel or due to

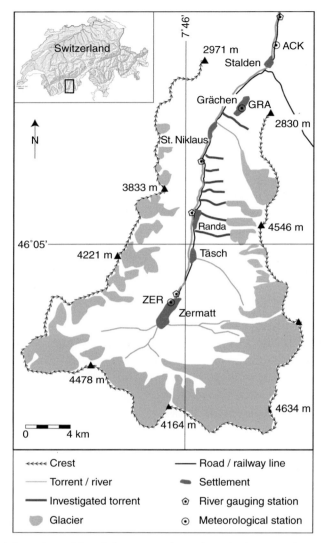

Figure 16.1 Location of the eight torrents investigated, river gauging stations and meteorological stations in the Zermatt valley.
Source: after [8].

failures at the rock glacier fronts during exceptional water input. Initiation is favoured by a liquefaction mechanism similar to that described for shallow landslides [28]. The wetting typically occurs during rainstorms, but debris flows at high-elevation sites also occur when sediment shear resistance is reduced by the melting of ice particles [29], by snow melting and/or a combination of both.

Debris flows propagate in steep channels (22–33°; mean 27.6°) and deposit in the transit zones and cones located in the valley floor (1200–1400 m asl). At

Table 16.1 *Geomorphic properties of the investigated torrents in the Zermatt valley (after [8]).*

Article III.	Ritigraben	Grosse Graben	Bielzug	Fallzug	Geisstriftbach	Birchbach	Dorfbach	Wildibach
Glacier	x				(x)	x	x	x
Glaciated area (km2)					0.08	3.4	2.1	2.4
Periglacial processes		x	x	x	x	x	x	x
Elevation catchment area (m asl)	3136 to 2600	3178 to 1900	3192 to 2100	3350 to 1900	4035 to 2100	4545 to 2000	4479 to 2000	4545 to 2100
Exposition	West	West	West	West	West	West	West	West
Rock	Gneiss	Gneiss	Gneiss	Gneiss	Gneiss	Gneiss	Gneiss	Gneiss
Catchment area (km^2)	0.8	1.5	1.5	2.1	4.3	7.1	5.6	7.7
Size cone (ha)	47	48	3.7	26	13	27	63	46
Elevation cone (m asl)	1460–1800	1200–1560	1230–1320	1250–1420	1260–1360	1300–1440	1400–1590	1420–1540

Figure 16.2 (a) Ritigraben system from its source to the confluence (catchment area: 1.36 km², channel lengths: 3.5 km). (b) Detailed view of the debris-flow cone (32 ha) and its forest.
Source: after [30].

Ritigraben (Figure 16.2), the cone is situated on a structural terrace (1460–1800 m asl; [30]), far above the valley floor. The proportion of debris-flow sediment originating from the initiation zones in the area typically is one order of magnitude smaller than the volume deposited on the cone [31]. However, where active-layer failures are the cause, up to one-third of the total debris-flow material is released from the rock glacier front [32].

Annual precipitation in the Zermatt valley is 589 mm in Grächen (1550 m asl) and 639 mm in Zermatt (1638 m asl; mean for the climate normal 1981–2010). Differences in annual precipitation totals and rainy days between the two stations are considerable and reflective of the proximity of the Zermatt meteorological station to the main Alpine divide. Mean annual air temperatures are 6 and 4.2 °C at the Grächen and Zermatt stations, respectively. January through April are driest, with a monthly mean of 30–40 mm precipitation (snow or rain), whereas October is wettest with 63 ± 52 mm.

The database on past debris flows has been constructed using archival and dendrogeomorphic records. The written archives on debris flows in the region are typically poor and incomplete, and lack information on volumes and triggers. In all eight catchments, debris-flow channels pass forested cones where vegetation has been impacted by debris flows. Woody vegetation damaged by debris flows or growing on deposits provides a means for dating and interpreting past events [26]. Impacts on trees will thereby depend on flow velocity, thickness and clast sizes. Effects can range from burial of lower trunks, tree tilting and stem corrosion to

Figure 16.3 Dendrogeomorphic reconstruction of debris-flow activity at Ritigraben since 1566; ten of the 124 events were known from archives. The dotted line shows the number of tree-ring cores available for analysis.
Source: after [16].

complete destruction of large forest areas [33,34]. As a result, damage in trees can be used to infer past process activity, both in temporal and spatial terms. The position of damage within a tree ring can be used to determine events with intra-seasonal precision [35].

The tree-ring database for the Zermatt valley contains 2467 *Larix decidua* and *Picea abies* trees [26,30] and covers AD 1570–2009; it contains 417 debris flows recorded in 226 different years. A total of 296 debris flows have been recorded after 1850. Decadal frequencies point to peaks in debris-flow activity after the end of the Little Ice Age and in 1920–1929. By contrast, activity was rather low during the most recent part of the record (2000–2009), when only 13 events have been recorded in the valley [7,26,30]. Figure 16.3 illustrates the temporal occurrence of debris flows at Ritigraben [16,30,36].

16.3 Periglacial debris flows in the Zermatt valley: triggers and thresholds

As a result of cooler mean air temperatures, debris-flow activity was mostly restricted to June through September during the nineteenth century, but tends to start in May and to cease in October since the start of pronounced warming in the 1980s [8]. The earliest event on record occurred on May 18, 1960, the latest on October 29, 1913. Over the past ~150 years, most debris flows were released in July and August (or JA; 60%), but events were also common in June and September (18% and 12%, respectively). Debris flows were rather scarce early and late in the debris-flow season, with 3% and 7% in May and October, respectively (Figure 16.4; [8]). Between 1864 and 1899, 76% of debris flows was triggered in July and August and no debris flows occurred in May and October. The first debris flow in May occurred in 1923 and the first in October for 1911. Since the 1970s, debris flows became more frequent early (May) and late (October) in the season, with 17% of all events, whereas debris flows have become less abundant in JA (51%).

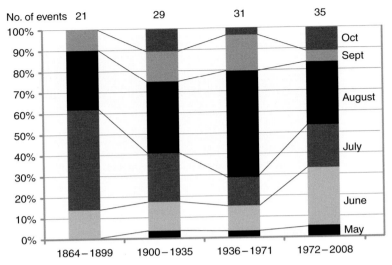

Figure 16.4 Temporal occurrence of debris flows between 1864 and 2008 in the Zermatt valley. Each bar represents 35 years and the percentage of debris flows occurring per month. Total number of debris flows per period is given on top. Whereas no debris flows occurred in May and October in the nineteenth century, 18% of all events now occur in these months.
Source: after [8].

Debris flows early in the season (May and June; or MJ) are generally triggered by low rainfall totals (<20 mm d^{-1}) as snowmelt adds considerable amounts of water to the system [7,31,37]. A majority of debris flows were released by short-lived rainfalls (≤ 1 day; mostly thunderstorms). This is particularly true for May through August, when 60% of all events fall into this category, whereas the occurrence of debris flows late in the season (September and October, SO) is more often related to longer-lasting advective rainfalls. Longer-lasting precipitation events (three-day events) were scarce in general and especially between May and August. In JA, less than 10% of all events were triggered by three-day rainfalls, but events become more crucial in SO when they are responsible for the release of one-third of the debris flows [8].

As the eight torrents in the valley have important sediment supply, the triggering of debris flows is mainly controlled by climatic factors (with water being the limiting factor). Coupling of debris-flow with daily rainfall data thus shows that events are triggered by precipitation >20 mm and that minimum rainfall thresholds are lower or sometimes even zero early in the season (MJ), suggesting that less water will be needed to achieve positive pore water pressure [32]. Debris flows are therefore mostly triggered by snowmelt or thunderstorms early in the season, whereas advective storms become important in late summer and fall [7].

16.4 Frequency–magnitude relations of debris flows: mirrors of the state of permafrost?

Debris-flow activity in a watershed is usually defined in terms of magnitude and frequency (M–F) [38,39]. While M–F relations have long formed the basis for risk assessment and engineering design in hydrology and hydraulics, only fragmentary data exists for debris flows. Several authors have derived M–F relationships for debris flows in the past [31,39,40], mostly based on stratigraphic techniques [41], lichenometry [42], LiDAR interpretation [43] or with empirical/statistical equations [44], but such data remain extremely scarce for high-elevation catchments.

Based on the tree-ring data presented above, Stoffel [31] reconstructed M–F relations of 62 debris flows of the Ritigraben since 1863. The magnitude of individual debris flows was classifed as *S, M, L, XL*, and derived from volumetric information of deposits, grain size distributions of boulders, and a series of debris-flow surrogates (snout elevations, tree survival, lateral spread of surges; Figure 16.5). *Class S* and *M* debris flows ($<5 \times 10^3$ m^3) encompass a typical size of events and have mean recurrence intervals of 5.4 and 7.4 years, respectively (Figure 16.6). *Class XL* events (10^4–5×10^4 m^3) are, in contrast, only identified in 1922, 1948 and 1993, and major erosional activity on the cone was restricted to the events of 1948 and 1993. *Class L* and *XL* debris flows were triggered by advective storms (rainfall >50 mm per event) in August and September, when the active layer of the rock glacier in the source area of debris flows is largest [32].

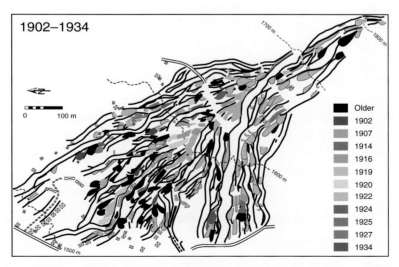

Figure 16.5 Deposition of debris-flow material on the cone between 1902 and 1934 as inferred from damage in trees. Deposits shown in black are dated, but are older than the time segment illustrated.
Source: after [30].

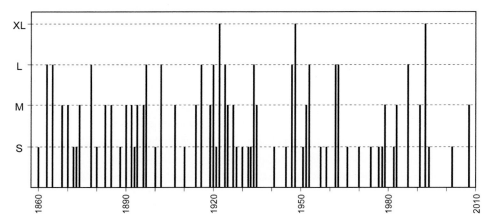

Figure 16.6 Reconstructed debris-flow magnitudes since 1863. Note the clustering of large events in the early twentieth century. The only *class XL* events on record occurred in September 1922, September 1948 and September 1993 and were triggered by long-lasting, advective rainfalls. Partial failures of the rock–glacier front were involved in the 1948 and 1993 debris flows, and probably also played a role during the 1922 event.
Source: after [31].

In view of the temperature evolution in the valley, one can demonstrate that climate exerted control on debris-flow volumes released from the source area of the Ritigraben torrent since 1863. The cooler temperatures at the end of the Little Ice Age and into the very early twentieth century have prevented the triggering of *class XL* events. Periods with warmer temperatures, by contrast, coincide with the occurrence of large events in September 1922, September 1948 and September 1993. The *class XL* events also concur with periods of above-average temperatures in the region and rock-glacier-related instability (mostly rockfalls) in other catchments in the valley, namely at Bielzug and Täschgufer [27], such that the linkage of enhanced permafrost thawing and mass wasting appears to be more than just a local phenomenon at Ritigraben. In contrast to other sites, however, the unprecedentented warming of the late 1990s and the twenty-first century did not cause enhanced debris-flow activity at Ritigraben.

16.5 The future of debris flows in the Zermatt valley: less, unless...

Based on point-based downscaling of climate scenarios for Grächen and Zermatt, for 2001–2050 and 2051–2100, the evolution of temperature and rainfalls above specific thresholds (10–50 mm) as well as the duration of precipitation events (1–3 days) reveal a drying tendency for future summers and more precipitation events during the shoulder seasons [44]. Projections agree with observations of Schmidli and Frei [45], who reported a shift in precipitation seasonality in the observational records. At the same time, we observe an increase in the occurrence of heavy (>40 mm) one-day precipitation events in the region. The drier conditions in future summers and the wetting of springs, falls and early winters are likely to have significant impacts on the behaviour of debris flows. One might expect a modification in debris-flow triggering from these high-elevation catchments, both in terms of frequency and magnitude. One might also expect only slight changes in the frequency of debris-flow events by the mid twenty-first century [44], but potentially an increase in the magnitude of debris flows due to larger sediment amounts delivered to the channels and an increase in extreme precipitation events. In the second half of the twenty-first century (Figure 16.7), the overall absolute number of days with conditions favourable for the release of rainfall-triggered debris flows is expected to decrease, especially in summer. The anticipated increase of liquid rainfalls during the shoulder seasons (March and April, November and December) is not expected to compensate for the decrease in future heavy summer rainfalls over two or three days [44].

In addition to changes in precipitation, permafrost stability will play a crucial role in future debris-flow activity. Increasing air temperatures can, on multi-annual timescales, impact on permafrost stability, even more so in steep terrain. Interpretation of satellite imagery from the Zermatt valley analysed with radar interferometry technology shows that five permafrost bodies – mostly rock glaciers – exhibited a significant acceleration of movement in the late 1990s (from a few decimetres to 2–5 m yr^{-1} [46]), such that more sediments are delivered to Grosse Grabe, Bielzug, Fallzug, Geisstriftbach, Dorfbach and Wildibach.

Instability and displacement rates have risen further recently to show movement rates without historical precedents. At Grabengufer (Dorfbach; Figure 16.8), increasing air and ice temperatures have favoured the multiplication of annual displacement rates from just a few decimetres in the past to 80 m yr^{-1} in 2010 [37,46]. While movements have been documented to be on the decline at the site since the 2010 crisis, rock glacier movements remain unusually high at >25 m yr^{-1} [46]. At Dirru (Geisstriftbach) and Gugla (Bielzug) rock glaciers, movement rates are somewhat less dramatic but still unusually high, with measured displacements of 10 and 5 m yr^{-1}, respectively.

As a consequence of the enhanced movement of these permafrost bodies and related slide and fall processes, increasingly large amounts of loose sediment are delivered into debris-flow systems. By way of example, an estimated total of 70 000–120 000 m^3 has been delivered from Grabengufer into Dorfbach [46], with a present-day annual increase in sediment input of 8000 m^3. At the same site, Kenner *et al.* and Willi *et al.* [48,49] used various sets of *in-situ* and remotely sensed data from summers 2010 and 2011 to map sediment transfer at

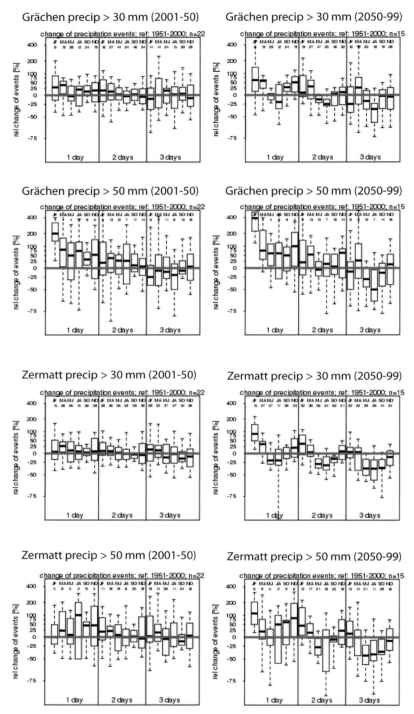

Figure 16.7 Changes in liquid precipitation events exceeding the 30 and 50 mm thresholds at Grächen and Zermatt. Left panels show 2001–2050 projections based on 22 downscaled RCMs, right panels give 2050–2099 projections obtained from 15 RCMs. Numbers in the top panel show the median number of events of the models in the reference period 1951–2000. Boxplots represent 10th, 25th, 50th, 75th, and 90th percentiles.
Source: after [44].

Figure 16.8 Dorfbach torrent with ice-covered Dom (left, 4545 m asl), Täschhorn (right, 4491 m asl) and the village of Randa (bottom, 1270 m asl). Recent instability at Grabengufer rock glacier (middle of photo) has delivered ample sediment to the Dorfach torrent and thus altered the frequency and magnitude of possible large debris flows. Catchments to the left and right of Dorfbach torrent (Birchbach and Wildibach) show fewer changes for the time being.
Source: Marcia Phillips, used with permission.

Grabengufer. Relocation of material was found to be substantial, especially on the rock glacier (6760 m^3 erosion) and gully (23 720 m^3 deposition) [47].

Sediment delivery into the torrential systems is smaller at Geisstriftbach, with a total of 20 000 m^3 and an annual addition of 8000 m^3, and only accounts for a few thousand cubic metres (and an annual input of 1500 m^3) at Bielzug (Figure 16.9). Annual sediment input are present-day values and are susceptible to changing significantly with time. In the case of Grabengufer, the frontal sector of the rock glacier has been disconnected from the main body, which likely contributed to reduced displacement rates as well as to the observed reduction in sediment delivery. The volume of the unstable rock glacier front has been estimated to be in the order of 45 000 m^3 and sediment is nowadays delivered mainly by continuous rockfalls and very small debris flows, but at much smaller rates than in 2010.

The massive delivery of rock glacier sediments into the debris-flow channels is likely to increase the frequency of debris flows despite the fact that the frequency of rainy days is expected to decline during future summers. In the case that debris flows are triggered by short-lived, moderate thunderstorms, magnitudes of individual debris flows should remain comparable to what has been observed in the past. In the case of intense water inputs into the system, be it in the form of extreme

Figure 16.9 Oversteepened front of Gugla rock glacier (Bielzug, yellow) during its most recent crisis in June 2013. Darker, wetted surfaces represent the release areas of smaller debris flows (blue lines). The southern part of the rock glacier front (right, red line) is moving faster than the reminder of the rock glacier. Gugla (Breithorn) is about 130 m wide and up to 40 m thick (yellow-dashed) at its front.

rainfalls or the rupture of water pockets, one may expect extreme events to be triggered from the source areas of Bielzug, Geisstriftbach and Dorfbach torrents.

The changes and process chains described herein are likely to represent situations for which, at best, only very limited or even no historical precedents exist. In addition to being of unprecedented intensity and magnitude, these rock-glacier instabilities and related debris flows would also affect a geographic context where both the demographic pressure and socio-economic development have been largely on the rise over the past few decades. In combination, the changes in the behaviour of natural processes and anthropogenic modifications in the valley are calling for new and innovative solutions to cope with climate change and related impacts in a sustainable way.

16.6 Anticipating events without historic precedence

One possible way of anticipating events without historical precedents is the physically based process modelling of debris flows, as it allows inclusion of changing and highly resolved terrain conditions [49], changing water and sediment inputs and the positioning of protective structures. For the torrents described herein, the numerical model RAMMS [50] has been used to test different combinations of water and sediment inputs and to define realistic scenarios of future debris flows, taking account of the changing conditions in the source areas and process chains.

At Dorfbach, modelling was realized for (1) the baseline scenario with conditions prior to rock glacier instability [51], (2) the new situation with more sediment, and (3) for the new situation but with protective measures along the debris-flow system. Model calibration was performed with a ten-year series of automated debris-flow observations [52] and earlier archival records [53]. Calibration as well as the model setup and different starting procedures are described in [51]. The hazard assessment of pre-rock glacier instability shows several locations where breakouts could have taken place. However, the central part of the village of Randa has not been affected by debris flows for at least 250 years [40]. Before the acceleration of rock glacier movements, event volumes of annually occurring debris flows in Dorfbach were <10 000 m^3 per event, often in several surges. In the absence of rock glacier instability, events with a return period of 100 years were in the order 15 000–20 000 m^3 for bedload events and 70 000–80 000 m^3 for debris flows.

Under the current situation of rock glacier instability and huge sediment transfers into the debris-flow system, [54] estimated the mean volume of debris flows with a return period of 100 years to be in the order of 180 000 m^3. Graf *et al.* [37] obtained event volumes for the same return period of 150 000 m^3, and expect events to occur in several surges after abundant rainfall and additional water from snowmelt. Peak discharge for the maximum surge was calculated as 625 m^3 s^{-1} based on estimations, extrapolation of observed and measured (smaller) events and empirical relationships (see Figure 16.11). The inclusion of data on rock glacier movements and sediment input [46] considerably improved the definition of event scenarios and an adaptation of event volumes. These data were complemented by (1) field-based data and (2) terrestrial and aerial survey on process initiation, sediment transfer, bulking mechanisms and deposition behaviour [47,48].

Intensity maps for different return periods and a preliminary hazard map are shown in Figure 16.10 for the new situation and in absence of additional protective measures. The illustration is for a low-probability event size, with the greenish layers showing zones with low- to high-intensity impacts. Potential flow paths for frequent to rare events from earlier studies are overlain, as are weak points along the torrent. Under the impression of ongoing instability and enhanced mass movements from the Grabengufer rock glacier as well as the accumulation of sediment in the transit zone, the positioning and extension of a lateral dam on the southern side of the torrent was evaluated and dimensioned with RAMMS. Modelling included not only the magnitudes of debris flows of different return periods, but also took account of changing geometric conditions, especially close to the fan apex, where major deposition has been included during scenario definition.

Figure 16.10 Colluvial fan of Dorfbach torrent with the village of Randa. Potential outbreak locations are indicated by arrows, potential flow paths for frequent to rare events are shown with dashed lines. The greenish layers define zones of low- to high-intensity impacts of low-probability debris-flow events. Intensities are derived from RAMMS.
Source: Orthophoto: swissimage 2014 swisstopo 5704000000.

16.7 Outlook and conclusions

Current understanding suggests that only moderate changes in the overall frequency of high-elevation debris-flow events from permafrost environments will occur over the short term, but that an increase in the overall magnitude of debris flows due to larger amounts of sediment delivered to the channels and an increase in extreme precipitation events is possible. Locally, however, strong and unprecedented changes in the frequency and magnitude of debris flows have recently been observed in relation to complex process coupling, associated with combined changes in climate, glacier, permafrost and sediment availability. Such local aggravations of general trends are likely to be observed in the future as well. More generally, the absolute number of days with conditions favourable for the release of debris flows will likely decrease, especially in summer, and the anticipated increase of days with heavy rainfalls during the shoulder seasons (March, April, November, December) is not expected to compensate for the decrease in future

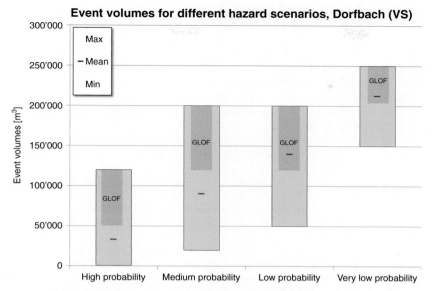

Figure 16.11 Bandwidth of event volumes from different assessments over the last 15 years for different hazard scenarios at Dorfbach. Whereas frequent events range in a narrow bandwidth, average and rare events span over a very wide range. This is due to different scenarios for event triggering, taking account of potential glacial lake outburst floods in addition to snowmelt and rainfall triggered events and actual changes in the catchment's sediment budget.

heavy summer rainfalls. The volume of entrained debris will crucially depend on the stability or acceleration of rock-glacier bodies.

The volume of entrained debris from the source areas tends to be larger in summer and autumn when the active layer of the permafrost bodies is largest and allows for larger volumes of sediment to be mobilized, but the situation has been shown to depend also on the stability and climate change-related accelerations of rock-glacier bodies. Along with the occurrence of more extreme precipitation events, these rock-glacier instabilities could lead to debris flows without historic precedents in the future.

References

1. M Borga, M Stoffel, L Marchi, F Marra, M Jakob, Hydrogeomorphic response to extreme rainfall in headwater systems: flash floods and debris flows. *Journal of Hydrology*, **518** (2014), 194–205.
2. M Jakob, O Hungr, *Debris Flow Hazard and Related Phenomena* (Heidelberg, Berlin, New York: Springer, 2005).
3. SN Jonkman, Global perspectives on loss of human life caused by floods. *Natural Hazards*, **34**: 2 (2005), 151–175.
4. M Jakob, D Stein, M Ulmi, Vulnerability of buildings to debris-flow impact. *Natural Hazards*, **60** : 2 (2012), 241–261.

5. GF Wieczorek, T Glade, Climatic factors influencing occurrence of debris flows. In: *Debris-flow Hazards and Related Phenomena*, ed. M Jakob, O Hungr (Chichester: Springer, 2005), pp. 325–362.

6. R Worni, C Huggel, M Stoffel, Glacier lakes in the Indian Himalayas: glacier lake inventory, on-site assessment and modeling of critical glacier lakes. *Science of the Total Environment*, **468–469** (2013), S71–S84.

7. M Stoffel, M Bollschweiler, M Beniston, Rainfall characteristics for periglacial debris flows in the Swiss Alps: past incidences – potential future evolutions. *Climatic Change*, **105** (2011), 263–280.

8. M Schneuwly-Bollschweiler, M Stoffel, Hydrometeorological triggers of periglacial debris flows in the Zermatt Valley (Switzerland) since 1864. *Journal of Geophysical Research*, **117** (2012), F02033.

9. PAGES 2k Consortium, Continental-scale temperature variability during the past two millennia. *Nature Geoscience*, **6** (2013), 339–346.

10. IPCC, *Summary for Policymakers: Working Group I Contribution to the IPCC Fifth Assessment Report Climate Change 2013: The Physical Science Basis* (Cambridge: Cambridge University Press, 2013).

11. IPCC, *Managing the Risks of Extreme Events and Disasters to Advance Climate Change Adaptation: A Special Report of Working Groups I and II of the Intergovernmental Panel on Climate Change* (Cambridge: Cambridge University Press, 2012).

12. A Gobiet, S Kotlarski, M Beniston, G Heinrich, J Rajczak, M Stoffel, 21st century climate change in the European Alps: a review. *Science for the Total Environment*, **493** (2014), 1138–1151.

13. G Lendernik, E Meijgarrd, Increase in hourly precipitation extremes beyond expectations from temperature changes. *Nature Geoscience*, **1** (2008), doi: 10.1038/ngeo262.

14. GA Meehl, C Covey, T Delworth, *et al.*, The WCRP CMIP3 multimodel dataset. *Bulletin of the American Meteorological Society*, **88** (2007): 1383–1394.

15. J Rajczak, P Pall, C Schär, Projections of extreme precipitation events in regional climate simulations for Europe and the Alpine Region. *Journal of Geophysical Research*, **118** (2013), 1–17.

16. M Stoffel, M Beniston, On the incidence of debris flows from the early Little Ice Age to a future greenhouse climate: a case study from the Swiss Alps. *Geophysical Research Letters*, **33** (2006), L16404.

17. P Pall, MR Allenand, DA Stone, Testing the Clausius–Clapeyron constraint on changes in extreme precipitation under CO_2 warming. *Climate Dynamics*, **28** :4 (2007), 351–363.

18. N Salzmann, J Noetzli, C Hauck, *et al.*, RCM-based ground surface temperature scenarios in high-mountain topography and their uncertainty ranges. *Journal of Geophysical Research*, **112** (2007), F02S12.

19. C Harris, LU Arenson, HH Christiansen, *et al.*, Permafrost and climate in Europe: monitoring and modelling thermal, geomorphological and geotechnical responses. *Earth-Science Reviews*, **92** (2009), 117–171.

20. M Stoffel, C Huggel, Effects of climate change on mass movements in mountain environments. *Progress in Physical Geography*, **36** (2012), 421–439.

21. S Gruber, W Haeberli, Permafrost in steep bedrock slopes and its temperature-related destabilization following climate change. *Journal of Geophysical Research*, **112** (2007), F02S18.

22. C Baroni, S Armiraglio, R Gentili, A Carton, Landform–vegetation units for investigating the dynamics and geomorphologic evolution of alpine composite debris cones (Valle dell'Avio, Adamello Group, Italy). *Geomorphology*, **84** (2007), 59–79.

23. M Keiler, J Knight, S Harrison, Climate change and geomorphological hazards in the eastern European Alps. *Philosophical Transactions of the Royal Society A*, **28** (2010), 2461–2479.

24. K Sattler, M Keiler, A Zischg, L Schrott, On the connection between debris-flow activity and permafrost degradation: a case study from the Schnalstal, South Tyrolean Alps, Italy. *Permafrost and Periglacial Processes*, **22** (2011), 254–265.

25. M Floris, A D'Alpaos, C Squarzoni, R Genevois, M Marani, Recent changes in rainfall characteristics and their influence on thresholds for debris flow triggering in the Dolomitic area of Cortina d'Ampezzo, north-eastern Italian Alps. *Natural Hazards and Earth System Sciences*, **10** (2010), 571–580.

26. M Bollschweiler, M Stoffel, Changes and trends in debris-flow frequency since AD 1850: results from the Swiss Alps. *The Holocene*, **20** (2010), 907–916.

27. M Stoffel, DM Schneuwly, M Bollschweiler, *et al.*, Analyzing rockfall activity (1600–2002) in a protection forest: a case study using dendrogeomorphology. *Geomorphology*, **68** (2005), 224–241.

28. RW Fleming, SD Ellen, MA, Algus, Transformation of dilative and contractive landslide debris into debris flows: an example from Marin county, California. *Engineering Geology*, **27**, (1989), 201–223.

29. L Arenson, S Springman, Mathematical descriptions for the behaviour of ice-rich frozen soils at temperatures close to 0 C. *Canadian Geotechnical Journal*, **42**: 2 (2005), 431–442.

30. M Stoffel, D Conus, MA Grichting, I Lièvre, G Maître, Unraveling the patterns of late Holocene debris-flow activity on a cone in the central Swiss Alps: chronology, environment and implications for the future. *Global and Planetary Change*, **60** (2008), 222–234

31. M Stoffel, Magnitude–frequency relationships of debris flows: a case study based on field surveys and tree-ring records. *Geomorphology*, **116** (2010), 67–76.

32. R Lugon, M Stoffel, Rock-glacier dynamics and magnitude–frequency relations of debris flows in a high-elevation watershed: Ritigraben, Swiss Alps. *Global and Planetary Change*, **73** (2010), 202–210.

33. M Stoffel, M Bollschweiler, Tree-ring analysis in natural hazards research – an overview. *Natural Hazards and Earth System Sciences*, **8** (2008), 187–202.

34. M Stoffel, C Corona, Dendroecological dating of geomorphic disturbance in trees. *Tree-Ring Research*, **70** (2014), 3–20.

35. M Bollschweiler, M Stoffel, MD Schneuwly, K Bourqui, Traumatic resin ducts in *Larix decidua* stems impacted by debris flows. *Tree Physiology*, **28** (2008), 255–263.

36. M Stoffel, I Lièvre, D Conus, *et al.*, 400 years of debris flow activity and triggering weather conditions: Ritigraben VS, Switzerland. *Arctic Antarctic and Alpine Research*, **37** (2005), 387–395.

37. C Graf, Y Deubelbeiss, Y Bühler, *et al.*, Gefahrenkartierung Mattertal: Grundlagenbeschaffung und numerische Modellierung von Murgängen. In: *Mattertal-ein Tal in Bewegung*, ed. C. Graf (Birmensdorf: Eidgenössische Forschungsanstalt WSL, 2013), pp. 85–112.

38. H Van Steijn, Debris-flow magnitude–frequency relationships for mountainous regions of central and northwest Europe. *Geomorphology*, **15** (1996), 259–273.

39. O Hungr, S McDougall, M Wise, M Cullen, Magnitude–frequency relationships of debris flows and debris avalanches in relation to slope relief. *Geomorphology*, **96** (2008), 355–365.

40. M Zimmermann, P Mani, H Romang, Magnitude–frequency aspects of alpine debris flows. *Eclogae Geologicae Helvetiae*, **90** (1997), 415–420.

41. TC Blair, Sedimentology of the debris-flow dominated Warm Spring Canyon alluvial fan, Death Valley, California. *Sedimentology*, **46** (1999), 941–957.
42. MM Helsen, PJM Koop, H van Steijn, Magnitude–frequency relationship for debris flows on the fan of the Chalance torrent, Valgaudemar (French Alps). *Earth Surface Processes and Landforms*, **27** (2002), 1299–1307.
43. C Scheidl, D Rickenmann, M Chiari, The use of airborne LiDAR data for the analysis of debris flow events in Switzerland. *Natural Hazards and Earth System Sciences*, **8** (2008), 1113–1127.
44. M Stoffel, T Mendlik, M Schneuwly-Bollschweiler, A Gobiet, Possible impacts of climate change on debris-flow activity in the Swiss Alps. *Climatic Change*, **122** (2014), 141–155.
45. J Schmidli, C Frei, Trends of heavy precipitation and wet and dry spells in Switzerland during the 20th century. *International Journal of Climatology*, **25** (2005), 753–771.
46. R Delaloye, S Morard, C Barboux, *et al.*, Rapidly moving rock glaciers in Mattertal. In: *Mattertal – ein Tal in Bewegung*, ed. C. Graf (Birmensdorf: Eidgenössische Forschungsanstalt WSL, 2013), pp. 21–31.
47. Y Bühler, C Graf, Sediment transfer mapping in a high-alpine catchment using airborne LiDAR. In: *Mattertal – ein Tal in Bewegung*, ed. C. Graf (Birmensdorf: Eidgenössische Forschungsanstalt WSL, 2013), pp. 113–124.
48. R Kenner, Y Bühler, R Delaloye, C Ginzler, M Phillips, Monitoring of high alpine mass movements combining laser scanning with digital airborne photogrammetry. *Geomorphology*, **206** (2014), 492–504.
49. C Willi, Y Deubelbeiss, C Graf, M Keiler, Methods for detecting stream bed surface changes in a mountain torrent: a comparison. *Geographica Helvetica*, (2015): in press.
50. M Christen, J Kowalski, P Bartelt, RAMMS: numerical simulation of dense snow avalanches in three-dimensional terrain. *Cold Regions Science and Technology*, **63**: 1–2 (2010), 1–14.
51. Y Deubelbeiss, C Graf, Two different starting conditions in numerical debris-flow models: case study at Dorfbach, Randa (Valais, Switzerland). In: *Mattertal – ein Tal in Bewegung*, ed. C. Graf (Birmensdorf: Eidgenössische Forschungsanstalt WSL, 2013), pp. 125–138.
52. C Graf, BW McArdell, *Die Murgangbeobachtungsstation Randa. FAN Agenda (Bulletin of Fachleute Naturgefahren Schweiz)*, **1** (2005).
53. M Zimmermann, Murgänge im Dorfbach von Randa (VS). *Beurteilung von Massnahmen. wasser, energie und luft*, **86**: 1–2 (1994), 17–21.
54. A Stocker, Geschiebeabschätzung im Dorf-und Wildibach mit den Geschiebeabschätzverfahren Gertsch und SEDEX. In: *Mattertal – ein Tal in Bewegung*, ed. C. Graf (Birmensdorf: Eidgenössische Forschungsanstalt WSL, 2013), pp. 75–83.

17

Contextualizing conflict

Vital waters and competing values in glaciated environments

ADAM FRENCH, JAVIERA BARANDIARÁN, AND COSTANZA RAMPINI

17.1 Introduction

In many regions, glaciers have long served as important freshwater reserves, providing vital meltwater during dry seasons and droughts to buffer fluctuations in water availability. Yet it is only recently that the metaphor of glaciers as natural water towers has gained prominence in the public imagination. Accompanying the rise of this symbolism are compelling narratives linking glacier recession with future water scarcity and resource conflict. Such thinking is hardly surprising given that in mountain ranges like the Himalayas and the Andes accelerating glacier decline is predicted to cause reductions in water supplies critical to a significant portion of the world's population. In some settings, conflicts over water supplies beneath melting glaciers have already arisen, at times before declines in glacial runoff have begun.

In light of these conflicts and concerns over future water scarcity linked to glacier recession, this chapter examines the evolving dynamics of water disputes in several glaciated environments in order to understand why they have emerged and what is at stake in each case. The chapter begins by briefly reviewing glacier recession's contemporary and predicted impacts on water availability, highlighting the complexity of these effects. We then problematize simplifying discourses that assume ongoing glacial recession and linked water scarcity will automatically drive environmental conflict. Our critique is rooted in theories from political ecology that treat resource access and control as the outcomes of socio-natural processes embedded in political economic systems. We complement this politicized analysis with insights from research focused on the dynamics of historical water conflicts and with critical perspectives from Science and Technology Studies (STS), suggesting that accepted forms of knowledge and expertise and the

The High-Mountain Cryosphere, ed. Christian Huggel, Mark Carey, John J. Clague and Andreas Kääb. Published by Cambridge University Press. © Cambridge University Press 2015.

regulatory regimes they influence also result from social and political processes. We then apply these theoretical perspectives to the empirical analysis of several case studies of conflicts in the Andean and Himalayan regions.

Complicating deterministic explanations of absolute scarcity as the key driver of water conflict, this chapter argues for research that interrogates the conjunctures of social and environmental factors under which disputes occur. The examination of the empirical cases illustrates that these conflicts are not driven by resource scarcity per se but, instead, by incompatible strategies of resource use linked to divergent values over glaciers and the water they provide as well as by political economic and techno-scientific processes occurring at a range of scales. We also highlight how such factors influence institutional and regulatory arrangements for the governance of glaciers and downstream water supplies. The chapter concludes with a synthesizing discussion of important themes emerging from the cases and several recommendations for future research.

17.2 Critical approaches to resource conflicts, scarcity, and governance

Water scarcity has emerged as one of the preeminent challenges of the twenty-first century, considered by some as the "single greatest threat to food security, human health, and natural ecosystems" [1:29]. While the causes of growing water scarcity are complex and heterogeneous, the rapid melting of most of the planet's glaciers is frequently identified as a threat to global water supplies and a potential driver of future hydrologic crises [2–5]. Studies from diverse mountain regions including the Himalayas [6–8], Central Asia [9], the Andes [10–14], and the Rockies [15], all suggest significant impacts of glacial recession on water supplies, though much uncertainty remains about the severity and timing of these effects, given limited historical data as well as difficulties in accurately modeling and predicting factors such as future precipitation [8,16,17]. Nevertheless, despite challenges in predicting future water availability, it is clear that as glacial masses melt away, their contributions to stream flows – especially critical in some regions during annual dry seasons and inter-annual droughts – eventually reach a threshold known as "peak water," after which they gradually decline until no longer influencing the hydrologic regime [10,18].

While long-term glacier wastage raises important concerns for the future, changes in water supplies are already being observed in glaciated settings on both sides of the peak-water transition. In many regions, impacts have occurred during annual dry seasons (a pattern predicted to continue where markedly seasonal precipitation regimes predominate, e.g. in monsoonal climates), and in some cases technological adaptations have helped to buffer the effects. For example, in Peru's Santa River, flows have decreased for several decades (~10% reduction since

1954) as glacial cover has declined [10] but water use has nevertheless intensified in the region's agricultural and energy sectors due to infrastructure developments [12,19] (though these sectors are predicted to face significant declines in the future) (e.g., [20]). Meanwhile, along some highland tributaries of the Santa watershed, residents report obvious declines in dry-season water supplies in recent decades, though the impacts on livelihoods are less clear [21,22]. In several communities of Nepal's Khumbu region, residents report less winter water availability in recent years but as a result of precipitation changes rather than glacial decline, as these highland populations do not rely upon glacial-fed sources [23]. In many regions, glacier change is also associated with natural hazards that threaten water supply systems and many other facets of society (e.g., [24]).

Given these observed and predicted impacts, concerns over future water scarcity are justifiable in the context of the rapid and accelerating decline of the world's glacial "water towers." Yet this scarcity, we argue, must be understood as the outcome of not only physical changes in water availability, but also of complex social relations that influence how and where water flows (also see Chapter 13 in this volume). The critical approach of political ecology (PE) provides a useful lens for analyzing the conjunctures of these biophysical and social factors. PE coalesced around scholarship examining the multi-scale environmental and political economic factors involved in coproducing problems like resource degradation, famine, and natural hazards in the Global South, focusing explicitly on what Blakie and Brookfield characterized as "the constantly shifting dialectic between society and land-based resources" (e.g., [25,26,27:17]). In diverse empirical contexts, political ecologists have challenged overly simplistic and reductionist explanations (i.e., blaming population growth or poverty) that fail to account for social factors like politics, power dynamics, and differential resource entitlements in producing environmental problems and conflicts (e.g., [28,29]).

Research in PE has illustrated that although "scarce" resources are central to many conflicts, the Malthusian link between population growth, resource scarcity, and conflict suggested by perspectives from the environmental security literature (e.g., [30]) is little more than a totalizing discourse that naturalizes scarcity and obscures the politics of allocation as well as the relations of power by which these politics are produced (e.g., [31,32,33]). These insights increasingly extend beyond critical academic scholarship as well. For example, in 2006 the UN Human Development Report rejected that the "water crisis is about absolute shortages of physical supply," arguing instead that the "roots of the crisis in water can be traced to poverty, inequality, and unequal power relationships, as well as flawed water management policies that exacerbate scarcity" [34:V].

While such critiques underscore the shortcomings of deterministic accounts of conflicts driven by a naturalized vision of scarcity, important reasons remain for considering water as a potential locus of socio-environmental crisis and conflict. As a substance vital to life and productive activity, water demand is omnipresent; yet the distribution of water is highly uneven both spatially and socially [1,34]. Moreover, society's hydrologic demand is rapidly increasing in the context of population growth, rising affluence, and modernization and urbanization – processes that often impact supply through the degradation of water quality [35,36]. Due to its material characteristics and complex social imbrications, "water ignores political boundaries, evades institutional classification, and eludes legal generalizations" in ways that link diverse stakeholders and their hydrologic needs across geographic, political, and cultural divides while significantly complicating water governance [37:243]. In its varied uses, water is valued distinctly across the dimensions of society connected by its flows [38], and these environmental, cultural, spiritual, and economic valuations often prove incommensurable.

These complexities of water's availability, valuation, and management contribute to its capacity to engender conflict, fueling fears over a future in which growing water scarcity leads to ever more serious and intractable crises. Yet scholarship reviewing the history of international water conflict refutes claims that water has been a major catalyst for wars in the past and presents a compelling argument for why it has instead fostered enduring international cooperation [39]. This research, however, also proposes an inverse relationship between geographic scale and the intensity of water conflict, suggesting that tensions over water between sectors and interest groups within national boundaries generate more conflict than those between nations. Looking towards the future, this work emphasizes the negative impact that declining water quantity and quality may have on political stability at the national and sub-national level, and, in congruence with work in PE, highlights how the increasing commoditization of water is becoming an important source of conflicts (e.g., [37,40]).

Regardless of the specific drivers of water conflicts or the scales at which they emerge, the role of politics in the governance of water disputes is critical [37,38,41]. These political processes extend well beyond the central state, encompassing actors from different levels of government, civil society, and the private sector, as well as the activities and influences of an array of advocacy networks and techno-scientific and legal practices and frameworks [42,43]. Different forms of "institutions" – ranging from state-led organizations to legally sanctioned regulatory processes – connect the various actors and spheres of influence involved in environmental governance [44–46]. But just as diverse social values shape conflicting conceptions of how water should be used, values also deeply influence how formal institutions govern water.

Uncovering how particular values affect governance decisions can be challenging, especially when seemingly apolitical technical and scientific criteria guide formal processes. In such contexts, critical perspectives from STS are useful for examining the social relations that influence how legal, scientific, and technical practices gain (and lose) credibility and contribute to reinforcing and privileging certain values while undermining others [47]. STS scholarship, for example, illustrates how the expert-led creation of environmental impact assessments (EIAs) is a socio-political process where who participates, what information is mobilized, and how criticism is produced and managed all shape resource valuation and use [48]. Given that glaciers are highly mediated by scientific and technical experts and their tools of assessment, we draw upon STS perspectives to understand the values inherent in these forms of expertise as well as how such values contribute to the privileging of specific conceptions of risk and opportunity and of social and environmental justice more broadly [49].

Through the case studies that follow, this chapter begins to address a gap in the literature on the cryosphere by contributing to our understanding of contemporary water conflicts in glaciated environments in the Andean and Himalayan regions. By drawing implicitly upon the PE and STS approaches described above in our analyses of the cases, we highlight the central role of social values (political, economic, techno-scientific, etc.) in structuring resource access and influencing the risks and opportunities that result for particular groups. In so doing, these studies link theoretical insights, critical historical and legal analysis, and the results of empirical fieldwork in efforts to understand the interplay of diverse biophysical and social processes in each case and to highlight and move beyond the limitations of overly deterministic and naturalized explanations of conflict in these dynamic environments.

17.3 Contemporary conflicts in glaciated environments

17.3.1 Pascua Lama, Chile

Pascua Lama is an open-pit gold and silver mining project located in the Andes on the Chilean–Argentine border (3800–5200 m). The proposed mine, reported as the world's largest bi-national mining project, is owned by Barrick Gold Corporation of Canada and contains proven and probable reserves reported at 17.9 million ounces of gold and 676 million ounces of silver [50]. Pascua Lama's development has been highly controversial for more than a decade, with the project's impacts on both glaciers and downstream water supplies the central points of contention for local residents and environmental activists.

Although explorations at the mine began in the 1980s, Barrick submitted the project's first EIA in 2000, and in 2004 the company submitted a second EIA for project expansion. Both EIAs were eventually approved, and the project moved forward until in April 2013 – after violations of the project's environmental permit – Chilean authorities imposed a $16 million fine and suspended project construction until a larger water treatment system is completed [51]. In mid-2014, construction remains paralyzed and Pascua Lama faces an uncertain future – a new Environmental Tribunal has ordered a technical review of the 2013 fine as well as the project's permit [52,53]. In this section, we provide an overview of the Pascua Lama project, and especially of its contested EIA process, that examines its central actors and their conflicting values and visions of development.

Organized resistance to Pascua Lama emerged during the project's initial EIA scoping phase in 2000, when advocates reviewing the EIA decided that local concerns over impacts on water supplies and glaciers were not suitably addressed. This EIA described how Barrick would relocate several small glaciers located in the proposed mine pit, but some residents living downstream in the Huasco Valley worried about the long-term effects of these actions. Glacial meltwater and sustainable development went hand-in-hand for many of these locals, who remembered how the region's rivers had kept flowing through a seven-year drought in the 1960s thanks to the "perpetual banks" of frozen water high in the Andes. Yet, some actors at Barrick, in the national government, and in the scientific community ridiculed these local values of glaciers. One scientist suggested that what local residents considered "pristine, pure, untouched waters" were really "a toxic stew of acidity and metals," and Barrick's CEO, Peter Munk, stated that he assumed people were upset because glaciers evoke the "pristine" and "pure" [54,55]. Activists, however, denied such romanticized views and insisted that the company's proposal to move the glaciers was "an insult to [their] intelligence," which came in the wake of a long list of broken promises for job opportunities for local youth and water for agriculture [55].

The different sides' conflicting visions of development and distinct values over the region's glaciers were clearly incompatible. In an article titled "A Valley is Dying," one activist described the hopelessness of local residents rooted in a belief that "the environment is subordinated to the interest of capital to maximize its profits, without the state applying any effective regulation. ... All justified by one economic and ideological explanation, as repetitive as it is false, that 'a country with US$5,000 annual per capita income can not afford to steward the environment.'" [56]. This complaint resonated with critiques of Chile's environmental policy regime emphasizing the regulatory system's market-enabling character [57–59]. Complicating the situation was the fact that Chilean glaciers existed in a legal blind-spot – the 1982 Water Code that establishes private property rights over water omitted glaciers entirely [57]. Since 1997, when EIAs became obligatory for large industrial projects, a project's environmental permit alone defines the legal obligations regarding construction and operation standards, and as a result EIAs are central to environmental management as well as to many socio-environmental conflicts. Despite the local concerns over the Pascua Lama EIA, the Chilean government's National Environmental Commission (CONAMA) approved the EIA and granted a permit in 2001, but community advocates continued to resist.

By 2004, when Barrick began a second EIA for Pascua Lama in the context of its plans to expand the project, public resistance both inside and outside the Huasco Valley had increased significantly. The local Irrigators' Association had become particularly concerned about the effects of declining hydrologic supplies and water quality on the viability of a new irrigation dam. This powerful group, which included a range of irrigators from multinational corporations to the local landed elite, was typically pro-development and pro-investment. Yet under pressure from residents concerned that Pascua Lama would "kill" their valley, the Association hired a local environmental consulting firm to provide an independent evaluation of the project's impacts [55]. Local activists opposing the project had also recruited a pair of glaciologists; and CONAMA, which in the second EIA process felt it necessary to have their own expert, hired another glaciologist. The Chilean Water Agency also sent an expert to assess the project's impacts on the glaciers and downstream hydrology [55,60].

Thus, while Pascua Lama's first EIA process suffered from a lack of scientific expertise in glaciology and insufficient attention to glaciers and downstream hydrology in general, the second process involved numerous experts and emphasized the proposed mine's impacts on glaciers. On one hand, this attention exposed serious shortcomings in the first EIA, and led to a unanimous decision that the small glaciers – recast diminutively as *glaciaretes* or relict ice in the new EIA – could not be preserved through relocation [61]. On the other hand, the scientists agreed that the demise of the relict glaciers located within the mine's pit would cause only minor volumetric changes in long-term water supplies, but many also felt that the mine's overall impact on the glaciers of the area and on water quality as well as quantity were not sufficiently addressed [61]. Several denounced that historical impacts from mining activity in the region had greatly contributed to the *glaciaretes*' degraded condition; CONAMA, however, ignored such accusations [55,60].

Eventually, CONAMA approved the second EIA, but the resulting environmental permit declared glaciers as "national patrimony" and "water reserves" and ordered Barrick to change the location of its pit to avoid destroying the *glaciaretes* [61]. By many measures resistance to Pascua Lama was successful: CONAMA had imposed more stringent water management measures, and Congress passed a law requiring that all EIAs address glacier impacts while also creating a Glacier Management Unit within the National Water Agency to inventory glaciers. Nevertheless, some project opponents were disappointed with the outcome, feeling it illustrated the ongoing cooptation of the technical realm of assessment and regulation by political and economic factors [60]. This distrust of the EIA process at Pascua Lama reflects Chilean society's more general doubts about the efficacy of the country's environmental regulatory regime and its propensity to document and compensate rather than prevent impacts from natural resource extraction [59]. Moreover, for many opponents, the EIA's approval illustrated the government's privileging of extractive industry and transnational capital over the values and livelihoods of Huasco Valley residents [55].

In a broader context, the Pascua Lama conflict greatly improved public awareness of Chile's glaciers and their vulnerability to extractive industries. Yet legislation that would have more comprehensively protected glaciers failed in Chile's congress, upholding a policy regime that "continues to eschew planning and zoning mechanisms … in favour of a case-by-case approach based on procedural compliance with legal requirements" [59:892–893]. In contrast, after significant political brokering, neighboring Argentina approved legislation protecting glaciers in 2010, in part due to resistance to Pascua Lama. The Argentine law, in addition to recognizing the strategic value of glaciers as water resources and creating a national glacier inventory as occurred in Chile, also explicitly recognizes the value of glaciers to biodiversity and tourism and prohibits activities "that could affect their natural condition or function as strategic reserves" [62]. Although Barrick and other mining companies in Argentina received preliminary injunctions protecting them from the glacier law, the country's Supreme Court overturned the injunctions in mid-2012, requiring the mining sector's compliance with the law [63].

Argentina's glacier law and Chile's recent suspension of the Pascua Lama mine illustrate increased attention to protecting glaciers and pro-glacial environments, but some analysts warn this stance could significantly depress foreign direct investment. One expert, suggesting that Chile's suspension of Pascua Lama be viewed in light of the country's recent accession to the OECD and related requirements for more stringent environmental controls on industry, stated that the ruling was "a good signal to the world that in Chile there are controls, that there's a new institutionality that is working" [51]. Indeed, effective environmental regulatory institutions are increasingly critical in a context of extensive growth of extractive industries coupled with growing access to glaciated environments and the resources they contain. Yet, whether these developments in Chile and Argentina signal a significant departure from policies that evaluate and compensate impacts toward a more precautionary approach that promotes inter-actor engagement and negotiation over competing values and development models will require continued analysis of regulatory performance and outcomes over time.

17.3.2 *Lake Parón, Peru*

Lake Parón is the largest of 830 inventoried glacial lakes in the Cordillera Blanca, the highest and most extensively glaciated mountain range in the world's tropical zone [64]. As a critical resource for local agriculturalists and urban residents as well as an important regulating reservoir for downstream hydropower, the lake became the center of a social conflict in 2008 when a coalition of local actors seized the lake's discharge infrastructure after nearly a decade of complaints over a foreign corporation's management of the lake's outflow. Efforts to resolve the conflict have lasted more than five years but have achieved only limited advances. In this section, we review the evolution of the case and discuss the conflicting values that have made it so intractable.

Located in a narrow glacial basin, Lake Parón sits behind a natural dam formed by the consolidated debris of a large moraine blocking the canyon. Despite its isolation at 4200 m, Lake Parón is connected to diverse landscapes and communities through its outflow, which forms the Parón–Llullán River. This river flows through the forests of Huascarán National Park and passes into a mosaic of agricultural lands, where its waters are diverted by a series of irrigation canals before joining the Santa River near the regional center of Caraz (pop. ~20 000) (see Figure 17.1). Along its roughly 20 km length, the Parón–Llullán River supplies water to thousands of farmers, and it is the principal source of potable water for the growing urban districts of Caraz. Maintained by both precipitation and glacial melt, the lake's waters are particularly critical during the tropical dry season (May–October) when little rain or snow falls in the region [65].

Although Lake Parón is a critical water source and an important cultural symbol of the region, its location upstream of population centers and vulnerable hydroelectric infrastructure in the Santa River valley has created enduring concerns over a potential glacier lake outburst flood (GLOF) [66,67]. After the lake had been filled to its brim in the early 1950s by avalanches and GLOFs higher up the canyon, risk-mitigation activities were finally undertaken at the lake beginning in the late 1960s [68]. Over almost two decades, a drainage tunnel was drilled through the granite mountainside adjoining the lake, and its volume was finally lowered in the mid-1980s. With the lake drained, evaluations of the morainal dam's stability were conducted. These analyses suggested that by following specific precautions the lake could be used as a regulating reservoir for downstream hydropower production [68]. With a burgeoning need for energy in the Peruvian economy, flow-regulating hydraulic valves were installed on the tunnel intake in 1992, permitting manipulation of the lake's volume and discharge regime by the downstream hydroelectric facility (see Figure 17.1) [67].

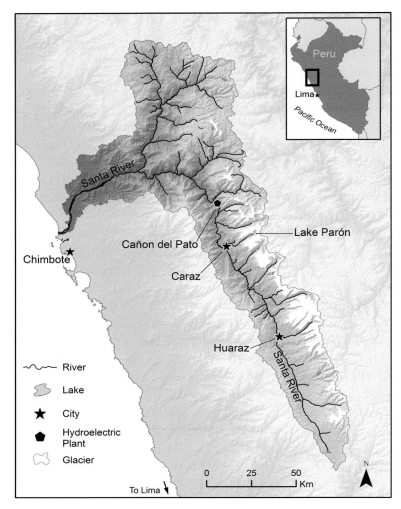

Figure 17.1 Lake Parón and the Santa River Watershed.

As a regulating reservoir, during the annual wet season Lake Parón is allowed to fill to a maximum secure level recommended by the engineers who evaluated the site's geophysical and seismic hazards. During the dry season, the lake's accumulated volume is then strategically discharged to generate electricity at the downstream hydropower plant. While the system helps satisfy peak energy requirements and also has the capacity to mitigate GLOF hazards and seasonal water scarcity, the manipulation of the Parón–Llullán River's discharge regime for energy production (a process known as hydro-peaking) generated significant problems for other downstream users [65].

Local complaints over hydro-peaking impacts began to mount around 2001, after Duke Energy – a large, private US-based corporation – had consolidated control of both Lake Parón's infrastructure and the downstream power plant in the context of the privatization of previously state-owned energy production facilities [67]. Initially, the grievances came from local irrigators who complained of excessive and unpredictable discharges that damaged irrigation infrastructure as well as a lack of sufficient flows for their water needs during some annual dry seasons [69]. Both residents and the local water provider also complained of increased turbidity in the potable water of Caraz and some locals noted detrimental effects on the tourism industry when the lake was drawn down to the level of its discharge tunnel [70,71]. By the mid-2000s, state entities like Huascarán National Park had formally acknowledged these complaints but were unable to force the completion of an EIA or influence the energy company's discharge practices in a way that satisfied local residents' demands. Finally, in mid-2008 a local coalition seized the discharge infrastructure at the lake, evicting Duke Energy's personnel from the site [65,67].

Under the national water law in force when local complaints were mounting against Duke, agricultural water use was prioritized over the demands of other sectors like industry and energy production. Duke's operation of Lake Parón's discharge infrastructure at times clearly contradicted this hierarchy of priorities. Nevertheless, the company's use of the lake was supported by its water license, which was originally granted to a state-owned energy company without consideration of how the license contradicted the national water law or the needs of other water users between Lake Parón and the downstream hydropower facility [67]. In fact, no EIA had been required for this license, and, although Duke had eventually been ordered to conduct a detailed EIA for its use of Lake Parón, the analysis had not been completed [65]. Thus, as the rights of local residents came into increasing conflict with the company's license, effective institutions were needed to mediate between these actors' distinct water uses in the context of contradictory legal regimes. Furthermore, there was a critical need for greater attention to hydro-ecological conditions in the Parón–Llullán watershed and incompatibilities between these conditions and the company's water license. In the absence of such factors, the conflict escalated, culminating in the infrastructure seizure.

Although the conflict has focused explicitly on corporate water-use practices and downstream impacts, glaciers play an important – though often overlooked – role in the case. First, GLOF threats contributed to initial decisions to drain the lake, which permitted technical studies supporting the installation of the infrastructure now at the center of the conflict [67]. Second, for many locals – especially in the context of growing predictions of future water scarcity linked to glacial recession – the lake and its surrounding glaciers represent long-term water security as well as important cultural symbols and tourist attractions, rather than mere sources of seasonal water supply [71]. Moreover, locals perceive a lack of water in the lake as the most significant source of risk, while Duke and state regulators have emphasized the risks associated with a lake level surpassing the "safe" maximum. Given a sustained lack of constructive dialogue and willingness to compromise between actors, there have been limited efforts to reconcile these conflicting visions of risk and the resulting proposals for lake management. Furthermore, since the occupation of the lake, concerns over GLOF hazards have become urgent on several occasions, when during annual wet seasons the lake's level exceeded its "safe" maximum. Although temporary management regimes have been negotiated to address these hazards, there has been little advancement in the design of a long-term strategy that balances users' competing values while accounting for current and future glacier recession and hydrologic change.

In conclusion, history illustrates that Duke's license at Lake Parón permits a discharge regime incompatible with other downstream water uses as well as with hydro-ecological conditions, which are complicated by ongoing climate change and glacial melt [10,65]. Thus far, conflict resolution efforts have focused on averting short-term GLOF risks through temporary management regimes imposed by state authorities. Yet enduring differences between how these authorities, Duke, and local residents perceive the value of the lake's water and associated risks challenge these strategies, requiring sustained negotiation and compromise between actors. Current progress towards the formalization of a multi-sectoral management committee for the Lake Parón watershed may eventually provide an institutional space for this negotiation. We argue, however, that for long-term conflict resolution to advance it will be crucial to develop a more flexible and adaptive management regime that at once works to balance the distinct values and rights of the different users involved while addressing the complex and shifting hydrological dynamics of the watershed.

17.3.3 Brahmaputra Basin, China and India

Originating in the glaciers of southwestern Tibet, the Brahmaputra River is one of Asia's largest and most important watercourses, as well as a source of growing geopolitical and socio-environmental conflict [3,72,73]. From Chinese-controlled territory, the river flows east before turning south at its "great bend" to enter the Indian state of Arunachal Pradesh. Beyond Arunachal Pradesh, the river reaches the Indian state of Assam and then passes into Bangladesh, the Ganges delta, and the Bay of Bengal. Given the many geographic and political regions crossed by the river, its flows link a broad diversity of ecosystems, human actors, cultural values, and strategic interests. As a vital source of freshwater and sediment transport for some of the most climate-vulnerable populations of Asia and a principal site of hydropower development for both China and India, the river is critical to residents along its course as well as in distant urban centers. While competition between China and India over the river's flows is a frequently cited cause of potential international conflict, tensions between different social, political, and economic sectors within each country are also generating governance challenges. In this section, we briefly survey a range of multi-scale socio-environmental factors and inter-actor dynamics that produce competing claims to and values over the river's flows.

Throughout the Himalayan region, ongoing glacial recession has led to concerns over future water availability and linked threats to livelihood security and public health for millions of people [7,8]. In general, Himalayan glaciers and snowpack are retreating at rates comparable to glaciers in other regions of the world, and the Himalayan region as a whole is expected to reach peak water by mid-century [16]. Yet, overall, it remains unclear how climate change and glacier recession will impact water availability in Asia due to significant uncertainty regarding the role of future monsoonal precipitation in compensating for declines in the cryosphere [16,74]. Nevertheless, in the Brahmaputra basin – where snow and glacier melt from the Himalayas contributes an estimated 27% of total downstream discharge – climate change impacts on these sources are predicted to diminish river flows significantly during the dry season by the middle of this century, particularly at higher elevations [7]. Despite these predicted reductions, social demands for the Brahmaputra's flows currently overshadow the contemporary impacts of glacier recession in the river basin. These demands are increasing rapidly with both China and India mobilizing their significant engineering capacities to tap the river's water resources, especially its hydropower potential.

China's position as the upstream riparian on the Brahmaputra, in particular, is raising concerns in India in relation to both hydroelectric development and potential water diversions. China is currently building the 510 MW Zangmu dam – a project the country denied plans of until construction was underway – and government documents propose another ten dams on the river, three of which were officially announced in early 2013 [72,75,76]. There has also been talk of the world's largest hydroelectric project at the "great bend" of the Brahmaputra, though few specific details about this project have emerged [6,76,77]. While China has assured its neighbor that all the dams are run-of-the-river projects that will not affect downstream flows, India has continued to voice concerns about impacts on flow regimes and sediment transport. India also worries about the potential inclusion of the Brahmaputra in China's South–North Water Diversion project, though current indications suggest that the river will not be a part of this massive water-transfer scheme [78].

Meanwhile, Sino-Indian cooperation on the Brahmaputra is complicated by ongoing geopolitical disputes over the Tibet–Arunachal Pradesh border region and related limitations on trans-boundary data sharing and scientific investigation in this conflicted context, as well as by the lack of any formal water-sharing agreement or international river treaty for the Brahmaputra [79–81]. Although data sharing between the nations during the monsoon season has reportedly improved in recent years [82], China has continued to reject India's requests for a bilateral water commission or an institutionalized treaty to guide the management of the river, stating that current mechanisms are "adequate" despite complaints of China's lack of transparency in its development plans [83].

With China planning multiple mega-dams on the Brahmaputra, India is also fast-tracking the development of hydropower on the river and its tributaries. By late 2010, the government of Arunachal Pradesh had signed initial contracts with both private and public corporations for the construction of 132 dams along these waters – 92 are large projects producing more than 25 MW, while 38 are mega-dams above 100 MW in capacity [84,85]. Many of these projects are contested by local residents, and critics have illustrated the inadequacies of some project EIAs [84,86]. Government and hydropower company officials accuse Maoists of infiltrating anti-dam protests and attribute local opposition to a lack of understanding of the development benefits that will ensue for Northeast India [85]. Yet, much local resistance centers around direct risks and impacts, including changes in flow regimes, increased seismic hazards and erosion, and reduced sediment transport that locals claim severely threaten the integrity of ecosystems supporting their livelihoods and cultures [84,85]. For example, anti-dam activists have argued that the water impoundment and hydro-peaking practices of the Lower Subansiri hydroelectric project on a major tributary of the Brahmaputra – one of Northeast India's most contested projects – will submerge lands and drastically change daily flow regimes with dramatic impacts on river morphology, biodiversity, and human livelihoods [84–86]. Such impacts seem probable under predictions that river flows below the Lower Subansiri dam in the non-monsoon period will be altered from their historical dry-season average of ~400 m^3 s^{-1} to 6 m^3 s^{-1} for all but four hours a day when hydro-peaking will raise flow rates to 2560 m^3 s^{-1} [84].

In addition to conflicts between local residents and dam proponents, hydropower development in Northeast India is causing tensions between Arunachal Pradesh and the downstream state of Assam, which critics suggest will see little direct benefit of hydropower development through flood control or subsidized energy while being the worst-exposed to negative impacts [85]. In some cases, including Lower Subansiri, projects initially envisioned by the Brahmaputra River

Board – a government authority formed to develop the basin's resources – were multipurpose dams designed to provide irrigation, flood mitigation, and erosion control in addition to energy production. Yet during the last decade's push to hasten project development and attract private investments in the hydropower sector, multipurpose functionality has often been sacrificed for run-of-the-river projects that will produce more energy and faster returns on investments [6,84–86]. While these changes have alleviated some tensions around land submergence in Arunachal Pradesh, they deprive downstream Assamese populations of benefits like flood and erosion control [85].

Some critics also argue that as a result of planned energy transfers from the Northeastern Region, both Arunachal Pradesh and Assam will suffer significant negative impacts from hydropower development while India's distant urban centers receive the principal benefits [85]. For example, a 6000 MW transmission system from Assam to Agra associated with the Lower Subansiri and Kameng projects is already underway [82], and more transmission lines are likely to be built as additional hydropower projects are commissioned. This transfer of energy, facilitated by the liberalization of the hydropower sector in India, is evidence of the increasing influence of powerful urban interests and distant values in reshaping the risks and opportunities of local residents.

The evolving dynamics described in this overview highlight only some of the complex management challenges confronting this important trans-boundary river system. Connecting Asia's most powerful and populous nations, diverse public and private corporations, and some of the planet's most vulnerable populations, the flows of the Brahmaputra are critical to an array of actors across geographic and political scales. These actors' competing claims to and values over the river's water are generating new hydrologic demands as well as greater potential for future water conflict. Although both contemporary and long-term impacts of climate change and glacial decline on water supplies in this basin are uncertain (and difficult to monitor and model in the context of limited data sharing and rapid development), predicted reductions in river discharge by the middle of this century linked to glacial decline will likely stress already tense social and political dynamics [7]. Under these conditions, crafting international institutions capable of promoting more collaborative and transparent management of the Brahmaputra across geo-political and cultural divides remains one of the region's greatest challenges.

17.4 Discussion and conclusion

Rather than providing directly comparable cases, the preceding examples of contemporary water resource conflicts in glaciated environments illustrate the diversity and complexity of socio-environmental dynamics in these rapidly changing contexts. In this section, we discuss several key issues that emerge across the cases: the role of water scarcity in the emergence of conflict, the ways in which sectoral interests and incompatible values over water resources generate distinct risks and opportunities for different actors, the role of social resistance and conflict in promoting adaptive resource governance, and the need for negotiated and flexible resource management regimes and allocation mechanisms under the uncertainties of global change.

Together these cases present a complex picture of water scarcity's role in producing conflict. In none of the cases is absolute scarcity the principal driver of conflict; rather, conflict development is influenced by historical experiences of temporary water shortage linked to climatic and social conditions as well as to the specter of future water scarcity. In the Chilean case, glacier runoff's buffering effect on local water supplies became clear during periods of drought, and, while the hydrological reserves actually stored in the *glaciaretes* at Pascua Lama may be minimal, the larger glaciated environment of the region and its role in sustaining hydrologic flows is deeply valued by local residents. In the Peruvian context, conflict-fomenting periods of water shortage for local irrigators were caused by management decisions at Lake Parón rather than by absolute water scarcity, though concerns of future scarcity purportedly influenced the decision to occupy the lake. Similarly, in the Brahmaputra basin, current tensions are driven by the direct impacts of unilateral water management strategies that affect diverse actors across social and geographic divides as well as by competing claims to future water supplies. Yet, while absolute scarcity is not a critical factor in these settings currently, it is important to note that future water reductions linked to glacial recession are likely to impact water availability in all of these contexts eventually. Thus while our analysis emphasizes the importance of avoiding environmentally deterministic, scarcity-driven explanations for conflict, we also underscore the critical need to understand and take account of shifting hydrological and biophysical conditions and their complex interplay with a broad array of social factors over time in conflict generation and persistence.

These cases also illustrate a variety of ways in which sectoral actors and their competing values influence the distribution of risks and opportunities in each context. At Pascua Lama, local concerns over long-term water availability are counterposed to values linked to global mineral prices and stakeholder dividends. At Lake Parón, a similar context of conflicting values emerges over local uses of the lake's outflow and the production of both corporate profit and energy for the national electricity sector. In the Brahmaputra watershed, the diverse values of nations, states, the private sector, and the general public are articulated across scales in complex and overlapping ways, sometimes directly and confrontationally – as is occurring, for example, between the energy sector and residents protesting dam projects like Lower Subansiri – and other times indirectly – as in cases involving inhabitants of India's Northeastern Region and the nation's distant cities and industries that stand to benefit from transferred hydropower. Through the evolution of each case, we see that the collision of these competing values does not typically lead directly to open conflict or violence, but rather to efforts to reconcile disagreements through formal institutions and processes, which also prove to be mediated by social relations and values. However, we also see that when such attempts at resolution fail to yield results in redistributing risk and reconciling divergent values over a vital resource like water, direct resistance and conflict often do occur.

While the nature of social resistance is obviously case-specific, in the examples examined here there is a common critique of the influence of economic and political objectives over the techno-scientific and legal-institutional processes structuring resource use and access. At Pascua Lama this has occurred largely in the formal context of public complaints and negotiations over the project's EIAs, which critics feel have been co-opted by political economic concerns and the powerful influence of industry. At Lake Parón, the failure of sustained requests for reform of Duke Energy's management regime eventually led to direct protest and property seizure as local residents felt state institutions were neglecting to enforce their own legal mandates. And in the Brahmaputra basin, China's seemingly unilateral management of the upper river as well as regional and sectoral disputes within India are generating geopolitical tensions and civil-society protests linked to a prioritization of national development agendas and hydropower production.

Each of these cases, and many other examples (e.g., [8,14]), highlight the importance of greater attention to the potential inadequacies of historical management regimes and water allocations in contexts undergoing rapid environmental change and shifting resource availability. Moreover, the increasing uncertainty of future conditions argues for more flexible and adaptive allocation mechanisms in general. Yet, altering existing arrangements or designing new resource management institutions can be exceptionally difficult, particularly in settings like river basins, where hydrologic flows often cross many jurisdictional boundaries. Moreover, in many instances, current arrangements support the interests of powerful economic actors well positioned to defend their current rights. Thus, while careful analyses of history stress that water management occurs through cooperation far more frequently than through conflict [37], we argue that it is also important to recognize that social resistance and conflict can be a critical catalyst for institutional change and the creation of spaces of negotiation and compromise that can promote adaptive resource governance [46,65].

In conclusion, we offer two general recommendations for future research concerning water conflicts in glaciated environments. First, in seeking explanations for water conflicts, analyses should consider the complex conjunctures of environmental and social factors and processes occurring across relevant geographic and social scales. This approach emphasizes analysis of the interacting processes of climate change (with direct impacts such as glacier recession) and globalization (with its myriad impacts on cultural, economic, and political conditions and processes) as well as the synergies and feedbacks that often exist between these concurrent aspects of global change [87]. Second, we recommend critical attention to the details and efficacy of existing water management institutions and regulatory regimes and to their role(s) in generating and perpetuating water conflicts. Specifically, we suggest a focus on how these arrangements distribute risks and benefits for different actors, if and how they hold legitimacy for different actors over time, and if and how current and future hydrologic change and social demand for water may challenge these arrangements. Progress on these two analytical fronts, we argue, will allow us to better understand and contextualize the complicated causes of water conflicts and their persistence, thereby avoiding deterministic and obfuscating claims of absolute scarcity's predominance as a driver of conflict. Moreover, this analytical approach has potential to support the negotiation of diverse values and growing demands over water as well as the crafting of resource governance strategies more capable of adapting to the interacting processes of contemporary global change. Rather than absolute resource scarcity, these value-driven negotiations and adaptations, we believe, are the preeminent challenge of water governance in the twenty-first century.

Acknowledgments

Adam French is grateful for support from the Fulbright-Hays Program, the Inter-American Foundation, the Pacific Rim Research Program, and the Ciriacy-Wantrup Postdoctoral Fellowship at the University of California, Berkeley. Javiera Barandiaran is grateful for support from the National Science Foundation (Award No. 0965524). Costanza Rampini is grateful for support from the EPA STAR Fellowship. We also thank the many individuals and organizations in Chile (Barandiaran), Peru (French), and India (Rampini) who have generously supported our field research, as well as this volume's organizers and two anonymous reviewers who provided comments on an earlier draft of this chapter.

References

1. D Seckler, R Barker, U Amarasinghe, Water scarcity in the twenty-first century. *International Journal of Water Resources Development*, **15**: 1–2 (1999), 29–42.
2. TP Barnett, JC Adam, DP Lettenmaier, Potential impacts of a warming climate on water availability in snow-dominated regions. *Nature*, **438**: 7066 (2005), 303–309.

3. Strategic Futures Group and Defense Intelligence Agency, *Global Water Security*, Intelligence Community Assessment White Paper (2012).

4. IPCC, *Climate Change 2007: Impacts, Adaptation and Vulnerability. Working Group II Contribution to the Intergovernmental Panel on Climate Change Fourth Assessment Report* (Cambridge: Cambridge University Press, 2007).

5. BP Kaltenborn, C Nelleman, II Vistnes, *High Mountain Glaciers and Climate Change: Challenges to Human Livelihoods and Adaptation*, (Arendal: United Nations Environment Programme, 2010).

6. B Crow, N Sing, The management of international rivers as demands grow and supplies tighten: India, China, Nepal, Pakistan, Bangladesh. *India Review*, **8**: 3 (2009), 306–339.

7. WW Immerzeel, LP van Beek, MF Bierkens, Climate change will affect the Asian water towers. *Science*, **328**: 5984 (2010), 1382–1385.

8. National Research Council. *Himalayan Glaciers: Climate Change, Water Resources, and Water Security* (Washington, DC: National Academies Press, 2012).

9. T Siegfried, T Bernauer, R Guiennet, *et al.*, Will climate change exacerbate water stress in Central Asia? *Climatic Change*, **112**: 3–4 (2012), 881–899.

10. M Baraer, B Mark, J McKenzie, *et al.*, Glacier recession and water resources in Peru's Cordillera Blanca. *Journal of Glaciology*, **58**: 207 (2012), 134–150.

11. RS Bradley, M Vuille, HF Diaz, W Vergara, Climate change: threats to water supplies in the Tropical Andes. *Science*, **312**: 5781 (2006), 1755–1756.

12. J Bury, BG Mark, M Carey, *et al.*, New geographies of water and climate change in Peru: coupled natural and social transformations in the Santa River watershed. *Annals of the Association of American Geographers*, **103**: 2 (2013), 363–374.

13. I Juen, G Kaser, C Georges, Modelling observed and future runoff from a glacierized tropical catchment (Cordillera Blanca, Peru). *Global and Planetary Change*, **59**: 1–4 (2007), 37–48.

14. R Rhoades, X Rios, J Ochoa, Mama Cotacachi: history, local perceptions, and social impacts of climate change and glacier retreat in the Ecuadorian Andes. In *Darkening Peaks*, ed. B. Orlove, E. Wiegandt, B. Luckman (Berkeley, CA: University of California, 2008).

15. DW Schindler, WF Donahue, An impending water crisis in Canada's western prairie provinces. *Proceedings of the National Academy of Sciences*, **103**:19 (2006), 7210–7216.

16. W Immerzeel, F Pellicciotti, M Bierkens, Rising river flows throughout the twenty-first century in two Himalayan glacierized watersheds. *Nature Geoscience*, **6**: 9 (2013), 742–745.

17. PCD Milly, J Betancourt, M Falkenmark, *et al.*, Stationarity is dead: whither water management. *Science*, **319** (2008), 573–574.

18. L Braun, M Weber, M Schulz, Consequences of climate change for runoff from Alpine regions. *Annals of Glaciology*, **31**: 1 (2000), 19–25.

19. M Carey, M Baraer, B Mark, *et al.*, Toward hydro-social modeling: merging human variables and the social sciences with climate–glacier runoff models (Santa River, Peru). *Journal of Hydrology*, **518** (2013), 60–70.

20. W Vergara, A Deeb, A Valencia, *et al.*, Economic impacts of rapid glacier retreat in the Andes. *EOS*, **88**: 25 (2007), 261–268.

21. J Bury, B Mark, J McKenzie, *et al.*, Glacier recession and human vulnerability in the Yanamarey watershed of the Cordillera Blanca, Peru. *Climatic Change*, **105** (2011), 179–206.

22. B Mark, J Bury, J McKenzie, A French, M Baraer, Climate change and Tropical Andean glacier recession: evaluating hydrologic changes and livelihood vulnerability

in the Cordillera Blanca, Peru. *Annals of the Association of American Geographers*, **100**: 4 (2010), 794–805.

23. G McDowel, J Ford, B Lehner, L Berrang-Ford, A Sherpa, Climate-related hydrological change and human vulnerability in remote mountain regions: a case study from Khumbu, Nepal. *Regional Environmental Change*, **13**: 2 (2013), 299–310.

24. E Hegglin, C Huggel, An integrated assessment of vulnerability to glacial hazards: a case study in the Cordillera Blanca, Peru. *Mountain Research and Development*, **28**: 3 (2008), 299–309.

25. M Watts, *Silent Violence: Food, Famine, & Peasantry in Northern Nigeria.* (Berkeley, CA: University of California, 1983).

26. P Blaikie, *The Political Economy of Soil Erosion in Developing Countries* (Harlow: Longman, 1985).

27. P Blaikie, H Brookfield, *Land Degradation and Society* (London: Methuen, 1987).

28. R Peet, M Watts, *Liberation Ecologies: Environment, Development, Social Movements* (London: Routledge, 1996).

29. K Zimmerer, T Bassett, *Political Ecology: An Integrative Approach to Geography and Environment-Development Studies* (New York: Guilford Press, 2003).

30. TF Homer-Dixon, Environmental scarcities and violent conflict: evidence from cases. *International Security*, **19**: 1 (1994), 5–40.

31. NL Peluso, M Watts, *Violent Environments.* (Ithaca, NY: Cornell University Press, 2001).

32. M Kaika, Constructing scarcity and sensationalising water politics: 170 days that shook Athens. *Antipode*, **35**: 5 (2003), 919–954.

33. L Mehta, (ed.), *The Limits to Scarcity: Contesting the Politics of Allocation* (London: Earthscan, 2010).

34. United Nations Development Programme, *Beyond Scarcity: Power, Poverty and the Global Water Crisis* (New York: UNDP, 2006).

35. V Srinivasan, E Lambin, S Gorelick, B Thompson, S Rozelle, The nature and causes of the global water crisis: syndromes from a meta-analysis of coupled human–water studies. *Water Resources Research*, **48**: 10 (2012). DOI: 10.1029/2011WR011087

36. CJ Vorosmarty, P McIntyre, MO Gessner, *et al.*, Global threats to human water security and river biodiversity. *Nature*, **467**: 7315 (2010), 555–561.

37. AT Wolf, Shared waters: conflict and cooperation. *Annual Review of Environment and Resources*, **32** (2007), 241–269.

38. B Orlove, SC Caton, Water sustainability: anthropological approaches and prospects. *Annual Review of Anthropology*, **39** (2010), 401–415.

39. AT Wolf, Conflict and cooperation along international waterways. *Water Policy*, **1**: 2 (1998), 251–265.

40. T Perreault, From the Guerra Del Agua to the Guerra Del Gas: resource governance, neoliberalism and popular protest in Bolivia. *Antipode*, **38**: 1 (2006), 150–172.

41. J Agnew, Waterpower: politics and the geography of water provision. *Annals of the Association of American Geographers*, **101**: 3 (2011), 463–476.

42. M Goldman, *Imperial Nature: The World Bank and Struggles for Social Justice in the Age of Globalization* (New Haven, CT: Yale University Press, 2005).

43. M Lemos, A Agrawal, Environmental governance. *Annual Review of Environmental Resources*, **31** (2006), 297–325.

44. E Ostrom. *Governing the Commons: The Evolution of Institutions for Collective Action* (New York: Cambridge University Press, 1990).

45. E Brondizio, E Ostrom, O Young, Connectivity and the governance of multilevel social-ecological systems: the role of social capital. *Annual Review of Environment and Resources*, **34** (2009), 253–278.

46. OR Young, Institutional dynamics: resilience, vulnerability and adaptation in environmental and resource regimes. *Global Environmental Change*, **20**: 3 (2010), 378–385.
47. A Iles, Identifying environmental health risks in consumer products: non-governmental organizations and civic epistemologies. *Public Understanding of Science*, **16**: 4 (2007), 371–391.
48. CA Miller, Resisting empire: globalism, relocalization, and the politics of knowledge. In *Earthly Politics: Local and Global in Environmental Governance*, ed. S Jasanoff, M Martello (Cambridge, MA: MIT Press, 2004), pp. 81–102.
49. T Forsyth, Climate justice is not just ice. *Geoforum*, **54** (2014), 230–232.
50. Barrick, www.barrick.com/operations/projects/pascua-lama/default.aspx, accessed January 18, 2014.
51. LA Henao, Chile fines Canada's Barrick Gold Corp. $16m for environmental violations. The Associated Press, 2013. http://globalnews.ca/news/587893/chile-fines-canadas-barrick-gold-corp-16m-for-environmental-violations.
52. S Garrido, Proyecto Pascua Lama cumplió un año paralizado. *BioBio Chile* (April 9, 2014).
53. D Medina, SMA acata fallo que revoca millonaria multa contra Pascua Lama y no recurrirá a la Suprema. *Biobio Chile* (March 19, 2014).
54. C Castillo, Pour tout l'or des Andes. Documentary film (France/Chile, 2010).
55. Anonymous. Informant interviews with scientists (consultants and government): 06/22/2011, 07/01/2011, 07/09/2011, 07/29/2011; activists: 06/22/2011, 06/30/2011, 07/25/2011; and personnel from Irrigators' Association and Barrick: 07/01/2011, 07/22/2011, 07/27/2011. Barandiaran J. 2011.
56. M Ossa, Un Valle se Muere. *Pastoral Popular*, (September–October, 2001).
57. C Bauer, *Against the Current: Privatization, Water Markets, and the State in Chile* (Boston, MA: Kluwer Academic Publishers, 1998).
58. J Budds, Contested H2O: science, policy and politics in water resources management in Chile. *Geoforum*, **40**: 3 (2009), 418–430.
59. D Tecklin, C Bauer, M Prieto, Making environmental law for the market: the emergence, character, and implications of Chile's environmental regime. *Environmental Politics*, **20**: 6 (2011), 879–898.
60. J Barandiaran, Regulatory science in a developing state: environmental politics in Chile, 1980–2010. Doctoral Dissertation, University of California, Berkeley, 2013.
61. *Pascua Lama Environmental Permit #024* (Chile, 2006).
62. *Régimen de Presupuestos Mínimos para la Preservación de los Glaciares y del Ambiente Periglacial* (Argentina, 2010).
63. D Rey, Argentine court rejects Barrick gold on glaciers. The Associated Press (2013) http://bigstory.ap.org/article/argentine-court-rejects-barrick-gold-glaciers-1.
64. ANA. *Inventario de Glaciares Cordillera Blanca* (Lima: Autoridad Nacional del Agua del Peru, 2011).
65. A French, Water is life: climate change, globalization, and adaptive resource governance in Peru's Santa River Watershed. Doctoral Dissertation, University of California, Santa Cruz, 2013.
66. M Carey, *In the Shadow of Melting Glaciers: Climate Change and Andean Society* (New York: Oxford University Press, 2010).
67. M Carey, A French, E O'Brien, Unintended effects of technology on climate change adaptation: an historical analysis of water conflicts below Andean glaciers. *The Journal of Historical Geography*, **38** (2012), 181–191.
68. S&Z. *Desague Laguna Paron: Estudio del Comportamiento del Dique* (Lima: S&Z Consultores Asociados S.A., 1986).

69. Parón-Llullán. *Oficio #003–2002-CRCY/CZ.* (Caraz, 2002).
70. EPS Chavin. Press release: pronouncement to authorities and general public against Duke's management (2007).
71. Parón-Llullán. Interviews with members of the Parón-Llullán Irrigators Association (2009–12). French, A. (Caraz, Peru).
72. P Malhotra-Arora, Sino-Indian water wars? In *Water Resource Conflicts and International Security*, ed. DK Vajpeyi (Lanham, md: Lexington Books, 2012).
73. DK Vajpeyi, Introduction. In *Water Resource Conflicts and International Security*, ed. DK Vajpeyi (Lanham, md: Lexington Books, 2012).
74. TF Stocker, D Qin, GK Platnerr, *et al.* (eds.) *Climate Change 2014: Impacts, Adaptation, and Vulnerability. Contribution of Working Group II to the Fifth Assessment Report of the Intergovernmental Panel on Climate Change.* (Cambridge and New York: Cambridge University Press, 2014).
75. A Krishnan, China gives go-ahead for three new Brahmaputra dams. *The Hindu* (2013). www.thehindu.com/news/international/china-gives-goahead-for-three-new-brahma putra-dams/article4358195.ece?ref=relatedNews.
76. Y Yong, Hydropower challenges for the Brahmaputra: a Chinese perspective. In *Brahmaputra: Towards Unity*, ed. I Hilton (n.p.: thethirdpole.net, 2014).
77. J Watts, Chinese engineers propose world's biggest hydro-electric project in Tibet. *Guardian* (2010). www.theguardian.com/environment/2010/may/24/chinese-hydroen gineers-propose-tibet-dam.
78. A Krishnan, Brahmaputra waters will not be diverted, indicates China. *The Hindu* (2011). www.thehindu.com/news/international/brahmaputra-waters-will-not-be-diverted-indicates-china/article2103736.ece?ref=relatedNews.
79. J Bandyopadhyay, N Ghosh, Holistic engineering and hydro-diplomacy in the Ganges–Brahmaputra–Meghna Basin. *Economic and Political Weekly*, **45** (2009), 50–60.
80. R Bhambri, T Bolch, Glacier mapping: a review with special reference to the Indian Himalayas. *Progress in Physical Geography*, **33**: 5 (2009), 672–704.
81. B Chellaney, The Sino-Indian water divide. project syndicate (August 3, 2009). www.project-syndicate.org/commentary/the-sino-indian-water-divide.
82. R D'Souza, Damming politics: India, China, and a trans-border river. India in transition (2013). http://casi.sas.upenn.edu/iit/dsouza.
83. PTI, China spikes India's proposal for joint mechanism on Brahmaputra. *The Hindu*, (2013). www.thehindu.com/news/national/china-spikes-indias-proposal-for-joint-mechanism-on-brahmaputra/article4627285.ece?ref=relatedNews.
84. N Vagholikar, PJ Das, *Damming Northeast India.* (Pune/Guwahati/New Delhi: Kalpavriksh, Aaranyak and ActionAid India, 2010).
85. S Baruah, Whose river is it anyway? Political economy of hydropower in the Eastern Himalayas. *Economic and Political Weekly*, **XLVII**: 29 (2012):41–52.
86. S Dharmadhikary, *Mountains of Concrete: Dam Building in the Himalayas.* (Berkeley, CA: International Rivers, 2008).
87. K O'Brien, R Leichenko, Double exposure: assessing the impacts of climate change within the context of economic globalization. *Global Environmental Change*, **10**: 3 (2000), 221–232.

Part IV
Conclusions

18

Synthesis and conclusions

The future of high-mountain cryospheric research

MARK CAREY, CHRISTIAN HUGGEL, JOHN J. CLAGUE, AND ANDREAS KÄÄB

18.1 Introduction

Ice is an incredibly complex substance – magical in some ways, deadly in others, but always difficult to study and understand. As Mariana Gosnell explains

It is more brittle than glass. It can flow like molasses. It can support the weight of a C-5A transport plane. A child hopping on one leg can break through it. It can last 20,000 years. It can vanish in seconds. It can carve granite. It can trace the line of a windowpane scratch. It can kill peach buds. It can preserve mammoths for centuries, peas for months, human hearts for hours.

[1]

Beyond its incredible physical properties, ice plays a central role in societies world-wide. Millions of people live in mountainous areas, and hundreds of millions more depend on mountain resources such as water from glaciers and gold, silver, and other minerals from icy environments. Others utilize mountainous terrain for activities such as tourism and recreation. Some define national and regional identities based on the nearby mountains. In high-mountain regions, where people live close to glaciers, they depend on snowmelt for water and energy, they carry out pilgrimages to venerated glacier-encased peaks, they ski and hike, or suffer the consequences when permafrost thaws and slopes become more unstable. For scientists, glaciers represent key laboratories for climate knowledge, with ice cores providing detailed climatic data going back 800 000 years [2,3].

Global change and high-mountain hazards are altering – and in many cases threatening, as this book has illustrated – these and many other high-mountain activities, livelihoods, cultural values, and economies that influence numerous stakeholders within and beyond mountains. Climate change has also transformed

The High-Mountain Cryosphere, ed. Christian Huggel, Mark Carey, John J. Clague and Andreas Kääb. Published by Cambridge University Press. © Cambridge University Press 2015.

glaciers into key icons of global warming: they represent clear and visible signs of the biophysical impacts of climate change and the cultural consequences for societies at high elevations and high latitudes [4–6]. These consequences can be catastrophic, not simply symbolic. The possibility of glacial lake outburst floods (GLOFs) and the loss of glacier runoff for water supplies are tangible results of global climate change that generate significant risks for high-mountain populations. Additionally, ice-clad volcanoes can produce deadly lahars, while ice-related debris flows, erosion from thawing permafrost, snow avalanches, and unconsolidated sediment pose risks to people in mountains on every continent except Australia. The complexity of the cryosphere coupled with other ever-increasing risks for high-mountain populations have heightened popular attention on the cryosphere and expanded research significantly in recent years among both natural and social scientists [7–10].

This book advances research on cryospheric risks through its cross-disciplinary approach to coupled natural–human systems in the world's high-mountain regions. Yet, as this conclusion suggests, there are still critical gaps to fill, new topics to explore, and critical concepts to better understand. Several areas that future cryospheric research could productively advance, as outlined below, include: (1) interacting spatial and temporal scales; (2) locally based studies or downscaling; (3) social differentiation and politics among affected populations; (4) expanded global coverage to all high-mountain ranges; (5) more precise understanding of drivers and triggers underlying biophysical and human processes, which includes analyzing complex coupled surface and subsurface processes of the cryosphere; (6) the sharing of local indigenous knowledge to augment the natural sciences and help societies adapt to cryospheric transformation and global environmental change; and (7) more research by interdisciplinary teams, including the incorporation of the humanities in cryospheric research.

18.2 Scale

One critical issue that emerges from the chapters in this book is the need to analyze both temporal and spatial scales on much deeper levels than research has dealt with to date. Two aspects of scale stand out as needing additional research and conceptualization: (1) the interaction of multiple scales, such as local and global forces; and (2) the integration of temporal scales into the analysis of high-mountain human–cryospheric systems. Many of the authors in this volume discuss the interaction of forces and drivers at multiple scales. The interaction of global atmospheric circulation with mountain ranges in specific countries and glaciers with, for example, particular topographies represent this intersection of global, regional, and local dynamics. The relationship between mining and mountain glaciers offers another example, whereby global climatic forces affecting glacier behavior interact with international political-economic trends and consumer demands that affect mining company strategies, national laws affecting mining company access, and regional demographics and livelihoods that can trigger local responses to mining companies. The interaction of these variables acting at various local, regional, national, and global scales makes understanding cryospheric environments and processes very complex.

The challenge of intersecting scales is clearly evident with biophysical conditions, such as climate impacts on the cryosphere. Temperature and precipitation are among the most important climatic controls on impacts in high mountains, and elsewhere as well. Other climatic parameters such as radiation, humidity, or wind may also be important, for instance for glacier mass balance. Climate impacts manifest locally or regionally, but at the same time there are important large-scale circulation patterns and processes that influence local and regional climate, and thus the impacts on the ground. Additionally, high-mountain regions have complex topography, often with strong elevation gradients that translate into temperature and precipitation gradients. Locally, hence, there is strong variability of meteorological conditions and climate at different spatial scales. Mountain systems such as the Andes and Himalayas are also good examples of how mountain topography controls climate (and vice-versa), by forming topographic barriers to moisture flow, resulting in impressive precipitation gradients (e.g. the dry highlands of Tibet).

A major difficulty or challenge, therefore, is understanding the climate–mountain–cryosphere systems over different spatial and temporal scales, and based on that knowledge, to make reasonable future projections (Rangwala *et al.*, this volume). For instance, in tropical and sub-tropical high mountains such as the Andes or the Himalayas, the monsoon represents a dominant source of moisture, manifesting in seasonally sustained and occasionally high-intensity precipitation. Such seasonally clear and distinct precipitation patterns not only exert an important influence on the natural environment, but also on people and their local or regional economies. Societies have adapted their agriculture and livestock to the corresponding dry and humid seasons. For mid-latitude high-mountain regions such as in Western and Central Europe, onshore westerly winds coupled with the northward movement of warm water by the Gulf Stream are important drivers of climate, both in terms of temperature and precipitation. The North Atlantic Oscillation (NOA) is a measure of these flow patterns.

Another challenge of research on different spatial scales is that conditions across mountain ranges are inconsistent. Observed changes in temperature and precipitation are not uniform across the world's mountains during recent decades, nor in fact within mountain ranges. The Karakoram, for instance, has what researchers refer to as the Karakoram "anomaly" because of much slower glacier loss, or even occasionally glacier growth, than in Himalayan regions farther east. This variation is likely related to temperature and precipitation patterns and changes that are different than in the central and eastern parts of the Himalayas. The central and eastern parts of the Himalayas are under strong monsoon influence, while the Karakoram is linked more to westerly flow regimes.

Variation at smaller scales has also been observed, such as in the Swiss Alps, where research indicates that changes in extreme warm temperatures have been much stronger at high elevations than at lower elevations. Such detailed analyses based on high-quality, long-term observations and observation networks do not exist, however, in most mountain ranges, thereby revealing significant gaps in knowledge about these cryospheric dynamics in many high-mountain regions, from the Himalayas to Central Asia and much of the American cordilleras. In recent years it has been increasingly recognized that the lack of such basic data represents a limitation for adaptation to climate change [11]. Climate data sets complementary to meteorological station observations are therefore increasingly developed, available, and evaluated. Authors such as Salzmann *et al.* (this volume) indicate the potential of such data sets for research on high-mountain impacts and adaptation. Yet they also emphasize that caution is necessary when using these data sets because they may have considerable inaccuracies that could generate, at worst, mal-adaptation.

In addition to spatial scales that can present challenges to researchers, temporal scales also pose difficulties for understanding hazard-event triggers, for predicting future events, and for analyzing human perceptions, vulnerabilities, and responses. For example, present-day glacier and glacial lake changes are typically visible and thus, in principle, comparably easy to investigate and monitor. The processes leading to glacier- and permafrost-related rock slope instabilities, on the other hand, are hidden and often play out on long timescales. Knowledge about these processes has increased only recently, and much remains unknown or at least uncertain about how they act on long timescales (Krautblatter and Leith, this volume). Rock slope stability, for example, is influenced by a highly transient system of mechanical and thermal processes acting on ever-changing boundary conditions, as well as by geological, geotechnical, hydrological, and hydraulic conditions at depth. In the case of glacial lakes, reliable regional-scale detection of high hazard dispositions is difficult because of the above spatio-temporal inter-actions. Special focus should thus be, and is, on developing coordinated regional to local-scale monitoring systems coupled with early warning mechanisms. Instal-lation and maintenance of ground-based systems can be challenging in high mountains; further technological developments are needed to enhance effective warning for surrounding populations. At the same time, community preparation and responsiveness to early warning systems are often shaped, not only by present-day concerns about hazards, but by long-term historical cultural values, perceptions of mountains, spirituality, socio-economic conditions, and political contexts – all of which are powerfully shaped by history and never remain as a static snapshot in time or space. Compared to rock slopes, periglacial debris slopes adjust much faster and with higher event frequency to changing environmental conditions because of the much bigger role of ground ice, which cements the material (Stoffel and Graf, this volume). Again, the issue hinges on analysis of the timescales in which these processes play out.

A monumental challenge for rock slope instabilities and risk reduction in high mountains thus involves grappling with high-magnitude and low-frequency mass movements such as rock avalanches that originate from these complex and difficult-to-quantify subsurface processes. At the same time, downslope populations are constantly in flux, and new building codes, urban development, land use practices, and tourism can introduce new vulnerabilities, exposure, and risks through time. More studies of past hazard events and historical societal conditions would help illuminate processes in the present, although to date the work on these historical timescales is limited except in places with long-term monitoring and scientific research, such as the European Alps. At the same time, historical events must be treated with caution: they can be misleading, given that systems are so dynamic. Climate change can disturb glacier and permafrost equilibrium and can thus shift hazard zones beyond historical limits. Human settlements and activities increasingly extend into threatened zones, which obviously heighten local vulnerability. Without more analysis of past events and the ways in which multiple drivers interact and trigger hazards or increase human vulnerability, it also becomes challenging to predict future disasters. This is true not only for debris flows but also for glacier runoff and the availability of water resources in the future. Historical data alone are not sufficient in dynamic human–cryospheric systems for hazard assessments; data on past events must be combined with new observations, analysis, and modeling approaches. Depending on locations, substantial future hydrologic change is projected to occur several decades in the future, a scale that, for most human societies and policy makers, will not generally yield action or stimulate adaptation. Moreover, the discourse invoked in local struggles for water often refers to the long term by pointing to future water deficiencies and merging that with present-day concerns about political agendas, environmental rights, and social justice (French *et al.*, this volume). These complex intersections of temporal scales – as well as the interaction of human and natural systems – need further investigation.

18.3 Regional studies

The challenges of scale reveal the difficulty of applying research results across time and space, which points to another need: more local studies and site-specific research, including research in what Korup and Dunning (this volume) refer to as the "forgotten mountains" – regions outside of the European Alps that have generally been much less extensively studied. Glacier shrinkage is occurring worldwide and transforming cryosphere environments in many ways. But those changes have different impacts in different places, thereby requiring local and regional studies to understand how cryospheric change plays out in particular places, such as for glacier hydrology and downstream impacts. In Switzerland, important hydrologic impacts from shrinking glaciers are projected mostly at the meso-scale, especially after around 2050, with summer drought potentially becoming a serious problem. In the Himalayas, a challenge will be to adapt to intra-annual changes of water resources and extremes. In North America, glaciers and snowpack are important for late summer and autumn stream flow. Mark *et al.* (this volume) conclude that climate–glacier–water issues need to be analyzed at the catchment scale. Deciphering the point of "peak water," a point after which glacier runoff from a shrinking glacier declines, depends on various catchment character-istics such as the percentage of glacierized area, the hypsography of the glacier, and the rate of climate change and alterations in temperature and precipitation that affect glacier mass balance and runoff.

Without regional specificity, however, mis-statements can emerge or, in the case of glacier runoff, exaggerations can occur about the downstream hydrologic impacts of glacier retreat, as Mark *et al.* (this volume) note. The US National Academy of Sciences recently identified common overstatements about Himalayan glacier retreat, such as the claim by many individuals and environmental NGOs that more than a billion people are supposedly going thirsty due to glacier shrinkage [12]. More regionally specific and catchment-based studies are needed to understand potential effects of decreasing glacier runoff because the total annual contribution of glaciers to downstream water supplies is influenced by many local and regional factors, including the monsoon (see French *et al.*, this volume). Generally, the farther one travels downstream from glaciers, the smaller is the influence of glacier melt on overall river runoff. Yet this trend can be complicated when human populations and water-use practices are considered: a hydroelectric plant far downstream, or a critical irrigation canal located in the lowlands, could play vital roles at regional and national levels and yet be profoundly affected by even minor fluctuations in dry-season river flow rates. Understanding the effects of changing glacier and snow cover requires analysis of the specific hydrology of specific watersheds, as well as the downstream human societies. Those societies may depend on hydropower for a significant portion of their energy, such as in the Andean region, or they may rely on glacier runoff for local and regional agricul-ture, whether for export-oriented production or household subsistence. Given the regional variations in climatic, topographic, and biophysical changes, and given the differences in socio-economic and cultural factors for regional populations, the emphasis on some mountain ranges and not others indicates significant shortcomings in knowledge about the high-mountain cryosphere. Some areas, such as Peru and Nepal, have been the site of a steadily growing number of studies. Yet in these same Andean and Himalayan mountains, Ecuador and Tibet, respectively, seem to have received less attention. A non-scientific analysis of the literature suggests that the Caucasus and Rwenzori Mountains have even fewer cryospheric studies, highlighting major gaps of understanding and creating a need for more regionally and locally based studies of high-mountain cryosphere–society dynamics.

18.4 Affected populations and social differentiation

Another significant area for future research is differences within affected local populations living near high-mountain ice. This research is required to understand the socio-economics and politics of who is most and least affected by cryospheric changes. Research to date, including that in this volume, has examined local populations to show how they are influenced by hydrologic change, glacier hazards, and slope instability. Whether it is decreased glacier runoff in the Andes, livelihood changes in the Himalayas, or mining operations adjacent to glaciers in Chile, studies in this book show that cryospheric changes can significantly affect local and regional populations. Often, however, the social science research on the cryosphere identifies broadly construed "affected" or "local" populations without uncovering how these local societies are divided by race, ethnicity, gender, class, age, or geographical location. In turn, scholarship misses how these aspects of social relations influence the degree to which someone is vulnerable to cryospheric changes, or is resilient to these changes, or lacks adaptive capacity. Little is thus known about which portion of the population is most affected by changing hazards due to melting ice on volcanoes, or by the reduction of glacier runoff. Research also demonstrates that vulnerability in high-mountain regions due to cryospheric changes does not necessarily follow typical patterns of vulnerability in which the poorest populations are most vulnerable. In the Peruvian Andes, for example, the populations most exposed to GLOFs have historically been the wealthiest populations living in cities founded by sixteenth-century Spanish colonizers that are located on alluvial fans and cling to glacier-fed riverbanks. The poorer rural indigenous populations living high on hillsides are much less exposed to GLOFs and avalanches, even though at first glance this group might appear more vulnerable given their race and class [13]. Diemberger *et al.* (this volume) also focus on various groups affected by precipitation (snow) changes in the Himalayas, showing how some groups, such as pastoralists and farmers, are more profoundly affected.

In many other studies of cryospheric changes, however, local populations are broadly lumped together in homogeneous populations: the local people, the vulnerable populations, those living in hazard zones, the high-mountain indigenous groups, among others. While useful for understanding human vulnerability and resilience, research should enhance efforts to consider issues such as gender or generational implications. Are men and women differently affected by glacier retreat, volcanic hazards, or increased permafrost thaw? How does ethnicity affect vulnerability to cryospheric change? Which social groups are most likely to follow government building codes, or have the political power to lobby successfully for disaster prevention projects, such as glacial lake lowering in Bhutan and Nepal? Given land use and demographic shifts that often favor emigration from rural mountain areas by men and younger generations, more studies should examine the role of gender and age on risk, vulnerability, adaptation, and resilience, drawing on experiences from regions other than high mountains. The chapters in this volume begin to address these issues, such as in the case of Jurt *et al.* (this volume) and French *et al.* (this volume). Future research could produce more nuanced and sophisticated analyses of social relations and their relationship to changes in the cryosphere.

18.5 Triggers and drivers

Understanding the triggers of cryospheric hazards or human vulnerability and risk are incredibly complex and, as this book and synthesis chapter demonstrate, are affected by multiple intersecting forces operating on different spatial and temporal scales. Additional research on these various triggers and drivers is essential, in large part because it continues to be nearly impossible to predict debris flows, GLOFs, jökulhlaups, peak water timing, and other cryospheric hazard events. High mountains are characterized by the energy associated with their high relief, typically coupled with the existence of snow and ice on and below the terrain surface. High-mountain relief gives rise to a number of horizontal and vertical climatic patterns that contrast massively with those found in lowlands. Surface and subsurface ice and snow are close to the melting point and thus sensitive to changes in boundary conditions, such as alterations in the surface energy balance as a consequence of climate changes. On the other hand, the latent heat content of ice requires a substantial amount of energy to transition from ice to water, giving high-mountain ice and snow bodies some thermal resistance to short-term variations in boundary conditions. Many environmental changes and resulting human risks are situated at the crossroads of these fundamental physical conditions in high mountains – the high gravitational energy, the relief, the sensitivity of ice to changes in boundary conditions, and, at the same time, the thermal inertia of cryospheric systems. These four physical factors manifest themselves clearly in cryosphere hazards, most prominently in glacial lake outbursts, slope instabilities related to glaciers and permafrost, periglacial debris flows, and high-mountain snow avalanches.

Floods from glacial lakes represent the most far-reaching high-mountain threat, posing risks to communities, hydroelectric installations, linear infrastructure, farmland, and housing far downstream, even into surrounding lowlands (Quincey and Carrivick, this volume). Substantial progress has been made in mapping, characterizing, and understanding the different types of glacial lakes and their potential outburst mechanisms, and, equally important, in disseminating this knowledge. Nevertheless, schemes for reliably ranking hazard potentials – necessary for prioritizing research or for taking preventative measures, or to forecast outburst floods – are still in early development. Given the potentially complex and often unforeseeable trigger mechanisms of lake outbursts, combined with the constant dynamic change and transient imbalance of cryospheric systems in high mountains, it remains questionable whether, or how far in advance, lake outbursts can be forecast. Consequently, it is still not possible to implement a highly reliable strategy for predicting outburst floods, even from the most unstable or dangerous individual glacial lakes. In recognition of these challenges a number of early warning systems have been developed and implemented for lake outbursts in mountain regions such as the Alps, Karakoram, or the Andes [14–16]. Furthermore, given the inherent problems and shortcomings in understanding precise trigger mechanisms, it is necessary for future research to focus more on understanding cascading systems with parameters that can be defined with different spatial and temporal resolutions, and on the sound coupling of such systems to adequate societal and technical responses to detected hazards.

Snow avalanche triggers are also difficult to detect and predict. Compared to glaciers and permafrost, snow is by far the most "nervous" interface between ground and atmosphere, reacting significantly at daily to seasonal timescales, and seasonally covering by far the largest area. Management of risks related to snow avalanches is in principle much further developed than for glacial and permafrost hazards, and in some ways it serves as an example of what risk management related to glaciers and permafrost in high mountains could accomplish, although event frequency and distribution are different for snow hazards. Yet snow-related hazards also respond to diverse forces and processes, from local meteorological and climatic conditions to large-scale circulation patterns, anthropogenic impacts on snow stability, and demographic and technological aspects of the exposed human populations (Fuchs *et al.*, this volume). More research on these drivers is essential to understand snow and other cryospheric hazards.

Human populations vulnerable to cryospheric hazards are also shaped by a variety of forces. Politics and laws shape the creation, implementation, and enforcement of building codes as well as urban development onto, or away from, alluvial fans, riparian zones, and avalanche shoots. Socio-economic factors exert significant influence on individuals' level of vulnerability and capacity to adapt and respond to new risks. It is also clear from abundant research that knowledge of risk does not necessarily trigger safe behavior. The dissemination of scientific knowledge about natural hazards and disaster prevention, therefore, does not yield populations free from risk. Risk, in other words, depends on a host of factors, including scientific knowledge, engineering practices, available technologies, cultural values, risk perceptions, communications and media reporting, politics, laws, and social relations, not to mention the biophysical forces affecting societies such as topography, hydrology, and tectonics. To date, however, most of these factors have not been sufficiently studied in integrative ways that ultimately reduce risk, and the studies that have been completed tend to concentrate in specific mountain ranges such as the European Alps.

Human migration is a classic case of multiple intersecting drivers affecting human vulnerability and responses to global environmental change, and it can thus be illuminating for cryospheric research. Migration is influenced by a host of drivers, with environmental change just one of them. Black and others identify five families of migration drivers: economic, political, social, demographic, and other environmental drivers [17]. Others point to more specific drivers of migration, such as education, access to services, access to health care, remittances, income decline, and income variability [18]. It thus is difficult to attribute migration to climate change alone. Climate change or cryospheric hazards may make environments more hazardous, can lead to conflicts over resources, and may negatively affect people's livelihoods or economic opportunities. Understanding migration in high mountains – as well as other human adaptations or responses to changes in the cryosphere and the development of new or more potent hazards – involves research on multiple drivers of change. To date, this exploration of multiple triggers of high-mountain human migration has been limited [19]. Researchers must be careful not to attribute societal impacts to cryospheric changes without first considering the role of multiple triggers of change. After all, further research is needed to measure or empirically demonstrate how changes in rainfall or temperature in mountains strengthen or weaken the various forces driving migration, especially income levels and income variability [18]. The same qualifications and limitations exist for the effects of changes in the high-mountain cryosphere on human societies.

18.6 Local knowledge integration

The value and use of local knowledge, indigenous knowledge, or traditional ecologic knowledge (TEK) in the analysis of high-mountain cryospheric change have received relatively little attention. In contrast, sea ice is one aspect of the cryosphere where TEK has been effectively utilized and more widely disseminated within and among indigenous populations; important initiatives have been undertaken to integrate sea-ice TEK and the natural sciences. The following provides some insights into the value of Arctic TEK and how it relates to scientific studies, as a possible pathway to explore TEK in high mountains and develop corresponding multi-disciplinary research that is sensitive to and helpful for local residents from widely divergent cultures and societies.

Local communities have extensive knowledge of sea ice, from its thinning and decreasing extent, to its consistency, color, shape, pileup in coastal areas, erosion rates, and ocean–ice dynamics. As Aporta *et al.* explain, Inuit peoples in the Arctic, who depend on sea ice for travel, hunting, fishing, and their culture, "understand the nuances and complexity of this dynamic environment through long-term use and occupancy, and thus Inuit sea ice experts have a great deal to contribute to collective knowledge of physical, human, and animal relationships with marine environments" [20]. In the Canadian Arctic, Inuit knowledge and scientific studies are consistent in highlighting the thinning of multi-year sea ice, the shortening of the sea ice season, and the declining extent of sea ice cover. Inuit experts report less predictability in the sea ice and more hazardous travel and hunting at ice edges [20]. The Inuvialuit indigenous community in Sachs Harbour, Northwest Territories, Canada, report a decrease in the quantity of multi-year and first-year sea ice, an increasing distance of multi-year ice from the shore, sea ice thinning, and significant changes in the times of ice breakup and freeze-up. They note particular variability and uncertainty in sea ice during the transition months of the year, when freeze-up and breakup occur. Residents detect changes in sea ice thickness and age by examining its texture and color. They note changes in off-shore ice because the bears and seals they previously hunted have become too far from shore to find. They have detected a change in the time of sea ice breakup through a cultural activity: from the 1950s to the 1990s, community members traveled to a July picnic site by crossing the ice. The ice, however, has been melted since the 1990s, and the Inuvialuit have since then used a boat to reach that same picnic site [21]. This kind of detailed data of dynamic sea ice conditions also exists – though it remains largely untapped or unstudied – in high-mountain environments, where local indigenous and non-indigenous peoples have lived for centuries or millennia and have generated sophisticated and extensive knowledge about the cryosphere, among other things.

In addition to the importance of TEK for increased understanding of the cryosphere, there are opportunities to integrate local knowledge with the natural sciences to further enhance awareness of distinct cultures and epistemological trajectories [22–23]. For example, the Siku-Inuit-Hila Project on sea ice and climate change in Barrow, Alaska, Clyde River, Nunavut, and Qaanaaq, Greenland, strives to create monitoring and information dissemination networks among local populations. This project also promotes dialogue between natural scientists and local indigenous ice experts so that each group can benefit from the other [24]. Through such partnerships, collaborative fieldwork, information sharing, co-authoring, and relationship building, knowledge about sea ice change has significantly expanded since the project began in 2007.

In the high-mountain cryosphere efforts of knowledge integration and local knowledge dissemination have not kept pace with all this research on Arctic TEK. There are hints, however, that such knowledge transfer and integration could be productive, provided TEK is utilized in culturally sensitive and responsible ways that recognize local people's intellectual property and cultural values. Anthropologist Julie Cruikshank has studied the relationship among local Alaskan indigenous peoples, geoscientists, and Romantic poets to show that, although their knowledge is presented quite differently – from oral tradition discussing sentient glaciers to scientific publications and poems – they all had similar aptitude for understanding late Little Ice Age glacier behavior [25]. In a study of six Tibetan villages in Yunnan province, many, but certainly not all, local residents observed warming temperatures, less snow, and shrinking glaciers, which are consistent with scientific interpretations [26]. In Peru's Cordillera Blanca, local residents and instrument-based scientific analysis both report increasingly rapid glacial recession, less snow in the upper watershed, and an increase of falling glacier "blocks" since the latter half of the twentieth century [27]. Local Cordillera Blanca residents also report that, despite interannual variability, there is a notable decline in water availability in recent years, which could be caused by climate change, glacier retreat, and/or shifting water management strategies [28–29]. In some cases in the Cordillera Blanca, indigenous perceptions of "enchanted lakes" may result from their detection of climate-induced glacier retreat and the formation of dangerous glacial lakes [13]. Although indigenous peoples inhabit high-mountain regions, reporting of high-mountain TEK related to dynamic cryospheric environments remains more limited than it is for the Arctic. There is much to learn about mountain glaciers, snow, permafrost, and ice from local populations. Future studies that explore local expertise and high-mountain TEK would add significantly to existing knowledge about cryosphere dynamics.

18.7 Interdisciplinary teams

A key takeaway message from the studies in this volume is that, given the complexity of human–cryospheric dynamics, there is urgent need for more cross-disciplinary studies to tackle these issues in the world's mountain ranges. Fuchs *et al.* (this volume) note that socio-economic changes such as deforestation, land use transitions, urban development, and demographic changes affect the likelihood and effects of snow avalanches. Mark *et al.* (this volume) discuss glacier runoff and aquifers, and, more broadly, the effects of glacier change on downstream hydrology. Yet they note how water security depends on socio-economic and political factors, not on the physical dynamics of climate and ice alone. Waitt *et al.* (this volume) explain that medieval priests saw Iceland's Hekla volcano as the portal to hell, and that its history of 16 eruptions in ten centuries has important implications, not only for human perceptions of volcanoes and glaciers, but also for how ice-clad volcanoes worldwide have affected surrounding populations, especially with catastrophic lahars. In all of these cases, and many others in this book and elsewhere, the issues of cryospheric changes and hazards in mountain regions involve deeply intertwined human–natural systems. Understanding the relationship between glacier behavior and downstream water management, or between volcanic activity and lahar impacts on communities, or between the differences in GLOF prevention in Switzerland versus Peru, requires in-depth analysis by social and physical scientists. Researchers have been calling for more studies on coupled natural–human systems or social–ecological systems for nearly two decades. Calls for a "new intellectual climate" to integrate the environmental social sciences and humanities into global change research, as well as pleas to insert the humanities into climate change research, have increased recently [30]. Yet, as is evident in this book, the humanities are not yet present at a desirable level, and studies focusing on rights, representations, perceptions, identities, faith, cruelty, love, trust, and fear should be strengthened in high-mountain regions.

While advances in coupled natural–human systems have occurred in the climate change community in the past decade, they remain still limited in the cryospheric research world. A recent call to conceptualize "socio-cryospheric systems" helps build on existing theories and relates this cross-disciplinary approach to ice, snow, and permafrost landscapes in particular [31]. This book also makes an important stride toward adopting this cross-disciplinary approach. Yet additional multi-disciplinary teams, which include researchers in the natural sciences, humanities, social sciences, and engineering, who are tackling cryosphere-related problems in integrated ways, should continue to be promoted. Cross-discipliniarity needs to involve this range of fields, and go beyond linking physical disciplines such as geology and hydrology. For relevant research with policy implications, studies will need to include specialists able to analyze society and dissect the socio-economic, political, cultural, and biophysical forces interacting through space and time.

References

1. M Gosnell, *Ice: The Nature, the History, and the Uses of an Astonishing Substance* (New York: Alfred A. Knopf, 2005).
2. R Alley, *The Two Mile Time Machine: Ice Cores, Abrupt Climate Change, and Our Future* (Princeton, NJ: Princeton University Press, 2000).
3. EPICA Community Members, Eight glacial cycles from an Antarctic ice core. *Nature*, **429** (2004), 623–628.
4. M Carey, The history of ice: how glaciers became an endangered species. *Environmental History*, **12**(3) (2007), 497–527.
5. A Kaijser, White ponchos dripping away? Glacier narratives in Bolivian climate change discourse. In *Deconstructing the Greenhouse Interpretive Approaches to Global Climate Governance*, ed. C Methmann, D Rothe, B Stephan (New York: Routledge, 2013), pp. 183–197.
6. B Orlove, E Wiegandt, BH Luckman, The place of glaciers in natural and cultural landscapes. In *Darkening Peaks: Glacial Retreat, Science, and Society*, ed. B Orlove, E Wiegandt, BH Luckman (Berkeley, CA: University of California Press, 2008), pp. 3–19.
7. K Gagné, MB Rasmussen, B Orlove, Glaciers and society: attributions, perceptions, and valuations. *WIRES Climate Change*, **5** (2014), 793–808.
8. W Haeberli, C Whiteman, eds. *Snow and Ice-Related Hazards, Risks, and Disasters* (Amsterdam: Elsevier, 2014).
9. B Orlove, E Wiegandt, B Luckman, eds. *Darkening Peaks: Glacier Retreat, Science, and Society* (Berkeley, CA: University of California Press, 2008).
10. C Harris, L Arenson, H Christiansen, *et al.*, Permafrost and climate in Europe: monitoring and modelling thermal, geomorphological and geotechnical responses. *Earth-Science Reviews*, **92**(3–4) (2009), 117–171.
11. N Salzmann, C Huggel, M Rohrer, M Stoffel, Data and knowledge gaps in glacier, snow and related runoff research: a climate change adaptation perspective. *Journal of Hydrology*, **518** (2014), 225–234.
12. National Academy of Sciences, Committee on Himalayan Glaciers, Hydrology, Climate Change, and Implications for Water Security, *Himalayan Glaciers: Climate Change, Water Resources, and Water Security* (Washington, DC: National Academy of Sciences, 2012).
13. M Carey, *In the Shadow of Melting Glaciers: Climate Change and Andean Society* (New York: Oxford University Press, 2010).
14. D Schneider, C Huggel, A Cochachin, S Guillén, J García, Mapping hazards from glacier lake outburst floods based on modelling of process cascades at Lake 513, Carhuaz, Peru. *Advances in Geosciences*, **35** (2014), 145–155.
15. C Haemmig, M Huss, H Keusen, *et al.*, Hazard assessment of glacial lake outburst floods from Kyagar glacier, Karakoram mountains, China. *Annals of Glaciology*, **55** (66) (2014), 34–44.
16. M Stoffel, C Huggel, Effects of climate change on mass movements in mountain environments. *Progress in Physical Geography*, **36** (2012), 421–439.
17. R Black, WN Adger, NW Arnell, S Dercon, A Geddes, D Thomas, The effect of environmental change on human migration. *Global Environmental Change*, **21** (2011), 3–11.
18. HB Lilleør, K Van den Broeck, Economic drivers of migration and climate change in LDCs. *Global Environmental Change*, **21**, (2011), 70–81.
19. DJ Wrathall, J Bury, M Carey, *et al.*, Migration admist climate rigidity traps: resource politics and social-ecological possibilism in Honduras and Peru. *Annals of Association of American Geographers*, **104**(2) (2014), 292–304.

20. C Aporta, DRF Taylor, GJ Laidler, Geographies of Inuit sea ice use: introduction. *Canadian Geographer*, **55**(1) (2011), 1–5.
21. T Nichols, F Berkes, D Jolly, NB Snow, and the community of Sachs Harbour, Climate change and sea ice: local observations from the Canadian Western Arctic. *Arctic*, **57**(1) (2004), 68–79.
22. I Krupnik, GC Ray, Pacific walruses, indigenous hunters, and climate change: bridging scientific and indigenous knowledge. *Deep-Sea Research Part II: Topical Studies in Oceanography*, **54**(23–26) (2007), 2946–2957.
23. GJ Laidler, Inuit and scientific perspectives on the relationship between sea ice and climate change: the ideal complement? *Climatic Change*, **78**(2–4) (2006), 407–444.
24. HP Huntington, S Gearheard, AR Mahoney, AK Salomon, Integrating traditional and scientific knowledge through collaborative natural science field research: identifying elements for success. *Arctic*, **64**(4) (2011), 437–445.
25. J Cruikshank, *Do Glaciers Listen? Local Knowledge, Colonial Encounters, and Social Imagination* (Vancouver, BC: University of British Columbia Press, 2005).
26. A Byg, J Salick, Local perspectives on a global phenomenon: climate change in eastern Tibetan villages. *Global Environmental Change: Human and Policy Dimensions*, **19**(2) (2009), 156–166.
27. J Bury, BG Mark, J McKenzie, *et al.*, Glacier recession and human vulnerability in the Yanamarey watershed of the Cordillera Blanca, Peru. *Climatic Change*, **105**(1–2) (2011), 179–206.
28. M Carey, A French, E O'Brien, Unintended effects of technology on climate change adaptation: an historical analysis of water conflicts below Andean glaciers. *Journal of Historical Geography*, **38**(2) (2012), 181–191.
29. M Baraer, BG Mark, J McKenzie, *et al.*, Glacier recession and water resources in Peru's Cordillera Blanca. *Journal of Glaciology*, **58**(207) (2012), 134–150.
30. N Castree, W Adams, J Barry, *et al.*, Changing the intellectual climate. *Nature Climate Change*, **4** (2014), 763–768.
31. M Carey, G McDowell, C Huggel, *et al.*, Integrated approaches to adaptation and disaster risk reduction in dynamic socio-cryospheric systems. In *Snow and Ice-Related Hazards, Risks, and Disasters*, ed. W Haeberli, C Whiteman (Amsterdam: Elsevier, 2014), pp. 219–261.

Index

Printed in the United States
By Bookmasters